Surveys in General Topology

Academic Press Rapid Manuscript Reproduction

Surveys in General Topology

edited by

George M. Reed

Institute for Medicine
and Mathematics
Ohio University
Athens, Ohio

ACADEMIC PRESS
A Subsidiary of Harcourt Brace Jovanovich, Publishers
New York London Toronto Sydney San Francisco 1980

ACADEMIC PRESS, INC.
111 Fifth Avenue, New York, New York 10003

United Kingdom Edition published by
ACADEMIC PRESS, INC. (LONDON) LTD.
24/28 Oval Road, London NW1 7DX

Library of Congress Cataloging in Publication Data
Main entry under title:

Surveys in general topology.

1. Topology–Addresses, essays, lectures.
I. Reed, George M.
QA611.15.S97 514 79-28483
ISBN 0-12-584960-5

PRINTED IN THE UNITED STATES OF AMERICA

80 81 82 83 9 8 7 6 5 4 3 2 1

This volume is dedicated to Helen Coast Hayes of Great Bend, Ohio, for the heritage of culture she represents, and for the generosity, intelligence, and enthusiasm with which this gracious lady transmits that heritage to younger generations.

Contents

Contributors

Numbers in parentheses indicate the pages on which the authors' contributions begin.

Dennis K. Burke (1), Department of Mathematics, Miami University, Oxford, Ohio 45056

W. Wistar Comfort (35), Department of Mathematics, Wesleyan University, Middletown, Connecticut 06457

Eric K. van Douwen (55), Institute for Medicine and Mathematics, Ohio University, Athens, Ohio 45701

Ryszard Engelking (131), Department of Mathematics, University of Warsaw, Warsaw, Poland

William G. Fleissner (163), Department of Mathematics, University of Pittsburgh, Pittsburgh, Pennsylvania 15213

Heikki J. K. Junnila (195), Department of Mathematics, University of Pittsburgh, Pittsburgh, Pennsylvania 15213

David J. Lutzer (247), Department of Mathematics, Texas Tech University, Lubbock, Texas 79409

Kiiti Morita (297), Department of Mathematics, Sophia University, Tokyo, Japan

Stelios Negrepontis (337), Department of Mathematics, Athens University, Athens, Greece

Peter J. Nyikos (367), Department of Mathematics, University of South Carolina, Columbia, South Carolina 29208

Teodor C. Przymusiński (399), Institute of Mathematics, Polish Academy of Sciences, Warsaw, Poland

Mary Ellen Rudin (431), Department of Mathematics, University of Wisconsin, Madison, Wisconsin 53706

Franklin D. Tall (445), Department of Mathematics, University of Toronto, Toronto, Ontario M5S 1A1, Canada

Michael L. Wage (483), Institute for Medicine and Mathematics, Ohio University, Athens, Ohio 45701

Howard H. Wicke (499), Department of Mathematics, Ohio University, Athens, Ohio 45701

John M. Worrell, Jr., (499), Institute for Medicine and Mathematics, Ohio University, Athens, Ohio 45701

Preface

During the academic year 1978–1979, the Institute for Medicine and Mathematics and the Department of Mathematics at Ohio University hosted a "Year of Topology." Activities for this occasion included several visiting positions for research topologists and advanced graduate students from other universities, a series of miniconferences and seminars throughout the year, and the annual spring international "rotating" topology conference held March 15–18, 1979 under the support of the National Science Foundation. The proceedings of the spring conference, as well as several papers presented during seminars, were published in the journal *Topology Proceedings,* a joint publication of the Institute for Medicine and Mathematics at Ohio University and the Department of Mathematics at Auburn University. In addition to this publication of recent research results, the editor felt that the concentration and diversity of research topologists brought together by these activities would provide an excellent opportunity for a collection of interacting survey articles in general topology, a collection that could provide focus and thoughtful perspective to current research and indicate promising directions for future study. Hence, the authors of this volume, most of whom took part in the year's activities at Ohio University, were invited to contribute such articles in their areas of specialization.

Articles in this volume may essentially be placed in the following four classifications: (1) Set Theory and Topology [Ultrafilters: An Interim Report (W. W. Comfort), Covering and Separation Properties of Box Products (E. K. van Douwen), Applications of Stationary Sets in Topology (W. G. Fleissner), S and L Spaces (M. E. Rudin), Large Cardinals for Topologists (F. D. Tall)]; (2) Set-Theoretic Topology and Analysis [Combinatorial Techniques in Functional Analysis (S. Negrepontis), Weakly Compact Subsets of Banach Spaces (M. L. Wage)]; (3) General Topological Structures [Closed Mappings (D. K. Burke), Three Covering Properties (H. J. K. Junnila), Ordered Topological Spaces (D. J. Lutzer), Order-Theoretic Base Axioms (P. J. Nyikos), Product Spaces (T. C. Przymusiński), The Development of Generalized Base of Countable Order Theory (H. H. Wicke and J. M. Worrell, Jr.)]; (4) Dimension Theory [Transfinite Dimension (R. Engelking), Dimension of General Topological Spaces (K. Morita)].

The editor strongly believes that the above articles will be of significant value both to researchers in general topology and to mathematicians outside the field who wish an overview of current topics and techniques. While the emphasis of the volume is on set-theoretic topology and its applications, the inclusion of articles on the classical study of dimension theory, including new considerations from shape theory, and others on the increasing infusion of set-theoretic and other topological techniques into functional analysis serve to present a broad sampling of ongoing related research in general topology. The editor, of course, does not claim to present an exhaustive overview of such topics. In particular, he regrets that planned surveys on cardinal invariance, continuua-theoretic constructions, generalized manifolds, and Moore spaces were omitted due to scheduling conflicts by the respective authors (including rather shamefacedly, the editor). However, the volume does contain a balanced presentation of some of the most interesting and productive topics in the area by undoubtly some of the best researchers. It provides sufficient evidence that general topology remains alive and well as a continually growing and interacting field of mathematical research.

Acknowledgments

The activities associated with the 1978–1979 "Year of Topology" at Ohio University, which inspired this volume, were made possible by the support of the National Science Foundation, the Ohio University Research Institute, and the Ohio University Fund, Inc. The helpful assistance of Don Norris, Chairman of the Department of Mathematics at Ohio University, in arranging these activities is also gratefully acknowledged.

The editor wishes to express his personal graditude to Gary and Elise Grabner, graduate students in the Department of Mathematics at Ohio University, and Stephanie S. Goldsberry, the typist for this volume. The successful completion of the volume owes much to the mathematical and diplomatic skills of the Grabners together with the incredible energy and patience of Mrs. Goldsberry. Finally, the editor wishes to thank Eric van Douwen for valuable proofreading and technical consultations.

CLOSED MAPPINGS

Dennis K. Burke
Miami University

1. *Introduction*

Problems involving the behavior of topological properties under the action of various kinds of continuous mappings, and the characterization of one class of topological spaces by another class via continuous mappings are central to general topology. Results of this type can often be applied to other problems which do not explicitly involve mappings so it may be beneficial to be aware of available results in mapping theory. In this note we survey some of the techniques and results in the area of closed mappings. The choice of material was guided by the author's interests and expertise, and does not include mention of several areas in which closed mappings play an important role such as dimension theory, multivalued mappings, and the theory of absolutes. Readers interested in other kinds of mappings such as open mappings, bi-quotient mappings, and others may wish to consult surveys and in depth articles from [G_2], [Mi_5], [MOS], [SMa].

For convenience, all spaces are assumed to be at least Hausdorff. The set of positive integers is denoted by N and an ordinal number denotes the set of smaller ordinals. Most of the topological properties discussed in this paper will not be defined--there should be sufficient references to guide the

ISBN: 0-12-584960-5

interested reader. The remainder of the paper is divided into sections discussing topics in: General properties of closed maps, Base axioms, Lasnev spaces and decompositions, σ-spaces, p-spaces, and related spaces, Covering properties, and Odds and ends.

Unless otherwise stated, a mapping from X to Y is a continuous onto function and may be denoted by $f: X \to Y$. A mapping $f: X \to Y$ is said to be a closed mapping if $f(F)$ is closed in Y for every closed subset $F \subset X$. A decomposition space $\mathcal{D}$ of closed subsets of X is upper semicontinuous if for any open $U \subset X$ and any $D \in \mathcal{D}$ with $D \subset U$ there is an open set $\mathcal{W}$ in $\mathcal{D}$ such that $D \subset \cup\mathcal{W} \subset U$. The natural mapping $\phi: X \to \mathcal{D}$ is closed if and only if $\mathcal{D}$ is upper semicontinuous. A closed mapping $f: X \to Y$ is said to be perfect (quasi-perfect) if each fiber $f^{-1}(y)$ is compact (countably compact). The term "proper map" has also been used in the literature for equivalent formulations of "perfect map". A mapping $f: X \to Y$ is said to be peripherally compact if the bdry $f^{-1}(y)$ (boundary of $f^{-1}(y)$) is compact for each $y \in Y$. It is well-known that if $f: X \to Y$ is a peripherally compact closed mapping (for X a T_1-space) there is a closed subspace $Z \subset X$ such that $f|_Z: Z \to Y$ is a perfect mapping. Hence, a closed-hereditary (preserved to closed subspaces) property P is preserved under a perfect mapping if and only if P is preserved under peripherally compact closed mappings. It is for this reason that we will not usually consider peripherally compact closed mappings as a special case.

2. *General Properties of closed mappings*

A mapping $f: X \to Y$ is a <u>k-mapping</u> if $f^{-1}(K)$ is compact in X for every compact $K \subset Y$. It is known that every perfect mapping is a k-mapping ([HI], [Ha]) and a k-mapping is perfect if the range space is a k-space [HI]. According to Whyburn [Wh], the k-space condition is necessary as Y must be a k-space if every k-mapping $f: X \to Y$ is perfect. The fact that perfect mappings preserve compactness in the inverse image direction plays an important role in many results about the inverse preservation of other topological properties under perfect mappings.

The presence of compactness also forces a "closed map" condition in several natural situations. For example, a mapping $f: X \to Y$ is clearly perfect whenever X is compact. If X is compact and Y is any space it is well-known that the projection map $\pi_2: X \times Y \to Y$ is closed and hence perfect. In general, some compactness condition on X is expected in order for π_2 to be closed since Mrówka has shown [Mr] that X must be compact if $\pi_2: X \times Y \to Y$ is closed for every space Y. Other authors ([H], [FF], $[Is_1]$) have weakened the compact condition on X in return for some restriction on Y. For example, the following is true.

Theorem 2.1 [FF], [K]. If X is countably compact and Y is sequential then $\pi_2: X \times Y \to Y$ is closed.

The fact, that the projection map $\pi_2: X \times Y \to Y$ is perfect when X is compact, is frequently applied in the following manner: If P is a topological property preserved in the inverse image direction of perfect mappings, then $X \times Y$ has property P whenever X is compact and Y has property P. It is apparently not so well known that a reverse argument can often be used because of the following result.

Theorem 2.2 [Mi_4], [N_1], [vdS]. If X is completely regular and Y is any space then there exists a perfect map f: X → Y if and only if X is homeomorphic to a closed subspace of Z × Y for some compact space Z.

Given the perfect mapping f: X → Y, any compactification of X can be chosen for the space Z in the above theorem, and the embedding map of X into Z × Y is defined by x → (x,f(x)). Theorem 2.2 gives the following formulation for the equivalence of preserving a topological property under inverse images of perfect mappings and preserving under products with a compact factor.

Theorem 2.3. If P denotes a closed hereditary property then P is inversely preserved by perfect mappings (with completely regular domain) if and only if X × Y has property P whenever X is compact and Y has property P.

The technique suggested by Theorem 2.3 can be applied to several topological properties such as paracompact, Lindelöf, σ-compact, subparacompact, locally compact, real compact, and many others.

Theorem 2.2 can be proved using the following result, which also allows a general technique for showing preservation of topological properties.

Theorem 2.4 [Ar_3], [Mi_4]. If X admits a perfect mapping onto a space Y and a one-to-one mapping into a space Z, then X is homeomorphic to a closed subspace of Y × Z.

This says:

Corollary 2.5 [Ar_3]. Suppose P denotes a topological property which is closed hereditary and preserved under finite products. If X admits a perfect mapping onto a space Y satisfying P and a one-to-one mapping onto a space Z satisfying P then X satisfies P.

Arhangel'skiĭ [Ar_3] calls a property (or class) P Dutch if P is closed hereditary and preserved under finite products; examples include: metrizable spaces, first countable spaces, spaces with a point-countable base, developable spaces and many others.

We continue with another general technique for showing preservation or inverse preservation of topological properties under perfect mappings. Henriksen and Isbell call a topological property P fitting if whenever $f: X \to Y$ is a perfect mapping then X has property P if and only if Y has property P. We have already mentioned several well-known fitting properties such as compactness, paracompactness, Lindelöf, and σ-compactness. (Other such properties will be given in later sections.) A completely regular space X is said to have property P at infinity if the Stone-Čech remainder $X^* = \beta X - X$ has property P. Quite often a property P at infinity for X translates to another familiar property Q for X. For example, X is compact at infinity if and only if X is locally compact, X is σ-compact at infinity if and only if X is Čech-complete, and X is Lindelöf at infinity if and only if X is of countable type [HI]. This concept of a property P at infinity is rendered useful by the following result.

Theorem 2.6 [HI]. For a completely regular space X, a property P is fitting if and only if P at infinity is fitting.

As a simple application, we see that the property of "countable type" is preserved in the image and preimage direction by a perfect map (with completely regular domain [HI]) by using the fact that the Lindelöf property is fitting.

A mapping $f: X \to Y$ is <u>irreducible</u> if the only closed set $F \subset X$ for which $f(F) = Y$ is $F = X$. It is frequently useful to have the irreducible property as an extra tool when working with mappings, and sometimes a given mapping can be restricted to a domain on which the map is irreducible. If $f: X \to Y$ is a perfect mapping, a Zorn's Lemma argument shows there is a closed subspace $Z \subset X$ such that $f|_Z: Z \to Y$ is an irreducible perfect mapping. Lašnev [L_1] has shown that if $f: X \to Y$ is a closed map, with X paracompact and Y Frechet, there exists a closed set $Z \subset X$ such that $f|_Z: Z \to Y$ is irreducible.

Given two closed mappings $f_1: X_1 \to Y_1$ and $f_2: X_2 \to Y_2$ it is natural to ask when the product map $f_1 \times f_2: X_1 \times X_2 \to Y_1 \times Y_2$ is closed. It is easy to verify that $f_1 \times f_2$ is closed if f_1 and f_2 are both perfect; in general, this perfect condition on f_1 and f_2 is not too strong as Bourbaki has shown [Bb] that f_1 is perfect if and only if $f_1 \times i_Z$ is closed for every space Z and identity map $i_Z: Z \to Z$. Michael gives an example in [Mi_6, Example 8.1] of a closed mapping $f: X \to Y$, with X locally compact separable metric, such that $f \times i_Z$ is not even a quotient mapping for any metric space Z which is not locally compact. Such a mapping f could not be bi-quotient [Mi_6].

We conclude this section with a word about the relationship between preservation under perfect maps and a closed locally finite sum theorem. A construction similar to that described below has been used by various authors but we have not found a suitable reference for its origin. Suppose a space X is covered by a locally finite collection of closed sets $\{A_\alpha : \alpha \in \Lambda\}$. Let $\sum_\alpha A_\alpha$ denote the free topological sum of the subspaces and let $\phi: \sum_\alpha A_\alpha \to X$ be the natural canonical map. It is easily verified that ϕ is a closed map with finite fibers; hence ϕ is perfect. If $\{A_\alpha : \alpha \in \Lambda\}$ is a hereditarily closure-preserving collection of closed sets then ϕ is still closed, but not perfect if $\{A_\alpha : \alpha \in \Lambda\}$ is not locally finite. This suggests the following sum theorem, which can be applied to a wide variety of topological properties. Examples include: metrizable, stratifiable, paracompact, metacompact, subparacompact, normal, collectionwise normal and many others.

Theorem 2.7. Suppose P is a topological property preserved under perfect mappings (closed mappings) and $\{A_\alpha : \alpha \in \Lambda\}$ is a locally finite (hereditarily closure preserving) closed cover of X with each A_α satisfying P. If the free topological sum satisfies P then so does X.

It should be noted that the canonical map ϕ may not be closed if $\{A_\alpha : \alpha \in \Lambda\}$ is simply a closure-preserving closed cover. Potoczny [P] has given an example of a nonparacompact space which has a closure-preserving cover of compact sets. In this case, the corresponding domain space (the free topological sum) would be paracompact, so ϕ could not be closed.

3. *Base axioms and perfect mappings*

One of the most well-known theorems concerning the preservation of base axioms under a perfect mapping is the following due to Morita, Hanai [MH], and Stone [S_n], with the (c) → (b) portion due to Vaĭnšteĭn [V].

Theorem 3.1. If f: X → Y is closed and X is metrizable the following are equivalent:

(a) Y is first countable.

(b) For each y ∈ Y, Bdry $f^{-1}(y)$ is compact.

(c) Y is metrizable.

Not only do we see that the image of a metric space, under a perfect mapping, is metrizable but that any closed f: X → Y preserving metrizability must have fibers with compact boundaries so that there would be a closed subset Z ⊂ X such that $f|_Z$: Z → Y is perfect. Other authors have weakened the metrizable condition on X in order to improve the (a) → (b) portion. We give one such result, due to Michael, and an application which shows that under fairly general conditions closed maps are compact covering.

Theorem 3.2 [Mi_2]. If X is paracompact, Y locally compact or first countable, and f: X → Y is closed then bdry $f^{-1}(y)$ is compact for every y ∈ Y.

Recall that a mapping f: X → Y is <u>compact covering</u> if every compact subset of Y is the image of a compact subset of X.

Corollary 3.3 [Mi_2]. If f is paracompact and f: X → Y is closed then f is compact covering.

These results can be sharpened somewhat, but any improvement quite often relies on the following, also due to Michael.

Theorem 3.4 [Mi_2] (see also [Ve]). Let f: X → Y be closed. If y ∈ Y is a q-point then every continuous real valued function on X is bounded on bdry $f^{-1}(y)$. Hence, if X is normal and Y is a q-space then every bdry $f^{-1}(y)$ is countably compact.

Michael also points out that the map $\pi_1: (\omega+1) \times \omega_1 \to \omega + 1$ is a closed map from a normal space onto a compact metric space and $\pi_1^{-1}(\omega) = \text{bdry } \pi_1^{-1}(\omega)$ is not compact.

There are several other important results that show certain base axioms are preserved under perfect mappings. Some of the positive and negative results are summarized in Theorems 3.5 and 3.6, where each base property is listed along with one or more authors and pertinent references, when available.

Theorem 3.5. The following base axioms are preserved under perfect mappings:

metrizable	Morita, Hanai, Stone	[MH], [Sn]
developable	Worrell	[Wo_1], see also [B_3]
uniform base	Worrell	[Wo_1], [Wo_2]
base of countable order	Worrell	[Wo_3]
point countable base	Filippov	[F_3], see also [BM]
σ-point finite base	Filippov	[F_3], [B_3]
stratifiable (under closed maps)	Borges	[Bo]
semi-stratifiable (closed maps)	Creede	[Cr]
γ-space	Nedev, Čoban	[NC]
quasi-metrizable	Kofner	
σ-Q	Aull, Kofner	[A]
σ-locally countable base	Burke	Corollary 3.10
primitive base	Burke	[B_4]
quasi-developable (θ-base)	Burke	[B_4]
γ-base	Chaber, Čoban, Nagami	[CCN], [WoW]

Theorem 3.6. The following base axioms are not generally preserved under perfect mappings.

σ-disjoint base	Heath-Reed	[B_3, Example 4.3]
weak uniform base	Slaughter	[S_2], [HL]
orthobase	Nyikos	
1st countable semimetrizable weak 1st countable symmetrizable Nagata space	Borges, Čoban, Lutzer, Vaughan	[Bo], [C], [Lu], [Vau_1] See Example 3.11

Question 3.7. Are any of the following base axioms preserved under a perfect mapping: θ-space [Ho], M_1-space [Ce], δθ-base [A_2], σ-minimal base [A_2].

Worthy of special note are the well-known and useful theorems by J. Worrell on the preservation of developable spaces and spaces with a base of countable order under perfect maps, and the result by Filippov showing that a point-countable base is preserved by bi-quotient s-mappings (hence by perfect mappings) [F_3], [F_4], [BM].

The primitive base and quasi-developable results from 3.5 are recent theorems by the author in [B_4] where it is shown that a primitive base is preserved under a perfect mapping, and quasi-developable spaces are characterized by:

Theorem 3.8. A space X is quasi-developable if and only if X has a primitve base and satisfies:

(*): If $\mathcal{U}$ is any well-ordered open cover of X there is an open refinement $\mathcal{G} = \bigcup_{n=1}^{\infty} \mathcal{G}_n$ of $\mathcal{U}$ such that if $x \in X$ there is some $n \in N$ with $St(x,\mathcal{G}_n) \subset F(x,\mathcal{G})$ and $ord(x,\mathcal{G}_n) = 1$ ($F(x,\mathcal{G})$ denotes the first element of $\mathcal{U}$ containing x.)

It is shown that property (*) is preserved under a perfect mapping [B_4], hence the quasi-developable property is preserved under a perfect mapping.

Other approaches to some of the results from Theorem 3.5 use characterizations of the base axioms which are more easily managed under perfect maps. For example, a characterization is given for spaces with a point-countable base in [BM] and for developable spaces in [B_3]; in both cases it is easily seen that the characterizations are preserved under a perfect mapping.

We also have the following unpublished result:

Theorem 3.9. A regular space X has a σ-locally countable base if and only if X has a σ-locally countable cover $\mathcal{P}$ such that if $x \in U \subset X$, with U open in X, there is a finite subcollection $\mathcal{F} \subset \mathcal{P}$ such that $y \in (\cup\mathcal{F})^\circ \subset (\cup\mathcal{F}) \subset U$.

Corollary 3.10. If X has a σ-locally countable base and $f: X \to Y$ is a perfect mapping onto a regular space Y then Y has a σ-locally countable base.

We include an example from [Lu] which shows some of the results in 3.6. Examples with similar results have been given in [Bo], [C], and [Vau_1].

Example 3.11. There is a Lindelöf semimetrizable space X, with a countable network, and a perfect mapping $f: X \to Y$ onto a space Y which is not any of the following: semimetrizable, 1st countable, symmetrizable, weak 1st countable, q-space, point-countable type.

Proof. Let X be the set of points in R × R with a topology described as follows: Points $(a,b) \in X$, with $b \neq 0$, have their usual neighborhoods. Points $(a,0) \in X$ have a "bow-tie" neighborhood base of sets of form:

$$\{(x,y) \in X: |x-a| < \frac{1}{n}, -|x-a| < ny < |x-a|\}.$$

Then X has the desired properties. To form the image space Y, let J be a closed bounded interval (of nonzero length) on the x-axis and let Y be the quotient space obtained by shrinking J to a point. The corresponding quotient map $f: X \to Y$ is perfect and it can be verified that Y has none of the properties listed above.

One should not expect base axioms to be inversely preserved by perfect maps (an arbitrary compact space would not generally satisfy any of the base axioms discussed in this section). However, with additional conditions on the domain and on the fibers of a closed (or perfect) map it is quite often possible to "pull back" certain base axioms. Results of this type are studied by Alkins and Slaughter in [AS] with a typical result given by:

Theorem 3.12. Let $f: X \to Y$ be a peripherally compact closed map from a regular space X. If Y and the fibers of f are developable, then X is a Moore space if and only if X has a G_δ-diagonal.

We conclude this section with a word about the interesting problem of whether M_1-spaces are preserved by perfect mappings. Borges and Lutzer have shown that the irreducible perfect image of an M_1-space is an M_1-space [BoL]. They point out that if $f: X \to Y$ is a perfect mapping from an M_1-space X, there is a closed subspace Z of X such that $f|_Z: Z \to Y$ is an irreducible perfect map. So if it can be shown that M_1-spaces are hereditary to closed subspaces then the perfect image problem is solved. Behavior under closed mappings is not known either--if M_1-spaces turn out to coincide with stratifiable spaces then M_1-spaces would be preserved under closed mappings. We note Slaughter has shown that the closed continuous image of a metric space is an M_1-space [S_1].

4. *Lašnev spaces and decompositions*

Spaces that are expressible as a closed (including continuous) image of a metric space are of special interest. Such spaces are called Lašnev spaces due to the work done by N. Lašnev in [L_1] and [L_2]. Lašnev proved the following "decomposition theorem".

Theorem 4.1 [L_1]. If X is a metric space and f: X → Y is a closed mapping then $Y = Y_0 \cup (\bigcup_{i=1}^{\infty} Y_i)$, where $f^{-1}(y)$ is compact for each $y \in Y_0$ and each Y_i, $i \in N$, is discrete.

Prior to Lašnev's theorem, Morita had proved a similar result for paracompact locally compact spaces, and since then quite a number of authors have strengthened Theorem 4.1 by weakening the condition on the domain space. To summarize these results we note that the conclusion of Theorem 4.1 is true for:

X a paracompact locally compact space	Morita	[M_3]
X a regular symmetrizable space	Arhangel'skiĭ	[Ar_7]
X a normal σ-space	Okuyama	[O_1]
X a normal stratifiable space	Stoltenberg	[St]
X a paracompact p-space	Filippov	[F_2]
X a metacompact p-space	Veličko	[Ve]

If "$f^{-1}(y)$ is countably compact" is substituted for "$f^{-1}(y)$ is compact" in 4.1 then the conclusion is true for:

X an SSM-space	Nagata	[N_2]
X a wM-space	Ishii	[I_2]
X a normal pre-σ-space	Suzuki	[Su]

The decomposition of Y in the above sense is more than just a curiosity. A decomposition is sometimes useful in metrizability theory, Atkins and Slaughter [AS] use such decompositions to "pull back" topological properties from Y to X under certain closed mappings f: X → Y, and Arhangel'skiĭ [Ar_4] uses Lašnev's Theorem 4.1 to show that every Lašnev space is submetrizable (has a weaker metric topology). As another application, the next corollary follows immediately from Theorem 4.1 and the Morita-Hanai-Stone Theorem 3.1.

Corollary 4.2 [L_1]. If the space Y is the image of a metric space of weight τ under a closed mapping then $Y = Y_0 \cup Z$, where Y_0 is metrizable and Z is σ-discrete with $|Z| \leq \tau$.

The subspace Y_0, in Corollary 4.2, may be quite "small" and, in fact, Lašnev gives an example in [L_1] of a Lašnev space Y where Y has no point of countable character. This space would not have a dense metrizable subspace. Stricklen [Sk_1] gives other examples of nowhere first countable Lašnev spaces, and proves there is a closed mapping from a metric space M onto a space with a dense nowhere first countable subspace if and only if M is nowhere locally compact. Contrast these examples with:

Theorem 4.3 [Van]. Every closed image of a complete metric space contains a dense completely metrizable subspace. Hence the image space must be a Baire space.

Because of the above result we fall to the temptation of calling the closed images of complete metric spaces, "complete Lašnev spaces". Van Doren also shows that for every complete Lašnev space Y there is a complete Lašnev space Z such that every nonempty open subspace of Z contains a copy of Y. This shows that a complete Lašnev space need not be a countable union of closed metrizable subspaces (see also Fitzpatrick, [Ft]). Stricklen [Sk_2] extends Van Doren's results by showing that every Lašnev space can be densely embedded in a Lašnev space whith has a dense completely metrizable subspace, and for every Lašnev space Y there is a Lašnev space Z, each nonempty open subset of which contains a copy of Y.

Every Lašnev space can be written as a union of c closed metrizable subspaces and (under MA + $\neg$CH) there is an example, by Van Doren, of a Lašnev space which cannot be written as a union of fewer than c closed metrizable subspaces [vDLPR].

Lašnev has given internal characterizations of Lašnev spaces and of complete Lašnev spaces. He shows that Y is a Lašnev space if and only if Y is a Fréchet space with a network consisting of an almost refining sequence of hereditarily closure-preserving coverings. See [L_2] for details.

In terms of more familiar topological properties, it is clear that every Lašnev space is a paracompact σ-space and, according to Slaughter [S_1] (see also [BoL]), every Lašnev space is M_1. That is, every Lašnev space has a σ-closure preserving open base.

Lašnev spaces are poorly behaved under products. It is known that if X and Y are nonmetrizable Lašnev spaces then X × Y is not a Lašnev space. In fact, if X and Y are Lašnev spaces then X × Y is a Lašnev space only when both X and Y are metrizable, or either X or Y is discrete ([Hr],[Ta]). This can be obtained from the following result.

Theorem 4.4 [Hr]. If Z and W are nondiscrete Lašnev spaces, then Z × W is metrizable if and only if it is Fréchet.

5. *σ-spaces, p-spaces, and related spaces*

Most of the results in this section are summarized in Table I, where the individual properties are listed along with an abbreviated statement indicating whether the given property is preserved or inversely preserved under the given mapping. Two open questions appear on this list.

Question 5.1. Are wΔ-spaces preserved under perfect mappings? Are strict p-spaces preserved under perfect mappings?

The question for strict p-spaces is related to the question of whether every strict p-space is θ-refinable, since Worrell has shown that the perfect image of a θ-refinable p-space is a p-space.

The theory of closed mappings is an integral part of the study of the topological properties in this section. Frolík [Fr_1] has shown that a completely regular space X is a paracompact Čech complete space if and only if X can be mapped onto a complete metric space by a perfect map. M-spaces are characterized as the quasi-perfect preimages of metric spaces

TABLE 1

Property	Image	Preimage
Čech complete	yes perfect [HI]	*yes perfect [HI]
p-space	no perfect [Wo_5]	*yes perfect [Ar_1]
paracompact p-space	yes perfect [F_1], [I_1]	*yes perfect [Ar_1]
metacompact p-space	yes perfect [Ve], [CCN]	*yes perfect
subparacompact p-space	yes perfect [CCN]	*yes perfect
θ-refinable p-space	yes perfect [Wo_5]	*yes perfect
strict p-space	? perfect	*yes perfect
monotonic p-space	yes perfect [CCN]	yes perfect [CCN]
regular λ_b, β_b	yes perfect [WoW]	yes perfect
regular λ_c, β_c	yes perfect [WoW]	yes quasi-perfect
quasi-complete	no perfect [G_1], [Wo_5]	yes quasi-perfect [G_1]
M-space	no perfect [M_2]	yes quasi-perfect [M_2]
normal M-space	yes quasi-perfect [I_1]	no perfect Example 6.3
M*-space	yes quasi-perfect [I_1]	yes quasi-perfect [I_1]
wM-space	yes quasi-perfect [I_2]	yes quasi-perfect [I_2]
wΔ-space	? perfect	yes quasi-perfect
σ-space	yes closed [SN]	no perfect
Σ-space	yes quasi-perfect [Na_1]	yes quasi-perfect [Na_1]
Σ*-space	yes closed [O_2]	no perfect [O_2]
Σ#-space	yes closed [Mi_3]	yes perfect [O_2]

*with completely regular domain

[M_1] and paracompact p-spaces (= paracompact M-spaces) are characterized as the perfect preimages of metric spaces

[Ar_1]. Using this characterization, J. Nagata [N_1] showed that paracompact p-spaces are exactly the closed subspaces of products $Z \times M$, of compact spaces Z with metric spaces M (this follows from Theorem 2.2). This prompted him to ask whether M-spaces could be characterized as the closed subspaces of products $Y \times M$, where Y is countably compact and M is a metric space. A negative answer was given by examples in [BvD] and [Ka].

The perfect inverse image (with completely regular domain) of a Moore space is a subparacompact p-space [B_1]. It is reasonable to expect that every subparacompact p-space could be mapped onto a Moore space by a perfect map (see [Ar_2]), but this is not the case as shown by examples in [B_2] and [Wo_4]. The perfect inverse images of developable spaces have been characterized in [Is_2] and [Pa], Isiwata asks in [Is_2] whether the property of being the perfect inverse image of a developable space is itself preserved under a perfect mapping.

It is interesting that two of the "core" properties from Table I, the p-spaces and the M-spaces, are not preserved under perfect mappings. Morita gives an example in [M_2] of a locally compact M-space X and a perfect mapping $f: X \to Y$ where Y is not an M-space, and Ishii [I_1] shows that the quasi-perfect image of an M-space is an M-space if the domain or range is normal. Morita and Rishel [MR] characterize the perfect images of M-spaces as the class of M*-spaces. They also give a characterization for the closed images of regular M-spaces; see [MR] for details.

The situation for p-spaces is similar. J. Worrell has given an example in [Wo_5] of a p-space X and a perfect mapping onto a space Y which is not a p-space, and several authors have noted that an additional covering condition on X does allow the preservation of the p-space condition. It would be interesting to characterize the perfect images of p-spaces; perhaps the answer will involve μ-spaces, as defined by Worrell and Wicke, since every p-space is a μ-space and completely regular μ-spaces (= completely regular space satisfying condition β_b) are preserved under perfect mappings [WoW].

6. *Covering properties*

The following beautiful result was proved by E. Michael in [Mi_1]:

Theorem 6.1. A regular space X is paracompact if and only if every open cover of X has a closure preserving refinement.

An immediate consequence of Theorem 6.1 is that the image of a paracompact space, under a closed mapping, is paracompact. This theorem is representative of several results in which covering properties are characterized using refinements (usually closure preserving type concepts) that are easily handled under closed mappings. A similar characterization of subparacompact spaces is given in [B_1], and Junnila [J] recently proved the following result which shows that θ-refinability is preserved under closed mappings.

Theorem 6.2. A space X is submetacompact (= θ-refinable) if and only if every directed open cover of X has a σ-closure preserving closed refinement.

The closed mapping results on covering properties are summarized in Table II. We have not provided references for the

TABLE II

Property	Image	Preimage
compact	yes continuous	yes perfect
Lindelöf	yes continuous	yes perfect
paracompact	yes closed [Mi_1]	yes perfect
metacompact	yes closed [Wo_2], [E]	yes perfect
subparacompact	yes closed [B_1]	yes perfect (with regular domain)
θ-refinable	yes closed [J]	yes perfect
δθ-refinable	? closed or perfect	yes perfect
weak θ-refinable	? closed or perfect	yes perfect
screenable	no perfect [B_3, Example 4.3]	yes perfect
mesocompact	yes perfect [Ma] ? closed	yes perfect
Property L	yes closed [D]	yes perfect
metalindelöf	? closed or perfect	yes perfect
paralindelöf	? closed or perfect	yes perfect
orthocompact	no perfect [B_5]	no perfect, Example 6.3
countably paracompact	yes perfect [HI] no closed, Example 6.4	yes perfect
countably subparacompact	yes closed	no perfect, Example 6.3
countably metacompact	yes closed	yes perfect

preimage results; most of these follow by a similar technique. In fact, many covering properties are inversely preserved by a closed mapping with Lindelöf fibers (or some other condition weaker than compactness).

The two negative results on inverse preservation are shown by the following example.

Example 6.3 [HI], [K_2], [Sc]. The projection mapping $\pi_1: \omega_1 \times (\omega_1 + 1) \to \omega_1$ is a perfect mapping onto a normal, countably subparacompact, orthocompact space and the domain space $\omega_1 \times (\omega_1 + 1)$ is not normal, not countably subparacompact and not orthocompact.

Proof. It is well known that $\omega_1 \times (\omega_1 + 1)$ is not normal. Results from [Sc] show that ω_1 is orthocompact and $\omega_1 \times (\omega_1 + 1)$ is not orthocompact. T. Kramer [K_2] has shown that ω_1 is countably subparacompact and $\omega_1 \times (\omega_1 + 1)$ is not countably subparacompact.

Zenor [Z_2] gives an example showing countable paracompactness is not preserved under closed mappings, and we provide such an example below, with completely regular domain and Hausdorff range. See [Mr] and [Z_2] for sufficient conditions to insure the preservation of countable paracompactness under closed mappings.

Example 6.4. There is a locally compact, countably paracompact space X and a closed map f: X → Y onto a space Y which is not countably paracompact.

Proof. For each $k \in N$, let $X_k = \omega_1 \times (\omega_1 + 1) \times \{k\}$ and let $\Delta_k = \{(\alpha,\alpha,k): \alpha \in \omega_1\}$. The space X is the topological sum of the spaces X_k, $k \in N$, and Y is the quotient space obtained from X by identifying all points in $\bigcup_{k=1}^{\infty} \Delta_k$ to a single point p. The corresponding quotient map is clearly closed and X is countably paracompact. To show Y is not countably paracompact, let $U_0 = \bigcup_{k=1}^{\infty} (\omega_1 \times \omega_1 \times \{k\})$, and for each k let

$U_k = \{(\alpha,\beta,k) \in X_k: \alpha \in \beta\}$; then $\{f(U_0), f(U_1), \ldots\}$ is a countable open cover of Y. If Y was countably paracompact there would be a precise locally finite open refinement $\{V_0, V_1, \ldots\}$ (i.e. $V_i \subset f(U_i)$). If W is an open neighborhood of p in Y such that W meets only finitely many V_i, there must be some $j \in N$ such that $W \cap V_j = \emptyset$. It follows that $f^{-1}(W) \cap X_j$ and $f^{-1}(V_j)$ are disjoint open sets in X_j about the closed sets Δ_j and $\omega_1 \times \{\omega_1\} \times \{j\}$ respectively, and this is known to be impossible. Hence Y is not countably paracompact.

Remark 6.5. A similar construction shows that if there exists a non-normal paralindelöf space Z then there is a paralindelöf space X and (allowing a nonregular range) a closed mapping onto a space which is not paralindelöf. The domain space would be a topological sum of an uncountable number of disjoint copies of Z.

An example showing that orthocompactness is not preserved under perfect mappings is given in [B_5], thus answering a question by B. Scott in [Sc]. The fibers of this map are infinite; it would be interesting to know whether orthocompactness is preserved under finite-to-one closed mappings.

7. *Odds and ends*

A summary of results on preservation and inverse preservation (under closed mappings), of certain properties not considered in earlier sections, is given in Table III. For our only remarks concerning the results in this table we refer the reader to an interesting discussion by J. Vaughan in [Vau_2] on the nonpreservation of the G_δ-diagonal, submetrizability, realcompactness, and property wD under

TABLE III

Property	Image	Preimage
point-countable type	no perfect [Bo], [C], [Lu], $[Vau_1]$ (Example 3.11)	yes perfect, $[Ar_6]$, $[Mi_5]$
countable type	yes perfect [HI], [C], $[Vau_3]$	yes perfect [HI]
k-space	yes closed	yes perfect $[Ar_6]$
q-space	no perfect [Bo], [C], [Lu], $[Vau_1]$ (Example 3.11)	yes quasi-perfect
sequential	yes closed (quotient)	no perfect
Fréchet	yes closed (pseudo-open)	no perfect
Baire space	yes closed irreducible [Aal] no perfect [AaL]	yes closed irreducible [AaL]
perfect (closed set G_δ)	yes closed	no perfect
G_δ-diagonal	no perfect $[B_6]$, $[S_2]$, [Po]	no perfect
submetrizable	no perfect $[S_2]$, [PO], $[Vau_2]$	no perfect
realcompact	no perfect (by Mrówka-see $[Vau_2]$)	yes perfect
normal realcompact	yes perfect $[Fr_2]$	
regular	yes perfect [HI]	yes perfect
completely regular	no perfect [HI]	no perfect [HI]
normal	yes closed	no perfect [HI]
collectionwise normal	yes closed	no perfect [HI] (Example 6.3)
expandable	yes quasi-perfect [Kr]	yes quasi-perfect [Kr]
scattered	yes perfect [Te]	no perfect (yes, if fibers scattered [Te])
point-countable p-base	? perfect	no perfect
Property D (and regular)	yes finite-to-one perfect $[Vau_2]$	yes quasi-perfect $[Vau_2]$

Table III continued

Property	Image	Preimage
Property wD	no finite-to-one perfect [Vau_2]	yes quasi-perfect [Vau_2]

finite-to-one perfect maps. The remainder of the section is devoted to a few isolated topics and questions.

The following result was proved independently by D. Burke and G. Gruenhage [Gr]:

Theorem 7.1. If X is a noncompact space in which all countable subsets have compact closure then X contains a perfect preimage of ω_1.

Theorem 7.1 is sometimes useful in questions of countably compactness versus compactness. Gruenhage [Gr] uses 7.1 to show that any regular countably compact space with an orthobase, or a base of subinfinite rank, is compact. Theorem 7.1 can be used to show:

Theorem 7.2 (MA + $\neg$CH). If X is a regular L-space and bX is any compactification of X then the remainder of bX - X contains a perfect preimage of ω_1.

In contrast to the perfect preimages of ω_1 we mention that perfect images of ω_1 are homeomorphic to ω_1 [RSJ]. See [RSJ] for other results on the perfect images of ordinals.

E. Michael and M.E. Rudin [MRu] recently gave a new proof of a result by Benyamini, Rudin, and Wage that every continuous image of an Eberlein compact is an Eberlein compact. If $c_0(\Gamma)$ denotes the Banach space of real-valued functions, vanishing at infinity on the set Γ, and τ_p the topology of pointwise convergence on $c_0(\Gamma)$, it is known that a compact space X is an Eberlein compact if and only if X is homeomorphic to a subset of some $(c_0(\Gamma), \tau_p)$. (See [MRu] for a discussion of this characterization and other topological characterizations of Eberlein compacts.) Michael and Rudin ask:

Question 7.3. If X is a normal space which is homeomorphic to a subset of some $(c_0(\Gamma), \tau_p)$, must every image of X under a perfect map be homeomorphic to a subset of some $(c_0(\Gamma), \tau_p)$? (M.E. Rudin has shown the answer is "yes", for X a <u>closed</u> subset of $(c_0(\Gamma), \tau_p)$.)

We conclude the paper with several questions:

Question 7.4. Is the class MOBI closed under perfect maps? (See [Ar_2] and [Na_2].)

Question 7.5. If X has a point-countable p-base and f: X → Y is perfect must Y have a point-countable p-base?

Question 7.6. [Cr]. Are the closed images of semimetrizable spaces characterized as the Fréchet semistratifiable spaces?

Question 7.7. [O1]. Is there a paracompact space, of point-countable type, which does not admit a perfect map onto a 1st countable space? (Example 3.1 in [B_2] and Example 7.3 in [O1] give nonparacompact examples.)

Question 7.8. [MOS]. Characterize those spaces Y such that every closed f: X → Y is countably bi-quotient.

Question 7.9. [vDvM]. If X is supercompact and f: X → Y is continuous (hence perfect) is Y supercompact? Is a closed

G_δ-subspace of a supercompact space also supercompact? A continuous image of a supercompact space?

Question 7.10. Is the perfect image of a continuously perfectly normal space also continuously perfectly normal? (See [Z].)

Question 7.11. [TV]. Are elastic spaces preserved under closed mappings?

References

[AaL] J.M. Aarts and D.J. Lutzer, *Completeness properties designed for recognizing Baire spaces*, Dissertationes Math. 116 (1974).

[Ar_1] A.V. Arhangel'skiĭ, *On a class of spaces containing all metric and all locally bicompact spaces*, Soviet Math. Dokl. 4 (1963), 1051-1055.

[Ar_2] ________, *Mappings and spaces*, Russian Math. Surveys 21 (1966), 115-162.

[Ar_3] ________, *Perfect images and injections*, Soviet Math. Dokl. 8 (1967), 1217-1220.

[Ar_4] ________, *The closed image of a metric space can be condensed onto a metric space*, Soviet Math. Dokl. 7 (1966), 1109-1112.

[Ar_5] ________, *On hereditary properties*, Gen. Top. Appl. 3 (1973), 39-46.

[Ar_6] ________, *Bicompact sets and the topology of spaces*, Trudy Moskov. Mat. Obsc. 13 (1965) (= Trans. Moscow Math. Soc. 1965, 1-62).

[Ar_7] ________, *Existence criterion of a biocompact element in a continuous decomposition. A theorem on the invariance of weight for open-closed finitely multiple mappings*, Dokl. Akad. Nauk. SSSR 166 (1966), 1263-1266 (= Soviet Math. Dokl. 7 (1966), 249-253).

[AS] J.M. Atkins and F.G. Slaughter, Jr., *On the metrizability of preimages of metric spaces*, Proc. Oklahoma Top. Conf. 1972, 13-22.

[A_1] C.E. Aull, *A survey paper on some base axioms*, Top. Proc. 3 (1978), 1-36.

[A_2] C.E. Aull, *Quasi-developments and $\delta\theta$-bases,* J. London Math. Soc. 9 (1974), 197-204.

[Bo] C.J.R. Borges, *On stratifiable spaces,* Pac. J. Math. 17 (1966), 1-16.

[BoL] C. Borges and D. Lutzer, *Characterizations and mappings of M_i-spaces,* Top. Conf. VPI, Springer-Verlag Lecture Notes in Math. no. 375, 34-40.

[Bb] N. Bourbaki, *Topologie Générale,* Hermann, 1961.

[B_1] D. Burke, *On subparacompact spaces,* Proc. Amer. Math. Soc. 23 (1969), 655-663.

[B_2] ________, *Subparacompact spaces,* Proc. Washington State University Top. Conf. 1970, 39-49.

[B_3] ________, *Preservation of certain base axioms under a perfect mapping,* Top. Proc. 1 (1976), 269-279.

[B_4] ________, *Spaces with a primitive base and perfect mappings,* preprint.

[B_5] ________, *Orthocompactness and perfect mappings,* preprint.

[B_6] ________, *A nondevelopable locally compact Hausdorff space with a G_δ-diagonal,* Gen. Top. Appl. 2 (1972), 287-291.

[BvD] D. Burke and E.K. van Douwen, *On countably compact extensions of normal locally compact M-spaces,* Set-Theoretic Topology, Academic Press, 1977, 81-89.

[BM] D. Burke and E. Michael, *A note on a theorem by V.V. Filippov,* Israel J. Math., 11 (1972), 394-397.

[Ce] J. Ceder, *Some generalizations of metric spaces,* Pac. J. Math. 11 (1961), 105-126.

[CCN] J. Chaber, M.M. Čoban and K. Nagami, *On monotonic generalizations of Moore spaces, Čech complete spaces, and p-spaces,* Fund. Math. 84 (1974), 107-119.

[C] M.M. Čoban, *Perfect mappings and spaces of countable type,* Vestnik Moskov. Univ. Ser. I. Mat. Meh. 22 (1967), 87-93.

[Cr] G. Creede, *Concerning semi-stratifiable spaces,* Pac. J. Math. 22 (1970), 47-54.

[D] S. Davis, *A cushioning-type weak covering property,* Pac. J. Math. 80 (1979).

[vDLPR] E.K. van Douwen, D.J. Lutzer, J. Pelant, and G.M. Reed, *On unions of metrizable subspaces,* to appear.

[vDvM] E.K. van Douwen and J. van Mill, *Supercompact spaces*, Rapport 46, Wiskundig Seminar. der Vrije Univ., Amsterdam (1976).

[E] R. Engelking, *General Topology*, Monografie Mathematyczne vol. 60, Polish Scientific Publ., Warsaw 1977.

[F_1] V.V. Filippov, *On the perfect image of a paracompact p-space*, Soviet Math. Dokl. 8 (1967), 1151-1153.

[F_2] ________, *On feathered paracompacts*, Soviet Math. Dokl. 9 (1968), 161-164.

[F_3] ________, *Preservation of the order of a base under a perfect mapping*, Soviet Math. Dok. 9 (1968), 1005-1007.

[F_4] ________, *Quotient spaces and multiplicity of a base*, Mat. Sb. 80 (1967), 521-532 (= Math. USSR Sb. 9 (1969), 487-496).

[Ft] B. Fitzpatrick, *Some topologically complete spaces*, Gen. Top. Appl. 1 (1971), 101-103.

[FF] I. Fleischer and S.P. Franklin, *On compactness and projections*, Report 67-20, Dept. of Math. Carnegie Inst. of Tech., 1967.

[Fr_1] Z. Frolík, *On the topological product of paracompact spaces*, Bull. Acad. Polon. Sci. Ser. Sci. Math. Astr. Phys. 8 (1960), 747-750.

[Fr_2] ________, *On almost realcompact spaces*, Bull. Acad. Polon. Sci. Ser. Sci. Math. Astr. Phys. 9 (1961), 247-250.

[G_1] R. Gittings, *Concerning quasi-complete spaces*, Gen. Top. Appl. 6 (1976), 73-89.

[G_2] ________, *Open mapping theory*, Set-Theoretic Topology, Academic Press, 1977, 141-191.

[Gr] G. Gruenhage, *Some results on spaces having an ortho-base or a base of subinfinite rank*, Top. Proc. 2 (1977), 151-159.

[Ha] E. Halfar, *Compact mappings*, Proc. Amer. Math. Soc. 8 (1957), 828-830.

[Hr] P.W. Harley, III, *Metrization of closed images of metric spaces*, TOPO 72, Springer-Verlag Lecture Notes in Math. 379, 188-191.

[H] S. Hanai, *Inverse images of closed mappings I*, Proc. Japan Acad. 37 (1961), 298-301.

[HL] R.W. Heath and W.F. Lindgren, *Weakly Uniform Bases*, Houston J. Math. 2 (1976), 85-90.

[HI] M. Henriksen and J.R. Isbell, *Some properties of compactifications*, Duke Math. J. 25 (1958), 83-105.

[Ho] R.E. Hodel, *Spaces defined by sequences of open covers which guarantee that certain sequences have cluster points*, Duke Math. J. 39 (1972), 253-263.

[I_1] T. Ishii, *On closed mappings and M-spaces, I, II*, Proc. Japan Acad. 43 (1967), 752-761.

[I_2] ________, *wM-spaces and closed maps*, Proc. Japan Acad. 46 (1970), 16-21.

[Is_1] T. Isiwata, *Normality and perfect mappings*, Proc. Japan Acad. 39 (1963), 95-97.

[Is_2] ________, *Inverse images of developable spaces*, Bull. Tokyo Gakugei Univ., Ser. IV 23 (1971), 11-21.

[J] H.J.K. Junnila, *On submetacompactness*, Top. Proc. 3 (1978).

[Ka] A. Kato, *Solutions of Morita's problems concerning countably-compactifications*, Gen. Top. Appl. 7 (1977), 77-87.

[Kr] L.J. Krajewski, *On expanding locally finite collections*, Can. J. Math. 23 (1971), 58-68.

[K_1] T. Kramer, *The product of two topological spaces*, Gen. Top. Appl. 6 (1976), 1-16.

[K_2] ________, *A note on countably subparacompact spaces*, Pac. J. Math. 46 (1973), 209-213.

[L_1] N. Lašnev, *Continuous decompositions and closed mappings of metric spaces*, Soviet Math. Dokl. 6 (1965), 1504-1506.

[L_2] ________, *Closed images of metric spaces*, Dokl. Akad. Nauk SSSR 170 (1966), 505-507 (= Soviet Math. Dokl. 7 (1966), 1219-1221).

[Lu] D.J. Lutzer, *Semimetrizable and stratifiable spaces*, Gen. Top. Appl. 1 (1971), 43-48.

[Ma] V.J. Mancuso, *Mesocompactness and related properties*, Pac. J. Math. 33 (1970), 345-355.

[Mr] H. Martin, *Product maps and countable paracompactness*, Can. J. Math. 24 (1972), 1187-1190.

[Mi_1] E. Michael, *Another note on paracompact spaces*, Proc. Amer. Math. Soc. 8 (1957), 822-828.

[Mi_2] ________, *A note on closed maps and compact sets*, Israel J. Math. 2 (1964), 173-176.

[Mi_3] E. Michael, *On Nagami's Σ-spaces and some related matters*, Proc. Washington State Univ. Top. Conf. 1970, 13-19.

[Mi_4] ________, *A theorem on perfect maps*, Proc. Amer. Math. Soc. 28 (1971), 633-634.

[Mi_5] ________, *A quintuple quotient quest*, Gen. Top. Appl. 2 (1972), 91-138.

[Mi_6] ________, *Bi-quotient map and cartesian products of quotient maps*, Annales de L'Institute Fourier 18 (1968), 287-302.

[MOS] E. Michael, R.C. Olson, and F. Siwiec, *A-spaces and countably bi-quotient maps*, Dissertationes Math. 133 (1976), 1-43.

[MRu] E. Michael and M.E. Rudin, *A note on Eberlein compacts*, Pac. J. Math. 72 (1977), 487-495.

[M_1] K. Morita, *Products of normal spaces with metric spaces*, Math. Annalen 154 (1964), 365-382.

[M_2] ________, *Some properties of M-spaces*, Proc. Japan Acad. 43 (1967), 869-872.

[M_3] ________, *On closed mappings*, Proc. Japan Acad. 32 (1956), 539-543.

[M_4] ________, *On closed mappings and dimension*, Proc. Japan Acad. 32 (1956), 161-165.

[MH] K. Morita and S. Hanai, *Closed mappings and metric spaces*, Proc. Japan Acad. 32 (1956), 10-14.

[MR] K. Morita and T. Rishel, *Results related to closed images of M-spaces, I, II*, Proc. Japan Acad. 47 (1971), 1004-1011.

[Mr] S. Mrówka, *Compactness and product spaces*, Coll. Math. 7 (1959), 19-22.

[Na_1] K. Nagami, *Σ-spaces*, Fund. Math. 55 (1969), 169-192.

[Na_2] ________, *Minimal class generated by open compact and perfect mappings*, Fund. Math. 78 (1973), 227-264.

[N_1] J. Nagata, *A note on M-spaces and topologically complete spaces*, Proc. Japan. Acad. 45 (1969), 541-543.

[N_2] ________, *On closed mappings of generalized metric spaces*, Proc. Japan Acad. 47 (1971), 181-184.

[NC] S.I. Nedev and M.M. Coban, *On the theory of O-metrizable spaces II*, Vestnik Moskow Univ. Ser. I Mat. Meh. 27 (1972), 10-17.

[Ol] R.C. Olson, *Bi-quotient maps, countably bi-sequential spaces, and related problems*, Gen. Top. Appl. 4 (1974), 1-28.

[O_1] A. Okuyama, *σ-spaces and closed mappings I, II*, Proc. Japan Acad. 44 (1968), 472-481.

[O_2] ________, *On a generalization of Σ-spaces*, Pac. J. Math. 42 (1972), 485-495.

[Pa] C.M. Pareek, *Moore spaces, semi-metric spaces and continuous mappings connected with them*, Can. J. Math. 24 (1972), 1033-1042.

[P] H. Potoczny, *A non-paracompact space which admits a closure-preserving cover of compact sets*, Proc. Amer. Math. Soc. 32 (1972), 309-311.

[Po] V. Popov, *A perfect map need not preserve a G_δ-diagonal*, Gen. Top. Appl. 7 (1977), 31-33.

[RSJ] M. Rajagopalan, T. Soundararajan, and D. Jakel, *On perfect images of ordinals*, TOPO 72, Springer-Verlag Lecture Notes in Math. 378, 228-232.

[Sc] B. Scott, *Toward a product theory for orthocompactness*, Studies in Topology, Academic Press, Inc. 1975, 517-537.

[SMa] F. Siwiec and V. Mancuso, *Relations among certain mappings and conditions for their equivalence*, Gen. Top. Appl. 1 (1971), 33-41.

[SN] F. Siwiec and J. Nagata, *A note on nets and metrization*, Proc. Japan Acad. 44 (1968), 623-627.

[S_1] F. Slaughter, *The closed image of a metrizable space is M_1*, Proc. Amer. Math. Soc. 37 (1973), 309-314.

[S_2] ________, *A note on perfect images of spaces having a G_δ-diagonal*, preprint.

[vdS] J. van der Slot, *Some properties related to compactness*, Math. Center Tracts 19, Amsterdam, 1966.

[St] R. Stoltenberg, *A note on stratifiable spaces*, Proc. Amer. Math. Soc. 23 (1969), 294-297.

[Sn] A.H. Stone, *Metrizability of decomposition spaces*, Duke Math. J. 17 (1950), 317-327.

[Sk_1] S.A. Stricklen, Jr., *Closed mappings of nowhere locally compact metric spaces*, Proc. Amer. Math. Soc. 68 (1978), 369-374.

[Sk_2] ________, *An embedding theorem for Lasnev spaces*, Gen. Top. Appl. 6 (1976), 145-152.

[Su] J. Suzuki, *On pre-σ-spaces*, Bull. Tokyo Gakugei Univ. Ser. IV Math. and Nat. Sci. 28 (1976), 22-32.

[Ta] Y. Tanaka, *A characterization for the product of closed images of metric spaces to be a k-space*, Proc. Amer. Math. Soc. 74 (1979), 166-170.

[Te] R. Telgársky, *C-scattered and paracompact spaces*, Fund. Math. 73 (1971), 59-74.

[V] I.A. Vaĭnšteĭn, *On closed mappings of metric spaces*, Dokl. Akad. Nauk SSSR 57 (1947), 319-321 (Russian).

[Vau_1] J.E. Vaughan, *Spaces of countable and point-countable type*, Trans. Amer. Math. Soc. 151 (1970), 341-351.

[Vau_2] ________, *Discrete sequences of points*, Top. Proc. 3 (1978), 237-266.

[Vau_3] ________, *Perfect mappings and spaces of countable type*, Can. J. Math. 22 (1970), 1208-1210.

[Van] K.R. Van Doren, *Closed continuous images of complete metric spaces*, Fund. Math. 80 (1973), 47-50.

[Ve] N.V. Veličko, *On p-spaces and their continuous maps*, Math. USSR Sbornik 19 (1973), 35-46.

[Wh] G.T. Whyburn, *Directed families of sets and closedness of functions*, Proc. Nat. Acad. Sci. USA 54 (1965), 688-692.

[Wo_1] J.M. Worrell, Jr., *Upper semicontinuous decompositions of developable spaces*, Proc. Amer. Math. Soc. 16 (1965), 485-490.

[Wo_2] ________, *The closed images of metacompact topological spaces*, Portugal. Math. 25 (1966), 175-179.

[Wo_3] ________, *Upper semicontinuous decompositions of spaces having bases of countable order*, Portugal. Math. 26 (1967), 493-504.

[Wo_4] ________, *On continuous mappings of metacompact Cech complete spaces*, Pac. J. Math. 30 (1969), 555-562.

[Wo_5] ________, *A perfect mapping not preserving the p-space property*, preprint.

[WoW] J.M. Worrell, Jr. and H.H. Wicke, *Perfect mappings and certain interior images of M-spaces*, Trans. Amer. Math. Soc. 181 (1973), 23-35.

[Z_1] P. Zenor, *Some continuous separation axioms*, Fund. Math. 90 (1976), 143-158.

[Z_2] P. Zenor, *On countable paracompactness and normality*, Prace. Mat. 13 (1969), 23-32.

ULTRAFILTERS: AN INTERIM REPORT

W.W. Comfort[1]
Wesleyan University

The adjective "interim" serves to describe a phenomenon or a hiatus which falls between two events. The events I have in mind are, first, the publication in 1974 of [CN_3], in which the authors attempted a comprehensive and reasonably complete survey of the theory of ultrafilters, and, secondly, the appearance of a serious survey improving that work and doing justice to the many results which have appeared in the field since 1974. I do not know if any mathematicians are now seriously contemplating this latter project (I believe that the authors of [CN_3] are not) and in any event I am not sure the time is yet quite ripe for the project. When and if a new and definitive treatment of the theory of ultrafilters is attempted, however, I do believe that the authors (whoever they are) will find there is a wealth of new material not available in [CN_3]. Let us see what some of that may be.

Because the present article is at best a preliminary assessment, not a formal, polished, comprehensive survey, I shall for the most part content myself with statements of theorems, leaving to the interested reader the task of contacting the authors for the preprints cited here.

---

[1]The author gratefully acknowledges support received from the National Science Foundation under grant NSF-MCS78-00912.

ISBN: 0-12-584960-5

I am indebted to Eric K. van Douwen for careful, detailed responses to several questions (both mathematical and historical) I posed to him about the content of this paper, and for additional helpful conversations and correspondence; he assumes no responsibility, however, for my choice of topics.

In this paper, "space" means "completely regular, Hausdorff space"; the Stone-Čech compactification of a space X is denoted βX, and $\beta X \backslash X$ is usually written X^*. The symbols α, γ and κ serve to denote infinite cardinals and, simultaneously, the discrete spaces of the corresponding cardinalities. The set $U(\alpha)$ of uniform ultrafilters on α is that subspace of α^* defined by

$$U(\alpha) = \{p \in \beta(\alpha): |A| = \alpha \text{ for all } A \in p\}.$$

1. *Shelah's P-point Theorem.*

It is remarkable how frequently mathematicians agree. As a group we seem to suffer from few controversies. Ask any ten interested mathematicians what single result concerning ultrafilters that has appeared in the past few years seems to them most important or striking and I'll bet at least eight will cite Shelah's Theorem. To put the result in context let us recall first a famous result of Walter Rudin [R]: Assuming CH, there is a P-point in ω^*--that is, a point p which is interior to every G_δ in ω^* which contains p or, equivalently, a point p with the property that for every sequence $\{A_n: n < \omega\} \subset p$ there is $A \in p$ such that $|A \backslash A_n| < \omega$ for each $n < \omega$. Since ω^* is an infinite, compact space and every compact space all of whose points are P-points is easily seen to be finite, this result settled neatly, at least assuming CH, the question whether ω^* was homogeneous (and *a fortiori* the question whether ω^* could admit a binary operation making it into a topological group). It was natural to conjecture in the face of Rudin's result that perhaps the existence of P-points in ω^* was equivalent to CH; this possibility was not laid to rest until 1968 when David Booth showed [Bo_1], [Bo_2], [Bo_3] that from Martin's axiom it follows that there are P-points in ω^* (in fact, non-selective P-points). The version of his result Booth first announced achieved wide popular currency: the system ZFC is equiconsistent with the system [ZFC + $2^\omega = \aleph_2$ + there are P-points in ω^*]. With this result available the perspective changed and the people I knew, if they remained interested in the general question at all, directed their efforts to attempts to prove in ZFC without any additional axioms that there are P-points in ω^*.

With this background the theorem of Shelah is seen as quite striking: It is consistent with ZFC that there are no P-points in ω^*. In the method of Cohen one uses a specially devised partially ordered set of "forcing conditions" to build a model of ZFC with a particular set of desirable properties; Shelah's proof was an order of magnitude more complicated in that the partially ordered set he used to define a model of ZFC in which ω^* has no P-points was itself defined by a forcing technique of Cohen type. The complete argument has not yet found its way into a textbook but the expository articles of Wimmers [Wi] and Mills [M], which contain some simplifications of the original arguments of Shelah, are reasonably accessible.

2. *Incomparable ultrafilters*

In remainder spaces of the form α^* there are several useful partial orders. The Rudin-Keisler partial order on $\beta(\alpha)$, here denoted $\leq$, is defined as follows: for ultrafilters $p,q \in \beta(\alpha)$ we write $p \leq q$ if there is $f \in \alpha^\alpha$ such that the Stone extension $\overline{f}: \beta(\alpha) \to \beta(\alpha)$ satisfies $\overline{f}(q) = p$. It is clear, since $|\alpha^\alpha| = 2^\alpha$, that $\leq$-below any ultrafilter q there are at most 2^α ultrafilters p, i.e. that $|\{p \in \beta(\alpha): p \leq q\}| \leq 2^\alpha$ for all $q \in \beta(\alpha)$; and Negrepontis and I [CN_2] showed that the set $\beta(\alpha)$ is $(2^\alpha)^+$-directed in the sense that if $\{p_\xi: \xi < 2^\alpha\}$ is any subset of $\beta(\alpha)$ then there is $q \in \beta(\alpha)$ such that $p_\xi \leq q$ for each $\xi < 2^\alpha$. A less trivial statement, which goes back to 1972 and is due to Kunen [K_1], is this: the set $\beta(\alpha)$ is not linearly ordered by $\leq$--i.e. there are Rudin-Keisler incomparable ultrafilters. In fact, writing $\alpha^\alpha \times \alpha \times \alpha = \{d_\eta: \eta < 2^\alpha\}$ and defining for $\xi < 2^\alpha$ an increasing 2^α-sequence $\{F_{\xi,\eta}: \eta < 2^\alpha\}$ of filters such that for $d_\eta = \langle f,\xi,\zeta\rangle$ with $\xi \neq \zeta$ there is $B \in F_{\xi,\eta}$ with $f^{-1}(B) \notin F_{\xi,\eta}$, and setting $p_\xi = \cup_{\eta<2^\alpha} F_{\xi,\eta}$, Kunen can define a set $\{p_\xi: \xi < 2^\alpha\}$ of Rudin-Keisler pairwise incomparable ultrafilters (details of this proof may be found, for example, in [CN_3] Theorem 10.4)). The number 2^α is evidently the maximal number of $\leq$-incomparable ultrafilters that can be defined by this argument but by more delicate reasoning (still based, however, on the existence of a large independent family of subsets of α) Shelah showed the following result.

Theorem [SR]. There is a set S of pairwise $\leq$-incomparable elements of $U(\alpha)$ such that $|S| = (2^\alpha)^+$.

It had been shown earlier ([CN_3], Corollary 10.15) that the following statements are equivalent: (a) $(2^\alpha)^+ < 2^{2^\alpha}$; and (b) every set of 2^{2^α} ultrafilters on α contains a subset of cardinality 2^{2^α} whose elements are pairwise $\leq$-incomparable.

Putting these together, we have the following statement from [SR].

Theorem. For $\alpha \geq \omega$ there is a set of cardinality 2^{2^α} of pairwise $\leq$-incomparable elements of $U(\alpha)$.

We should note that there is another recent, accessible proof that the order $\leq$ is not linear on ω, this one due to van Mill [vM_4] and using Kunen's independent matrices. An element of ω^*, in the terminology of van Mill, is an R-point if there is an open F_σ-set U of ω^* such that p is in the boundary of U but p is not in the closure of any subset of U of cardinality $<2^\omega$. The proof of van Mill's theorem as given in [vM_4] breaks into two parts as follows: (a) There are 2^{2^ω} R-points in ω^*; and (b) for every R-point p the set of elements of ω^* not $\leq$-comparable to p has cardinality 2^{2^ω}.

Kunen [K_3] has also shown, in ZFC without any special set-theoretic assumptions, that ω^* contains what he calls weak P-points. These are by definition points p not in the closure of any countable subset of $\omega^*\backslash\{p\}$. The weak P-points have proved adequate to some of the tasks for which P-points were previously enrolled. Using them, for example, Victor Saks has responded (positively) to a question I asked with Charles Waiveris with the following theorem: For every non-compact, realcompact (Tychonoff) space X there is a set $\{P_\xi: \xi < 2^{2^\omega}\}$ of pairwise non-homeomorphic countably compact subspaces of βX every two of which have intersection equal to X. (Waiveris [W_1], [W_2], [W_3], [CW] had shown that spaces P_ξ exist with pairwise intersection equal to X, and with each P_ξ extra countably compact in βX in the sense that every infinite subset of βX has an accumulation point in each P_ξ; but we had been unable to show that the spaces P_ξ might be chosen pairwise non-homeomorphic without additional restrictive assumptions.)

3. Other Special Points in ω^*

The Rudin-Frolík order on ω^*, here denoted $\sqsubseteq$, is defined as follows: $p \sqsubseteq q$ if there is a (continuous) function $f: \omega \to \beta(\omega)$, with $f[\omega]$ discrete, whose Stone extension $\overline{f}$ satisfies $\overline{f}(p) = q$. The order $\sqsubseteq$ is stronger than $\leq$ in the sense that if $p \sqsubseteq q$ then $p \leq q$. It is clear that within ω^* the $\sqsubseteq$-minimal points are those points p not in the closure of any countable discrete subset of $\omega^* \backslash \{p\}$; each of Kunen's weak P-points is such a point.

(a) It is natural to ask whether every point p in the closure of a countable subset of $\omega^* \backslash \{p\}$ is in the closure of a countable discrete subset of $\omega^* \backslash \{p\}$. This question has been answered in ZFC in the negative by van Mill [vM_3], who showed that the Gleason space of the Cantor set 2^ω, which does contain a point not the limit of any countable discrete set, can be embedded in ω^* as a weak P-set (i.e. as a set S such that if A is countable set in ω^* and $A \cap S = \emptyset$, then $(\text{cl } A) \cap S = \emptyset$). In fact, van Mill shows that every continuous image of ω^* with the countable chain condition has a Gleason space which embeds into ω^* as a weak P-set.

As a consequence of the existence of these points, van Mill can give alternative examples (appropriate quotient spaces of ω^*) of phenomena described earlier by van Douwen and C.F. Mills: (1) There is a compact space which is not an F-space in which each countable subspace is C*-embedded; and

(2) there is a compact space in which each countable discrete subspace is C*-embedded but not every countable subspace is C*-embedded.

(b) In connection with (a) we point out the following: If X is a compact space without isolated points of weight 2^ω, then Frankiewicz [F] shows, assuming MA, that there is $p \in X$ such that p is not in the closure of any countable, discrete subset of $X\backslash\{p\}$, provided X is extremally disconnected; and van Mill [vM_1], assuming CH, shows that such p exists if X is an F-space.

(c) Of course Kunen's weak P-points in Shelah's model of ω^* are not P-points. More generally Kunen showed, assuming CH or Martin's axiom in [K_2] and without any such assumption in [K_3], that in (every model of) ω^* there are 2^{2^ω} weak P-points that are not P-points. Subsequently the result was generalized by van Mill [vM_5]: If X is an infinite, compact F-space of weight 2^ω with no isolated point in which each non-empty G_δ has non-empty interior, then there are 2^{2^ω} points of X which are weak P-points but not P-points. (It is well-known that ω^* satisfies the hypotheses required of X.) Whether the hypothesis $w(X) \leq 2^\omega$ is extraneous is unknown. Van Mill [vM_5] conjectures that it is, and he has derived the truth of his conjecture from the following statement:

(*) There is a compactification $B(\omega)$ of ω such that $B(\omega)\backslash\omega$ is a non-separable space with the countable chain condition.

Statement (*) follows from CH; van Mill has offered a prize for a proof (or disproof) of (*) in ZFC.

(d) It is a well-known result of Pospísil [Po], noted also by Juhász [J_1], [J_2] and Kunen [K_1], that there are points of $U(\alpha)$ whose local weight in $U(\alpha)$ is 2^α. The π-character of a point p in a space X, here denoted $\pi\chi(p)$, is the least cardinal number of a family $\mathcal{U}$ of non-empty open subsets of X (not necessarily neighborhoods of p) such that every neighborhood of p contains an element of $\mathcal{U}$. An argument of M.G. Bell [Be], using a family $\{A_\xi: \xi < 2^\omega\}$ of subsets of ω with $|\cap_{\xi\in F} A_\xi| = \omega$ for all finite $F \subset 2^\omega$ and with $|\cap_{\xi\in B} A_\xi| < \omega$ for all $B \subset 2^\omega$ such that $|B| = 2^\omega$, shows that there are 2^{2^ω} elements p of ω^* such that $\pi\chi(p) \geq cf(2^\omega)$; this result cannot be improved since Kunen has shown that it is consistent with ZFC that 2^ω is a singular cardinal and $\pi\chi(p) \leq cf(2^\omega)$ for all $p \in \omega^*$.

(e) A point p in X^* is called a remote point (of X!) if for every nowhere dense subset A of X one has $p \notin cl_{\beta X}A$. The concept was introduced, and the points in question shown to exist assuming the continuum hypothesis in a number of special cases, including for example, the case X = R, by Fine and Gillman [FG$_2$]. Unsettled for a number of years was the question whether remote points exist without CH.

Although there remain a few related open questions around the edges, it seems fair to say that this question has been satisfactorily solved. Chae and Smith [CS], for example, have shown that if X is any metric space such that $cl_X(X\backslash cl_X A)$ is not compact, where A is the set of isolated points of X, then X has remote points. By quite different methods, van Douwen [vD$_2$] showed that every non-pseudocompact space of countable π-weight has remote points. For non-pseudocompact spaces of (π-) weight $\aleph_2$ the conclusion can fail [vDvM$_1$], [vM$_2$]; the question whether $\omega \times \omega^*$ has a remote point is independent. It is unknown whether or not every nonpseudocompact space of

π-weight $\aleph_1$ has a remote point, but the paper [vM_2] of van Mill shows that many of them do, including those that are the product of at most ω^+ spaces of countable π-weight.

4. Subspaces of ω^*

(a) It is a familiar theorem of Efimov [E] that for $\alpha \geq \omega$ every extremally disconnected compact space X of weight $\leq 2^\alpha$ can be embedded into $\beta(\alpha)$. (Proof. Since X embeds into $[0,1]^{2^\alpha}$, and there is a continuous function f from $\beta(\alpha)$ onto $[0,1]^{2^\alpha}$, there is compact $K \subset \beta(\alpha)$ such that $f|K$ is an irreducible function from K onto X. It follows from the fact that X is extremally disconnected that $f|K$ is a homeomorphism.) Assuming CH, A. Louveau [L] has shown that the compact spaces which embed into ω^* are those which are totally disconnected F-spaces of weight $\leq 2^\omega$. And Woods [Wo] has shown, again assuming CH, that for a subspace A of a compact F-space X such that $|C(X)| = 2^\omega$ (e.g. for $X = \omega^*$), the following are equivalent: $|C^*(A)| = 2^\omega$; A is C*-embedded in X; A is weakly Lindelöf in the sense that every open cover of X has a countable subfamily whose union is dense in X. It follows, again assuming CH, that a space X is homeomorphic to a C*-embedded subspace of $\beta(\omega)$ if and only if X is an F-space such that $|C^*(X)| = 2^\omega$ and βX is zero-dimensional. Thus under CH the question whether a subspace of $\beta(\omega)$ is C*-embedded depends not on how the space sits inside $\beta(\omega)$ but on topological properties intrinsic to the subspace itself: If one of its homeomorphs into inside $\beta(\omega)$ is C*-embedded, they all are.

Let us note that the corresponding statement cannot be proved in ZFC without assuming CH. The following argument is due to van Douwen. There is in (every model in ZFC of) ω^* a discrete subspace of cardinality ω^+ that is not C*-embedded. Indeed it has been shown by Balcar and Simon [BS], and independently by Kunen and independently by Shelah, that there are discrete $D \subset \alpha^*$ with $|D| = 2^\alpha$ and $p \in \alpha^*$ such that $|D \backslash U| \leq \omega$ for every neighborhood U of p. (I am forced to wonder whether the author of the question in 2.4(b) of [C] was aware of the construction described in 4.11(c) of [C].)

(b) Fine and Gillman [FG_1] proved, assuming CH, that if $p \in \omega^*$ then $\beta(\omega^* \setminus \{p\}) \neq \omega^*$, but they left unsettled the question whether the same conclusion could be established in ZFC without additional set-theoretic assumptions. The answer is now known to be No: van Douwen [vD_4] has shown, using a result of Kunen concerning the non-existence of certain generalized gaps (in the sense of Hausdorff) in ω^*, that in the equi-consistent system [ZFC + Martin's axiom + ¬CH + there is no $(2^\omega, (2^\omega)^*)$-gap], the relation $\beta(\omega^* \backslash \{p\}) = \omega^*$ holds for every P_c-point of ω^*; in fact in this model, van Douwen shows with X the set of non-P_c-points of ω^* that $\beta X = \omega^*$.

(c) A discrete subset $\{p_\xi : \xi < \kappa\}$ of $U(\alpha)$ is said to be strongly discrete if there is a set $\{A_\xi : \xi < \kappa\}$ of subsets of α such that $A_\xi \in p_\xi$ for all $\xi < \alpha$ and $A_\xi \cap A_\zeta \neq \emptyset$ for $\xi < \zeta < \kappa$. We remark that for such sets $\{p_\xi : \xi < \kappa\}$ and $\{A_\xi : \xi < \kappa\}$, the family

$$\{(cl_{\beta(\alpha)} A_\xi) \cap U(\alpha) : \xi < \kappa\}$$

is a faithfully indexed family of non-empty pairwise disjoint open subsets of $U(\alpha)$. We note further that such a subset $\{p_\xi : \xi < \kappa\}$ of $U(\alpha)$ is C*-embedded in $U(\alpha)$. Indeed if f is a bounded function on $\{p_\xi : \xi < \kappa\}$ to R and if $\{A_\xi : \xi < \kappa\}$ is the given disjoint family of subsets of α (with $p_\xi \in cl_{\beta(\alpha)} U_\xi$ for $\xi < \kappa$), one defines g to have the constant value $f(p_\xi)$ on

A_ξ and one defines g arbitrarily elsewhere on α subject only to the condition that g remain bounded on α; then the Stone extension $\overline{g}$ of g satisfies $\overline{g}(p_\xi) = f(p_\xi)$ for $\xi < \kappa$.

It is not difficult to see that every (countable) discrete subset of ω^* is strongly discrete; it is known (cf. [GJ] (Problem 14N.5)) that in ω^*--indeed in every F-space--every countable subspace is C*-embedded. With this and the definitions above it is now natural to ask the following question, which to the best of my knowledge was first posed by Frolík in Prague in 1966: If A is a discrete subset of $U(\alpha)$ such that $|A| = \alpha$ and A is C*-embedded, must A be strongly discrete? Let us show, following an argument suggested by van Douwen (see also [CN_2] (Corollary 12.4) and [CH] (Theorem 6.2)), that the answer to this question is No for every cardinal $\alpha > \omega$.

Theorem (van Douwen). Let $\alpha \geq \kappa > \omega$. There is a discrete, C*-embedded subspace D of $U(\alpha)$ such that $|D| = \kappa$ and D is not strongly discrete.

Proof. Let f be a continuous function from $U(\alpha)$ onto $\{0,1\}^{2^\alpha}$ (see for example 6.2 of [CH] for the existence of such a function f) and let K be a compact subspace of $U(\alpha)$ such that $f|K$ is an irreducible function from K onto $\{0,1\}^{2^\alpha}$; since $\{0,1\}^{2^\alpha}$ satisfies the countable chain condition, the space K does also.

There is in $\{0,1\}^{2^\alpha}$ a copy of $\beta(\kappa)$. Let D be a subset of K such that $f[D] = \kappa$ and $f|D$ is a one-to-one function, and let $B(D)$ be the compactification of D given by $B(D) = cl_K D$. It is clear that the (discrete) space D is C*-embedded in $B(D)$ (and hence in $U(\alpha)$): For $g \in C^*(D)$ define $h \in C^*(\kappa)$ by

$h(f(p)) = g(p)$ and define $\tilde{g}$ on $B(D)$ by $\tilde{g}(q) = \overline{h}(f(q))$ (with $\overline{h}$ the Stone extension of h); then $\tilde{g}$ is a continuous, real-valued extension of g. It remains to show that D is not strongly discrete in $U(\alpha)$.

If D is strongly discrete then according to the remark above there is a family $\{U_\xi : \xi < \kappa\}$ of non-empty, pairwise disjoint, open subsets of $U(\alpha)$. Then $\{U_\xi \cap K : \xi < \kappa\}$ is a faithfully indexed family of non-empty, open subsets of K; this contradicts the conditions that $\kappa > \omega$ and K satisfies the countable chain condition. The proof is complete.

We note that for $\kappa = \alpha$ the discrete set D defined in the preceding proof has the property that none of its uncountable subsets are strongly discrete. Indeed if $\{p_\eta : \eta < \omega^+\}$ is a faithfully indexed subset of D then there is no family $\{A_\eta : \eta < \omega^+\}$ of subsets of α such that

$A_\eta \in p_\eta$ for $\eta < \omega^+$, and

$|A_\eta \cap A_{\eta'}| < \alpha$ for $\eta < \eta' < \omega^+$;

for from such a family, if it were to exist, one could again define the family $\{(cl_{\beta(\alpha)} A_\eta) \cap K : \eta < \omega^+\}$ and thus contradict the fact that K satisfies the countable chain condition.

(d) I cited in (a) above the result of Louveau [L] to the effect that (assuming CH) every compact zero-dimensional F-space X of weight $\leq 2^\omega$ embeds into ω^*. In fact, according to Frankiewicz and Mills [FM], X can be embedded as a P_c-set--i.e. as a set with the property that every intersection of $<2^\omega$ open subsets of ω^* containing X contains an open set containing X. (The idea of the proof of Frankiewicz and Mills is to choose a continuous function f from $(2^\omega)^*$ onto X, and to show that the identification space $(2^\omega)^* \cup_f X$ satisfies

Parovičenko's [Pa] characterization (assuming CH) of ω^*: It is a compact, zero-dimensional F-space Y without isolated points, such that $|C(Y)| = \omega^+$, in which every non-empty zero-set has non-empty interior.)

The conclusion in the theorem of Frankiewicz and Mills, and even in that of Louveau, cannot be achieved in ZFC. Van Douwen and van Mill [$vDvM_2$] have shown, assuming Martin's axiom and $2^\omega = \aleph_2$, that there is a compact F-space of weight 2^ω that cannot be embedded into any basically disconnected space.

Van Douwen and van Mill [$vDvM_3$] have also shown that the validity of Parovičenko's characterization of ω^* is equivalent to CH.

While citing recent developments concerning $\beta(\omega)$ and its subsets, let us list three conclusions derived by van Douwen from the hypothesis that ω^* has a P_c-point: (a) For all $p \in \omega^*$, the local weight of ω^* at p is 2^ω; (b) ω^* has 2^c P_c-points; and (c) every sequence of first-countable, compact spaces has a paracompact box-product. The first of these conclusions seems particularly surprising, since it shows that the nature of some one ultrafilter can influence the behavior of every ultrafilter. Van Douwen's method of proof is to find a consequence of the existence of a P_c-point which is known to imply all three of his conclusions; this consequence is the fact that in ω^ω, partially ordered in the usual way, no subset of cardinality $<2^\omega$ is cofinal.

For a short proof of Parovičenko's theorem that every compact space of weight $\leq\omega^+$ is the continuous image of ω^*, and for a topological characterization of ω^* similar to that of

Negrepontis [N] but simpler, see Błaszczyk and Szymański [BSz].

5. γ-points in U(α)

A point p of a space X is said to be a γ-point of X if there are γ pairwise disjoint, open subsets of X with p in the boundary of each. A family of subsets of α is said to be almost disjoint if each of the sets in the family has cardinality α, and the intersection of every two of the sets in the family has cardinality <α. It is known that there is on ω an almost disjoint family of subsets of cardinality 2^{ω}, and that there is on (arbitrary) $\alpha \geq \omega$ an almost disjoint family of subsets of cardinality α^+ (see for example [CN_3], 12.2 and 12.20, respectively, for proofs of these statements). A routine compactness argument then shows that there is in ω* a 2^{ω}-point, and that if $\alpha^+ = 2^{\alpha}$ then there is in U(α) a 2^{α}-point.

It is tempting to conjecture the strong, sweeping statement that, for every $\alpha \geq \omega$, every point of U(α) is a 2^{α}-point of U(α). Unfortunately, Baumgartner [Bm] has described a model of set theory in which there is on ω^+ no almost disjoint family of sets of cardinality $2^{(\omega^+)}$; it is easy to see that in this model no point of $U(\omega^+)$ is a $2^{(\omega^+)}$-point, so the conjecture must be drastically revised.

Prikry [Pr_2] has shown that if α is regular and $\alpha^+ = 2^{\alpha}$, then every point of U(α) is a 2^{α}-point; it is not known if the same conclusion holds for singular α, but Taylor [T] has shown that it is consistent with ZFC that every point of $U(\aleph_\omega)$ is a $2^{(\aleph_\omega)}$-point.

While Prikry's work [Pr_1], [Pr_2] and the results of [CH] leave several loose ends (see below), it is nice to be able to record the definitive success achieved lately by Balcar and Vojtáš [BV]: Every point of ω^* is a 2^ω-point. (The weaker conclusion that every point of ω^* is a ω-point, as well as a number of related results dealing with larger cardinals, had been given earlier by Szymański [Sz].)

Among the questions left unsolved in [CH], [He] and [BV] are these two: (a) If there is on α an almost disjoint family of sets of cardinality γ, must every element of $U(\alpha)$ be a γ-point? (b) Is every nowhere dense subset of ω^* a 2^ω-set?

6. *Products reduced by an ultrafilter*

The survey [CN_3] devoted substantial space to the structural, the topological, and the model-theoretic aspects of the theory of ultrafilters. I hope that in the preceding five sections of this informal paper I have indicated convincingly that there is plenty of action these days in the first two of these three aspects of the theory. As the result either of ignorance and incompetence on my part, or of reduced activity on the part of the wide mathematical community, or a combination of the two, I am much less aware of recent model-theoretic advances in topology in which ultrafilters have played a substantial role. The work of Bankston [Ba_1], [Ba_2], is relevant in this connection and potentially interesting to topologists; he notes himself ([Ba_2], page 261) that the formulation of the notion of reduced (ultra-) product in the context of general topology gives rise to "a vast untapped source of research problems, many of the type already encountered in the theory of box products."

It will follow from the definition in the next paragraph that every topological product reduced by a countably incomplete ultrafilter is a P-space. Accordingly we need no longer retain our standing hypothesis that the topological spaces we consider are completely regular, Hausdorff spaces; in what follows we do not assume any separation properties.

Let $\{X_\xi: \xi < \alpha\}$ be an α-sequence of spaces, and $p \in U(\alpha)$. The topological product of $\{X_\xi: \xi < \alpha\}$ reduced by p, here denoted $(\Pi_\xi X_\xi)/p$, is (by definition) the set $(\Pi_\xi X_\xi)/\sim$ with $\sim$ defined by $f \sim g$ if $\{\xi < \alpha: f(\xi) = g(\xi)\} \in p$; the topology is the coarsest in which each $(\Pi_\xi U_\xi)/p$, with U_ξ open in X_ξ, is open.

Among the results of [Ba_1] are these: (a) Every regular space has a paracompact ultrapower (see below); (b) ultraproducts of normal spaces need not be normal; and (c) every two regular perfect spaces have homeomorphic ultrapowers.

Now let us with Bankston say that a space X is α-open if every intersection of $<\alpha$ open subsets of X is open; and $(X)_\alpha$ is the set X with the topology whose open sets are unions of sets which are intersections of $<\alpha$ open subsets of X. If $\{X_\xi: \xi < \alpha\}$ is a set of regular spaces each with weight $\leq 2^\alpha$, and if the ultrafilter p is regular on α (i.e. if there is A p such that $|A| = \alpha$ and such that $\cap F = \emptyset$ for all $F \subset A$ with $|F| = \omega$), then $(\Pi_\xi X_\xi)/p$ is α^+-open, regular, and of weight $\leq 2^\alpha$; hence if in addition $\alpha^+ = 2^\alpha$, then $(\Pi_\xi X_\xi)/p$ is paracompact. Conversely we may note, using what is essentially an argument of A.H. Stone [St] as generalized by Borges [Br], that for regular, uncountable cardinals κ the space $(2^{(\kappa^+)})_\kappa$

is not normal. (More recently the same conclusion has been achieved for arbitrary $\kappa > \omega$; see [vD_1], [vD_5] for details and related results.) It then follows, since for every compact Hausdorff space X the space $(X)_{\kappa^+}$ embeds as a closed subspace of the ultrapower X^{κ}/p, that if $\alpha^+ < 2^{\alpha}$ then $(\Pi_{\xi}X_{\xi})/p$ with each $X = 2^{(\alpha^{++})}$ is a non-normal reduced product of an α-sequence of spaces each with weight $\leq 2^{\alpha}$. Thus we have the following result (from [Ba_2]): For $\alpha \geq \omega$ the condition $\alpha^+ = 2^{\alpha}$ is equivalent to each of the following $(3 \times 2 \times 2) = 12$ conditions: For every α-sequence $\{X_{\xi}: \xi < \alpha\}$ of regular/normal/compact Hausdorff spaces, and for every/some regular ultrafilter p on α, the reduced product $(\Pi_{\xi}X_{\xi})/p$ is paracompact/normal.

7. *Concluding remarks*

Among the many recent striking advances in the theory of ultrafilters and their applications which we have ignored in this survey are those relating to the algebraic structure of the set ω of non-negative integers and, more generally, of discrete semi-groups.

[The principal agents of progress in this connection have been Hindman and van Douwen. Let us cite just two results by each of these mathematicians. Here the functions + and $\cdot$ denote the natural extensions over $\beta(\omega)$ of the usual addition and multiplication in ω. Specifically, we write

$$A - n = \{k < \omega: k + n \in A\} \text{ for } A \subset \omega,\ n < \omega, \text{ and}$$

$$p + q = \{A \subset \omega: \{n < \omega: A - n \in p\} \in q\} \text{ for } p,q \in \beta(\omega);$$

and $p \cdot q$ is defined analogously. It is not difficult to see that the functions + and $\cdot$ are right-continuous on $\beta(\omega)$, in the sense that for $p \in \beta(\omega)$ the functions $q \to p + q$,

$q \to p \cdot q$ are continuous.

Theorem (Hindman [Hi]). If $p, q \in \omega^*$ then $p + q \neq p \cdot q$.

Corollary (Hindman [Hi]). + and $\cdot$ have no simultaneous idempotent in ω^*.

Now let + denote restriction to $\omega^* \times \omega^*$ of the function + defined above.

Theorem (van Douwen [vD_3]). $\{p \in \omega^*$: + is left-continuous at $\langle p,q \rangle$ for no $q \in \omega^*\}$ is dense in ω^*.

Theorem (van Douwen [vD_3]). The following are equivalent.

(a) There is a P-point in ω^*;

(b) $\{p \in \omega^*$: + is continuous at $\langle p,q \rangle$ for all $q \in \omega^*\}$ is dense in ω^*; and

(c) $\{p \in \omega^*$: there is $q \in \omega^*$ with + left-continuous at $\langle p,q \rangle\}$ is dense in ω^*.

Despite these various omissions it is clear that the theory of ultrafilters continues to florish. Especially in connection with the spaces $\beta(\omega)$ and ω^*, we are approaching the time when it will be appropriate for a serious, comprehensive study designed to codify the theory and bring it up to date; I hope that the present paper and its bibliography will be at least marginally useful to the architect of that project.

References

[BS] Balcar, Bohuslav and Petr Simon, *Convergent nets in the spaces of uniform ultrafilters*, preprint.

[BV] Balcar, Bohuslav and Peter Vojtáš, *Almost disjoint refinement of families of subsets of N*, preprint.

[Ba_1] Bankston, Paul, *Ultraproducts in topology*, Gen. Top. and Its Appl. 7 (1977), 283-308.

[Ba_2] Bankston, Paul, *Topological reduced products and the GCH*, Topology Proceedings 1 (1976), 261-267.

[Bm] Baumgartner, James E., *Results and independence proofs in combinatorial set theory*, Doctoral Dissertation, University of California (Berkeley), 1970.

[Be] Bell, Murray G., *Points in* $\beta\omega$-ω *of* π-*character c*, preprint.

[BSz] Błaszczyk, A. and A. Szymański, *Concerning Parovičenko's theorem*, preprint.

[Bo_1] Booth, D.D., *Some consistency questions in topology*, Notices Amer. Math. Soc. 15 (1968), 641 [Abstract].

[Bo_2] ________, *Countably indexed ultrafilters*, Doctoral Dissertation, University of Wisconsin, 1969.

[Bo_3] ________, *Ultrafilters on a countable set*, Annals of Math. Logic 2 (1970), 1-24.

[Br] Borges, C.J.R., *On a counterexample of A.H. Stone*, Quarterly J. Math. Oxford 20 (1969), 91-95.

[CS] Chae, Soo Bong and Jeffrey H. Smith, *Remote points and G-spaces*, preprint.

[C] Comfort, W.W., *Compactifications: recent results from several countries*, Topology Proceedings 2 (1977), 61-87.

[CH] Comfort, W.W. and Neil Hindman, *Refining families for ultrafilters*, Math. Zeitschrift 149 (1976), 189-199.

[CN_1] Comfort, W.W. and S. Negrepontis, *Homeomorphs of three subspaces of* $\beta N \backslash N$, Math. Zeitschrift 107 (1969), 53-58.

[CN_2] ________, *On families of large oscillation*, Fundamenta Math. 75 (1972), 275-290.

[CN_3] ________, *The Theory of Ultrafilters*, Grundlehren der math. Wissenschaften vol 211. Springer-Verlag. Berlin-Heidelberg-New York, 1974.

[CW] Comfort, W.W. and Charles Waiveris, *Intersections of countably compact subspaces of Stone-Čech compactifications*, preprint.

[vD_1] van Douwen, Eric K., *Another non-normal box product*, Gen. Top. and Its Appl. 7 (1977), 71-76.

[vD_2] ________, *Remote points*, Dissertationes Math., to appear.

[vD_3] van Douwen, E.K., *The Čech-Stone compactification of a discrete cancellative groupoid*, preprint.

[vD_4] ________, *There can be* $p \in \omega^*$ *with* $\beta(\omega^* \backslash \{p\}) = \omega^*$, preprint.

[vD_5] ________, *Covering and separation properties of box products*, in this volume.

[$vDvM_1$] van Douwen, E.K. and J. van Mill, *Spaces without remote points*, Pac. J. Math., to appear.

[$vDvM_2$] ________, *Subspaces of basically disconnected spaces, or quotients of countably complete Boolean algebras*, Trans. Amer. Math. Soc., to appear.

[$vDvM_3$] ________, *Parovičenko's characterization of* $\beta\omega$-ω *implies CH*, Proc. Amer. Math. Soc. 72 (1978), 538-541.

[E] Efimov, B.A., *Extremally disconnected compact spaces and absolutes*, Trans. Moscow Math. Soc. 23 (1970), 243-285.

[FG_1] Fine, N.J. and L. Gillman, *Extension of continuous functions on* βN, Bull. Amer. Math. Soc. 66 (1960), 376-381.

[FG_2] ________, *Remote points in* βR, Proc. Amer. Math. Soc. 13 (1962), 29-36.

[F] Frankiewicz, Ryszard, *Non-accessible points in extremally disconnected compact spaces*, Fundamenta Math., to appear.

[FM] Frankiewicz, Ryszard and Charles F. Mills, *More on nowhere dense closed P-sets*, Rapport 85, Wiskundig Seminarium, Free University of Amsterdam.

[GJ] Gillman, L. and Meyer Jerison, *Rings of Continuous Functions*, D. van Nostrand Co., Inc., Princeton-Toronto-London-New York, 1960.

[He] Hechler, S.H., *Generalizations of almost disjointness, c-sets, and the Baire number of* βN-N, Gen. Top. and Its Appl. 8 (1978), 93-110.

[Hi] Hindman, Neil, *Ultrafilters and combinatorial number theory*, Proc. March, 1979 Southern Illinois University Number Theory Conference, to appear.

[J_1] Juhász, I. *Remarks on a theorem of B. Pospíšil*, Commentationes Math. Univ. Carolinae 8 (1967), 231-247.

[J_2] ________, *On the character of points in* βN_m, In: *Contributions to Extension Theory of Topological Structures*, Proc. 1967 Berlin Symposium, edited by J. Flachsmeyer et al., pp. 139-140. VEB Deutscher Verlag der Wissenschaften, Berlin, 1969.

[K_1] Kunen, Kenneth, *Ultrafilters and independent sets*, Trans. Amer. Math. Soc. 172 (1972), 299-306.

[K_2] ________, *Some points in* βN, Math. Proc. Cambridge Philosophical Soc. 80 (1976), 385-3981

[K_3] ________, *Weak P-points in* N*, preprint.

[L] Louveau, Alain, *Caractérisation des sous-espaces compacts de* βN, Bull. Sci. Math. 97 (1973), 259-263.

[vM_1] van Mill, Jan, *A simple observation concerning the existence of non-limit points in small compact F-spaces*, Topological Structures 2, Mathematisch Centrum, Amsterdam, to appear.

[vM_2] ________, *More on remote points*, Rapport 91, Wiskundig Seminarium, Free University of Amsterdam.

[vM_3] ________, *Another RF-minimal point in* βω-ω, Rapport 104, Wiskundig Seminarium, Free University of Amsterdam.

[vM_4] ________, *A remark on the Rudin-Keisler order of ultrafilters*, Rapport 106, Wiskundig Seminarium, Free University of Amsterdam.

[vM_5] ________, *Weak P-points in compact F-spaces*, Preprint.

[M] Mills, Charles F., *An easier proof of the Shelah P-point independence theorem*, Rapport 78, Wiskundig Seminarium, Free University of Amsterdam.

[N] Negrepontis, S., *The Stone space of the saturated Boolean algebras*, Trans. Amer. Math. Soc. 141 (1969), 515-527.

[Pa] Parovičenko, I.I., *On a universal bicompactum of weight* ℵ, Doklady Akad. Nauk. SSSR 150 (1963), 36-39. [Russian, Eng. Trans.: Sov. Math. Dokl. 4 (1963), 592-95]

[Po] Pospíšil, Bedřich, *On bicompact spaces*, Publ. Fac. Sci. Univ. Masaryk 270 (1939), 3-16.

[Pr_1] Prikry, Karel, *Ultrafilters and almost disjoint sets*, General Topology and its Applications 4 (1974), 269-282.

[Pr_2] ________, *Ultrafilters and almost disjoint sets, II*, Bull. Amer. Math. Soc. 81 (1975), 209-212.

[R] Rudin, Walter, *Homogeneity problems in the theory of Čech compactifications*, Duke Math. J. 23 (1956), 409-419; 633.

[SR] Shelah, S. and M.E. Rudin, *Unordered types of ultrafilters*, Top. Proceedings 3 (1978), 199-204.

[St] Stone, A.H., *Paracompactness and product spaces*, Bull. Amer. Math. Soc. 54 (1948), 977-982.

[Sz] Szymański, Andrej, *On the existence of $\aleph_0$-points*, Proc. Amer. Math. Soc. 66 (1977), 128-130.

[T] Taylor, Alan D., *Regularity properties of ideals and ultrafilters*, Annals of Math. Logic 16 (1979), 33-55.

[W_1] Waiveris, Charles, *Intersections of countably compact spaces*, Notices Amer. Math. Soc. 26 (1979), A-230 [Abstract].

[W_2] ________, *Intersections of countably compact spaces*, Doctoral Dissertation, Wesleyan University, Connecticut, 1979.

[W_3] ________, *Intersections of countably compact subspaces of* βX, preprint

[Wi] Wimmers, E., *The Shelah P-point independence theorem*, preprint.

[Wo] Woods, Grant, *Characterizations of some C*-embedded subspaces of* βN, Pacific J. Math. 65 (1976), 573-379.

Added in proof by Eric K. van Douwen (October 1, 1979): It has been reported that Murray Bell has proved that (*), mentioned in 3(c), holds in ZFC.

COVERING AND SEPARATION PROPERTIES OF BOX PRODUCTS

Eric K. van Douwen
Institute for Medicine and Mathematics
Ohio University

ABSTRACT: We give a fairly complete survey, including proofs, of what is known about the question of when a box product of compact spaces is paracompact, and show how badly a box product of compact or metrizable spaces can fail to be normal. A side result is that the Tychonoff product of uncountably many infinite discrete spaces is not countably orthocompact.

1980 Math. Subj. Class.: 54B10, 54D15, 54D18, 54A35; 54E35, 54E65

1. INTRODUCTION

The simple minded way to generalize the formation of finite[1] products to infinite products leads to the box (= □) product rather than the Tychonoff product. The lower separation properties, up to and including complete regularity, are preserved by □-products, [Kn, 2.1,3.3], see Fact 6.3. Things become interesting, as so often, if we consider normality and paracompactness. Since a product of only two paracompact spaces can already badly fail to be normal, [So], we will consider only □-products of factors which are at least

[1]By common abuse of language, a product is called *finite* if it has finitely many factors.

ISBN: 0-12-584960-5

metrizable or compact[2], so that all finite subproducts are paracompact.

The first result in this area is Rudin's theorem that under *CH* a countable $\square$-product of σ-compact[3] locally compact metrizable spaces is paracompact, [Ru_1]. Later I showed that a countable $\square$-product of separable metrizable spaces need not be normal, [vD_1,§2], see §§7,12, and Kunen showed, in an early version of [Ku_1] that a countable $\square$-product of compact spaces need not be normal, see §§8,13. This shows that in Rudin's theorem some restriction on the factors is essential.

Rudin's theorem has been improved, see §10, but the improvements are incomplete: we know that *CH* can be weakened, and that the factors don't have to be metrizable, but it is still unknown if a countable $\square$-product is normal (or paracompact) in *ZFC* if all factors are $\omega+1$, or compact metrizable, or compact first countable, or compact and of weight at most ω_1. Also, it is unknown if an uncountable $\square$-product of infinite compact spaces can be normal or paracompact, under any additional axioms. There is some nontrivial information about $\square$-products which does not require additional axioms nor the restriction to countably many factors, see Theorem 11.1.

There are three main techniques for dealing with $\square$-products:

1. FINITE (!) PRODUCTS: The trick is factor a $\square$-product as a product of two sub-$\square$-products $\square_0$ and $\square_1$. This is done in three cases:

---

[2]We will completely ignore countable $\square$-product of ordinals (whether compact or not), except $\omega+1$, and refer to [ER], [Ku_2,§5], [Ru_2], [W_2].

[3]This condition is inessential, [vD_1,§1], see the proof of Proposition 9.8.

(a) countable □-products of metrizable spaces.

In this case $\square_0$ has one factor, and it is important that $\square_1$, when given the Tychonoff topology, is metrizable, [vD_1,§2], see §§7,12.

(b) countable □-products of compact spaces

In this case we treat the countable □-product by letting $\square_0$ be bigger and bigger finite—hence compact—subproducts, see §9.

(c) uncountable □-products

This case occurs in the proof that a □-product is pseudo-normal iff each countable sub-□-product is, see §11.

2. ∇-PRODUCTS. Each □-product has a certain quotient, called the nabla (=∇) product, which is not a real product. Its main features, due to Kunen, [Ku_2, §2,3.1], are

(a) a countable ∇-product is easier to work with than a countable □-product, and

(b) a countable □-product of compact spaces is paracompact iff the associated ∇-product is.

∇-products were used implicitly in the proof of Rudin's theorem, mentioned above. The proof is admittedly messy, [Ru_3, p. 56], precisely because they were not used explicitly.

[In §15 an idea, implicit in a proof of Roitman, [Ro_2, Thm. 5], will be used to show that (a) holds without "countable". Indeed, we show that under certain additional axioms uncountable ∇-products are paracompact. However, it will also be shown that this is useless since (b) is (consistently) false without "countable".]

3. G_δ-MODIFICATIONS AND POWERS OF 2: Kunen, in an early version of [Ku_2], pointed out that a countable □-product of

factors X has a closed subspace homeomorphic to the G_δ-modification of X, and used this if X is a Tychonoff product of factors 2. We will use this idea, and an obvious generalization, see Proposition 6.4, in §§8,13,15; it is used in all our negative results about □-products or ∇-products of compact spaces.

This survey is fairly complete, and contains results I announced in [vD$_3$], [vD$_4$], or obtained later. The proofs given here often differ in their outward appearance of the original proofs, yet are essentially the same. We use ¤ to denote begin and end of a proof, or the absence of a proof. We put short historical comments between ⟦ and ⟧ at the end of a proof. There are extensive cross references; and we have indicated in the references where literature, except for standard references, has been cited.

2. CONVENTIONS

An ordinal is the set of smaller ordinals, and a cardinal is an initial ordinal. We use the common conventions that ω denotes ω_0 and $\underline{c}$ denotes 2^ω, and use the classical convention that $\underline{f}$ denotes $2^{\underline{c}}$, the number of functions $\mathbb{R} \to \mathbb{R}$. κ, λ, ν denote *infinite* cardinals.

WE USUALLY SUPPRESS INDEX SETS: we either use k,n as indices, in which case the index set ω is tacitly understood, or use α,β,ξ,η as indices, in which case the index set ν[4] is tacitly understood. Other index sets will be used when necessary.

---

[4]ν = *nu* = *nu*mber of factors.

Cardinals get the discrete topology, and $\omega+1$ gets the order topology.

The set of functions $A \to B$ is denoted by ${}^{A}B$. For $X \subseteq A$, or $Y \subseteq B$ we use $f^{\to}X$, or $f^{\leftarrow}Y$, to denote the image, or inverse image, of X or Y, under f.

Unless otherwise indicated all spaces are assumed to be regular T_1.

3. VARIOUS SORTS OF PRODUCTS, THE G_ν-MODIFICATION

Given spaces X_α, a κ-*box* in the (set theoretic) product $\Pi_\alpha X_\alpha$ is a set of the form

$\Pi_\alpha U_\alpha$, with each U_α open in X_α, and $U_\alpha \neq X_\alpha$ for less than κ indices α.

The topology on $\Pi_\alpha X_\alpha$ which has the κ-boxes as a base is called the $<\kappa$-*box* topology, and $\Pi_\alpha X_\alpha$, when carrying this topology, is called the $<\kappa$-*box*-product of the X_α's, and is denoted by $<\kappa\text{-}\square_\alpha X_\alpha$. If $\kappa = \omega$ this simply yields the ordinary, or *Tychonoff* topology (and product); we will *not* use $\Pi_\alpha X_\alpha$ to denote $<\omega\text{-}\square_\alpha X_\alpha$: $\Pi_\alpha X_\alpha$ is used only to denote the set theoretic product, without topology. If $\kappa > \nu$ this yields the *box* topology (and product); in this case we omit the prefix $<\kappa$. We use $<\kappa\text{-}\square$-products only as tool to get information about $\square$-products and ∇-products.

We define an equivalence relation E on $\square_\alpha X_\alpha$ by

xEy if $x_\alpha \neq y_\alpha$ for less than ν indices α,

and denote the quotient space $\square_\alpha X_\alpha/E$ by $\nabla_\alpha X_\alpha$, the *nabla* $(=\nabla)$ product of the X_α's. Although $\nabla_\alpha X_\alpha$ is not a real product we will refer to the X_α's as the *factors* of $\nabla_\alpha X_\alpha$. Throughout this paper we will use q to denote the *quotient map*

$$q\colon \square_\alpha X_\alpha \to \nabla_\alpha X_\alpha.$$

3.1. REMARK: One can also consider the quotient $<\kappa\text{-}\Box_\alpha X_\alpha / E(\lambda)$, where

$$xE(\lambda)y \text{ if } x_\alpha \neq y_\alpha \text{ for less than } \lambda \text{ indices } \alpha$$

for $\lambda < \kappa$ (otherwise the quotient is indiscrete, [Kn, 1.9]), but we have no applications.

Often all factors X_α are the same space X, in this case we use

$^\nu X$ for $<\omega\text{-}\Box\text{-}X_\alpha$, (this double use of $^\nu X$ will not cause confusion),

$<\kappa\text{-}\Box^\nu X$ for $<\kappa\text{-}\Box_\alpha X_\alpha$,

$\Box^\nu X$ for $\Box_\alpha X_\alpha$,

$\nabla^\nu X$ for $\underline{\nabla}_\alpha X_\alpha$;

we also use

$G_\nu X$ for the G_ν-modification of X,

i.e. X, retopologized by taking the family of G_ν-sets as a base: here a subset of X is called a *G_ν-set* if it is the intersection of at most ν open sets of X. We emphasize that we only consider what in a more precise notation would be $G_{\leq\nu}(X)$, and not $G_{<\nu}(X)$.

4. WORKING WITH ∇-PRODUCTS

When working with a ∇-product $\nabla_\alpha X_\alpha$ we will frequently use elements of $\Box_\alpha X_\alpha$ as names for elements of $\nabla_\alpha X_\alpha$; in other words, we use the element x of $\Box_\alpha X_\alpha$ to denote the element $q(x)$ of $\nabla_\alpha X_\alpha$. Also, if each $A_\alpha \subseteq X_\alpha$, then we use $\nabla_\alpha A_\alpha$ to denote the set $q^{\rightarrow}\Pi_\alpha A_\alpha = \{q(x): x \in \Pi_\alpha A_\alpha\}$, even though the ∇-product of the A_α's and the image $q^{\rightarrow}\Pi_\alpha A_\alpha$ are formally different sets (unless each $A_\alpha = X_\alpha$).

The advantage of doing so is that it facilitates working with ∇-products: We have the following simple rules, which all follow from the first one

$x = y$ iff $x_\alpha = y_\alpha$ for almost all α (= with less than ν exceptions),

$x \in \nabla_\alpha A_\alpha$ iff $x_\alpha \in A_\alpha$ for almost all α,

$\nabla_\alpha A_\alpha = \nabla_\alpha B_\alpha$ iff $A_\alpha = B_\alpha$ for almost all α,

$\nabla_\alpha A_\alpha \subseteq \nabla_\alpha B_\alpha$ iff $A_\alpha \subseteq B_\alpha$ for almost all α,

also

$$\nabla_\alpha A_\alpha \cap \nabla_\alpha B_\alpha = \nabla_\alpha (A_\alpha \cap B_\alpha).$$

Hence ∇_α operates on sets in much the same way Π_α does. It also operates on spaces in much the same way $\Box_\alpha$ does: we will see in §6 that $q\colon \Box_\alpha X_\alpha \to \nabla_\alpha X_\alpha$ is open, hence that

$$\{\nabla_\alpha U_\alpha\colon \text{each } U_\alpha \text{ open in } X_\alpha\} \text{ is a base for } \nabla_\alpha X_\alpha.$$

We will also see that the image $q^{\to}\Pi_\alpha A_\alpha$ can be naturally identified with the ∇-product $\nabla_\alpha A_\alpha$, and that (after this identification) $\nabla_\alpha A_\alpha$, as subspace of $\nabla_\alpha X_\alpha$, coincides with $\nabla_\alpha A_\alpha$, as ∇-product of the A_α's; the analogous statement for $\Box_\alpha$ is obvious. ⟦The notation $\nabla_\alpha X_\alpha$ and this method of working with ∇-products were introduced in [vD$_3$].⟧

5. Some Covering and Separation Properties

In this paper we show that several $\Box$-products are not paracompact; the proofs show in fact that they badly fail to be normal. We find it convenient to have the following concepts available.

5.1. Definitions: The space X is called (*discretely*) *pseudo-normal* if every two disjoint closed sets, one of which is countable (and both of which are discrete) have disjoint neighborhoods.

The space X is called *paranormal*, or *metanormal* if for every indexed discrete sequence[5] $\langle F_n: n < \omega\rangle$ of closed sets there are open sets $\langle U_{n,k}: n,k < \omega\rangle$ with $F_n \subseteq U_{n,k}$ for all $n,k < \omega$, such that

$\cap_{n,k}\overline{U}_{n,k} = \emptyset$ (for paranormal), or $\cap_{n,k}U_{n,k} = \emptyset$ (for metanormal).

Pseudonormality (also called weak normality) is well known, discrete pseudonormality is essentially due to Tall, [T1], and is considered in [vD$_5$]. Paranormality and metanormality are introduced here for the only reason that certain □-products or products don't have this property. The relevance of the four weakenings of normality, just defined, for our study of □-products is that they occupy the bottom line of the following diagram, so by showing that a certain space doesn't have one of these properties one shows that it doesn't have several more interesting properties. All implications have straightforward proofs, with exception of the implications subparacompact ⇒ submetacompact, due to Burke, [Bu, 1.6], and countably submetacompact ⇒ countably metacompact, due to Gittings, [Gi, 2.2]; as Gittings' proof is indirect we include a direct proof below. The question of whether there are any implications missing in this diagram will be ignored, except for implications between properties on the bottom line: we will show that there are no more.

We now state and prove four simple results which will be used later. The first two of these show that certain results about countably paracompactness are results about paranormality (there is no essential change in the proof);

---

[5]i.e. each point has a neighborhood that intersects F_n for at most one n.

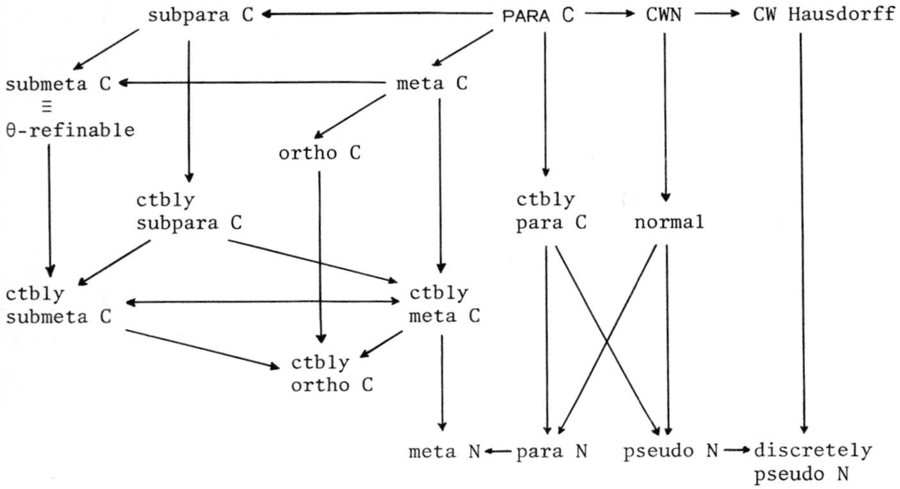

C=compact, ctbly=countably, CW=collectionwise, N=normal

the reason this is of interest is that similar results hold for normality, so we feel that we have found the (or at least a) reason for this similarity.

Our first result will be used to prove that ${}^{\omega}\omega \times \square^{\omega}(\omega+1)$ is not paranormal, Theorem 7.1.

5.2. THEOREM. If Y has a closed subset K which is not a regular G_{δ}[6], then $Z = \omega \times Y \cup \{\omega\} \times (Y - K)$ is a nonparanormal subspace of $(\omega + 1) \times Y$.

⌑ Let $\langle C_n : n < \omega \rangle$ be a decomposition of ω into infinite sets. Then $\langle C_n \times K : n < \omega \rangle$ is an indexed discrete sequence of closed sets in Z. Suppose $U_{n,k}$ is open in Z and includes

---

[6] We call K a *regular* G_{δ} of Y if it is the intersection of some countable family of closed neighborhoods.

$C_n \times K$, for all $n,k < \omega$. For $j,k < \omega$, if $j \in C_n$ for (unique) $n < \omega$, the set $V_{j,k} = \{y \in Y\colon \langle j,y\rangle \in U_{n,k}\}$ is open in Y and includes K. Since K is not a regular G_δ, we can find $q \in (\cap_{j,k} Cl_Y V_{j,k}) - K$. Then $\langle \omega,q\rangle \in \cap_{n,k} Cl_Z U_{n,k}$ since $\omega \in Cl_{\omega+1} C_n$ for each n.

⟦If X has a countable subset C with a limit point p, consider the analogous subspace $\tilde{Z} = C\times Y \cup \{p\}\times(Y - K)$ of $X \times Y$. Katětov has shown that $\tilde{Z}$ is not normal, [Ka, Cor. 1], and later Zenor has shown that $\tilde{Z}$ is not countably paracompct, [Z, Thm. 8], under the superfluous condition that C be relatively discrete. (Outline of a joint proof: Since $S = C \cup \{p\}$ is zero-dimensional, being countable, C admits a decomposition $\langle C_n\colon n < \omega\rangle$ with each C_n clopen in S. Then $\langle C_n \times K\colon n < \omega\rangle$ is an indexed discrete sequence in Z that admits no expansion to an indexed discrete open sequence.) I don't know if $X\times Y$ must have a non-paranormal subspace.⟧

(I am indebted to Jeff Cohen for pointing out an oversight.) ¤

Our second result will be used to prove that $\square^\omega(\underline{\mathrm{f}}2)$ is not paranormal, Theorem <u>8.18</u>.

5.3. THEOREM. If X is paranormal then $|D| < 2^{d(X)}$ for every closed discrete subset D of X.

¤ Let κ be the number of regularly closed sets. We will prove that $|D| < \kappa^\omega$ for each closed discrete subset D of X; since $\kappa \leq 2^{d(X)}$, hence $\kappa^\omega \leq 2^{d(X)}$, this proves our result. [Actually: $\kappa^\omega = \kappa$.]

Suppose there is a closed discrete D with $|D| = \kappa^\omega$. Then

$$\{\langle U_{n,k}\colon n,k < \omega\rangle\colon \text{each } U_{n,k} \text{ is regularly closed, and } \cap_{n,k} U_{n,k} = \emptyset\}$$

can be enumerated as $\langle\langle U_{x,n,k}: n,k < \omega\rangle: x \in D\rangle$. For each $n < \omega$ define

$F_n = \{x \in D: n = min\{i < \omega: x \notin \cap_k U_{x,i,k}\}$.

Clearly $\langle F_n: n < \omega\rangle$ is a decomposition of D, hence is a discrete sequence of closed sets. We built in that this sequence witnesses that X is not paranormal.

⟦Fleissner proved this for countably paracompact spaces, [F1, 2.4]; our proof is the same. Earlier Jones proved (but didn't formulate) for normal X the stronger result that $2^{|D|} \leq 2^{d(X)}$ for each closed discrete D in X, [Jo, Thm. 1]. It is not known whether or not this strengthening holds for countable paracompactness or for paranormality.⟧ ¤

Our third result gives a quick way to see why in a special case a nonparacompact box product must be nonnormal. It will be used to show that $\square^{\omega}(^{\underline{c}^+}2)$ is not normal and is not orthocompact, Theorem 8.16.

5.4. PROPOSITION: Let $\kappa^{\omega} = \kappa$, and let $\square$ denote $\square^{\omega}(^{\kappa}2)$.

(a) $\square$ is paracompact if (and only if) it is normal.

(b) $\square$ is metacompact if (and only if) it is orthocompact.

¤ Clearly $w(\square) \leq \kappa^{\omega} = \kappa$, hence $\square$, being zero-dimensional by Fact 6.3, can be embedded in $^{\kappa}2$, [En_1, 6.2.16]. Since $\square$ and $^{\kappa}2 \times \square$ are homeomorphic, it now suffices to remember that if X is a subspace of a compact space K, then

X is paracompact if (and only if) $X \times K$ is normal,

this is Tamano's theorem, [Tm,3.1], [En_1, 5.1.8], and then

X is metacompact if (and only if) $X \times K$ is orthocompact,

this is Junnila's analogue of Tamano's theorem, [Jn, 4.1]. ¤

We should point out that we have better results about $\square^{\omega}({}^{\kappa}2)$ than those obtained from 5.3 and 5.4: $\square^{\omega}({}^{\kappa}2)$ is neither paranormal nor orthocompact (the second part is due to Scott, [Sc_2, 7.0.1]) if $\kappa \geq \omega_2$ by Theorem 13.8, and is neither metanormal nor countably orthocompact if $\kappa \geq \underline{c}^{+}$ by Theorem 13.9. The weaker results have been included since I believe that their proofs are of interest.

Our fourth result will be used in the proof that ${}^{\omega_1}\omega$ and $\square^{\omega}({}^{\underline{c}^{+}}2)$ are not countably orthocompact.

5.5. PROPOSITION: Assume $\nu \geq \omega$, and that either X has an infinite compact subset or $\kappa = \omega$. Denote $<\kappa - \square^{\nu}X$ by $\square$.

(a) If $\square$ is normal, then $\square$ is countably paracompact.

(b) If $\square$ is countably orthocompact, then $\square$ is countably metacompact.

¤ If X has an infinite compact subset then $\square$ has a subspace homeomorphic to $b\omega \times \square$ for some compactification $b\omega$ of ω, since then $\square$ and $X \times \square$ are homeomorphic. If $|X| \leq 1$ there is nothing to prove. In the remaining case ${}^{\omega}X \times \square$ and $\square$ are homeomorphic and ${}^{\omega}X$ has a (necessarily closed) subspace homeomorphic to $\omega + 1$. Hence it suffices to recall that for each space Y and each compactification $b\omega$ of ω,

Y is countably paracompact if $Y \times b\omega$ is normal,

this is (essentially) part of Dowker's theorem, [Do, Thm. 4], [En_1, 5.2.8], and

Y is countably metacompact if $Y \times b\omega$ is countably orthocompact,

this is part of Scott's analogue of Dowker's theorem, [Sc_1, 1.3], [Sc_2], [Sc_3, 1.1]. ¤

5.6. REMARK: The facts

(1) Y is countably paracompact if $Y \times (\omega + 1)$ is normal, and

(2) Y is countably metacompact if $Y \times (\omega + 1)$ is countably orthocompact,

aren't very good analogues of each other since normality and countable paracompactness are incomparable, while countable orthocompactness is (properly) weaker than countable metacompactness. In this context we point out without proof another result involving paranormality:

(3) Y is countably paracompact if $Y \times (\omega + 1)$ is paranormal.

(Here the factor $\omega + 1$ can be replaced by any compactification $b\omega$ of ω for which ω admits a disjoint infinite family $\mathcal{D}$ such that $(\cap_{D \in \mathcal{D}} \overline{D}) \cap (b\omega - \omega) \neq \emptyset$. I didn't verify if this condition is essential.)

We finish this section with some left-overs.

5.7. EXAMPLES (a) A space that is pseudonormal but not metanormal.

(b) A space that is both metanormal and discretely pseudonormal, but is neither paranormal nor pseudonormal.

(c) A space that is paranormal but not discretely pseudonormal.

¤ (a): We will see that $\square^{\omega}(\overset{+}{\underline{c}}\, 2)$ is pseudonormal, Theorem 11.1, but is not metanormal, Theorem 13.9.

(b). Let $W = \omega_1$ and $\alpha W = \omega_1 + 1$ have the order topology. Then the subspace $B = (\omega + 1) \times \alpha W - \{\langle \omega, \omega_1 \rangle\}$ of $(\omega + 1) \times \alpha W$ (i.e. the Tychonoff plank without corner point) is as required: Since W is countably compact it is easy to see that

B is countably metacompact. Also, B is collectionwise Hausdorff since every closed discrete set is at most countable. But $\langle\{\langle n,\omega_1\rangle\}: n < \omega\rangle$ witnesses that B is not paranormal, and $\omega \times \{\omega_1\}$ and $\{\omega\} \times W$ witness that B is not pseudonormal.

(c) Let $p \in \beta\omega - \omega$, let $\alpha D = D \cup \{\infty\}$ be the one-point compactification of an uncountable discrete space D. Our example will be the subspace $C = ((\omega \cup \{p\}) \times \alpha D) - \{\langle p,\infty\rangle\}$ of $(\omega \cup \{p\}) \times \alpha D$. It is easy to see that $\omega \times \{\infty\}$ and $\{p\} \times D$ witness that C is not discretely pseudonormal.

Let $\langle F_n: n < \omega\rangle$ be an indexed discrete sequence of closed subsets of C. We want to show that there are open $U_{n,k}$'s with $F_n \subseteq U_{n,k}$, and $\bigcap_{n,k}\overline{U}_{n,k} = \emptyset$. We apply two simplifications: we may assume that no F_n has an isolated point of C, and it suffices to find an injection $s: \omega \to \omega$ and for each n an open set $U_n \supseteq F_{s(n)}$ such that $\bigcap_n\overline{U}_n = \emptyset$. For each n let

$A_n \subseteq \omega$ and $B_n \subseteq D$ be defined by $F_n = A_n \times \{\infty\} \cup \{p\} \times B_n$.

Since the A_n's are pairwise disjoint, and p is a free ultrafilter, we can find

$P \in p$ and an injection $s: \omega \to \omega$ such that $A_{s(n)} \cap P = \emptyset$ and $n \notin A_n$ for each n.

Then we define our promised U_n's by

$$U_n = A_{s(n)} \times \alpha D \cup ((P - \{n\}) \cup \{p\}) \times B_{s(n)},$$

since each $A_{s(n)} \times \alpha D$ is closed, we see that

$$\overline{U}_n \subseteq U_n \cup (P - \{n\}) \times \{\infty\}.$$

Then $\overline{U}_m \cap \overline{U}_n \subseteq (P - \{m,n\}) \times \{\infty\}$ whenever $m < n < \omega$, hence $\bigcap_n\overline{U}_n = \emptyset$. ¤

5.8. REMARK: The reader may have wondered why we consider $U_{n,k}$'s in the definition of paranormal and metanormal, instead of simply U_n's, especially after example C above. The reason is that we show (formally) more when we show that a certain space is not metanormal or paranormal as defined above, and that this does not affect the difficulty of the argument. (I have not checked if considering $U_{n,k}$'s instead of U_n's leads to a different concept.)

5.9. THEOREM: A space X is countably metacompact if (and only if) it is countably submetacompact.

¤ Let $\mathcal{U}$ be a countable open cover of X. Since X is countably submetacompact there is a sequence $\langle \mathcal{V}_k \rangle_k$ such that

(1) each $\mathcal{V}_k$ is an open cover of X that refines $\mathcal{U}$; and

(2) for each $x \in X$ there is $k < \omega$ with $\mathcal{V}_k$ point-finite at x.

Enumerate $\mathcal{U}$ as $\langle U_n \rangle_n$. For $n < \omega$ define

$$W_n = \{x \in U_n : \forall k \leq n \ \exists V \in \mathcal{V}_k \ [x \in V \text{ and } V \not\subseteq \cup_{j<n} U_j]\}$$

We claim that $\mathcal{W} = \{W_n : n < \omega\}$ is a point-finite open refinement of $\mathcal{U}$ that covers X. Since it is clear that $\mathcal{W}$ refines $\mathcal{U}$, we must check three facts:

$\mathcal{W}$ is point-finite: Let $x \in X$. There is k with $\mathcal{V}_k$ point-finite at x. Since $\mathcal{V}_k$ refines $\mathcal{U}$ there is $m < \omega$ such that $V \subseteq \cup_{i<m} U_i$ whenever $x \in V \in \mathcal{V}_k$. Then $x \notin W_n$ whenever $n \geq max\{k,m\}$.

$\mathcal{W}$ is open: Let $x \in W_n$. For $k \leq n$ pick $V_k \in \mathcal{V}_k$ with $x \in V_k$ and $V_k \not\subseteq \cup_{j<n} U_j$. Then $x \in \cap_{k \leq n} V_k \subseteq W_n$.

$\mathcal{W}$ covers X: Let $x \in X$. There is a (unique) $n < \omega$ with $x \in U_n - \cup_{j<n} U_j$. Since each $\mathcal{V}_k$ covers X, we have $x \in W_n$.

⟦Gittings' proof uses a characterization of countable metacompactness, [Gi,2.2].⟧ ¤

6. BASIC FACTS ABOUT □-PRODUCTS AND ∇-PRODUCTS

Much in this section is elementary and will be used without explicit reference.

6.1. FACT: (a) If each A_α is closed in X_α then $\Pi_\alpha X_\alpha$ is closed in $<\kappa - \square_\alpha X_\alpha$.

(b) If each A_α is open/closed in X_α then $\nabla_\alpha A_\alpha$ is open/closed in $\nabla_\alpha X_\alpha$.

(c) If κ is regular, and I is a decomposition of ν, then the natural bijection $<\kappa - \square_{\alpha\in\nu} X_\alpha \to <\kappa - \square_{I\in I}(<\kappa - \square_{\alpha\in I} X_\alpha)$ is a homeomorphism.

¤ We prove (b): Denote $\square_\alpha X_\alpha$ by $\square$. Let $A = q^{\leftarrow}\nabla_\alpha A_\alpha$, then

$$A = \{x \in \square\colon x_\alpha \in A_\alpha \text{ with exception of less than } \nu\ \alpha\text{'s}\}.$$

Clearly A is open if each A_α is. If each A_α is closed, and $x \notin A$, consider

$$E = \{\alpha \in \nu\colon x_\alpha \notin A_\alpha\} \text{ and } U = \{y \in \square\colon y_\alpha \notin A_\alpha \text{ for } \alpha \in E\}.$$

Then $|E| = \nu$, hence U is a neighborhood of x that misses A. Therefore A is closed. Since q is quotient this completes the proof.

⟦The fact that q is open was noted in [Kn, 1.8]. The fact that (c) is not generally true without the condition that κ be regular was overlooked in [Kn, 1.12].⟧ ¤

Part (b) of the following fact justifies our using $\nabla_\alpha X_\alpha$ to denote different objects. (Its precise meaning will, become clear from the proof.)

6.2. FACT: Let each A_α be a subspace of X_α.

(a) $\Pi_\alpha A_\alpha$, as subspace of $<\kappa - \square_\alpha X_\alpha$, coincides with $<\kappa - \square_\alpha A_\alpha$.

(b) The natural way to identify $\nabla_\alpha A_\alpha$, as ∇-product, with $\nabla_\alpha A_\alpha$, as subset of $\nabla_\alpha X_\alpha$, is a homeomorphism which respects the quotient map.

¤ If $\Pi_\alpha U_\alpha$ is a basic open set in $<\kappa - \Box_\alpha X_\alpha$, then $\Pi_\alpha(A_\alpha \cap U_\alpha)$ is a basic open set in $<\kappa - \Box_\alpha A_\alpha$. As $\Pi_\alpha(A_\alpha \cap U_\alpha) = \Pi_\alpha A_\alpha \cap \Pi_\alpha U_\alpha$, this proves (a). The proof of (b) will be similar; this illustrates that one can operate with ∇ as with a product.

For clarity we use q_A and q_X to denote the quotient maps $\Box_\alpha A_\alpha \to \nabla_\alpha A_\alpha$ and $\Box_\alpha X_\alpha \to \nabla_\alpha X_\alpha$, respectively. Since clearly $q_A(x) = q_A(y)$ iff $q_X(x) = q_X(y)$ for $x,y \in \Box_\alpha X_\alpha$, we can define a bijection $h\colon \nabla_\alpha A_\alpha \; (= q_A^{\rightarrow}\Box_\alpha A_\alpha) \to q_X^{\rightarrow}\Box_\alpha A_\alpha$ by

$$h(q_A(x)) = q_X(x), \qquad x \in \Box_\alpha A_\alpha.$$

Clearly

(1) $\quad h \circ q_A = q_X \restriction \Box_\alpha A_\alpha,$

so h respects the quotient maps as functions. But (1) only implies that h is continuous, [En$_1$, 2.4.1], and not that it is a homeomorphism, nor that it respects the quotient maps in the topological sense. To prove this we observe that

(2) $\quad q_X \restriction \Box_\alpha A_\alpha\colon \Box_\alpha A_\alpha \to q_X^{\rightarrow}\Box_\alpha A_\alpha$ is a quotient map, even is open.

[Here we use (a).] Indeed, if each U_α is open in X_α, then

$$q_X^{\rightarrow}\Pi_\alpha(A_\alpha \cap U_\alpha) = q_X^{\rightarrow}\Box_\alpha A_\alpha \cap q_X^{\rightarrow}\Pi_\alpha U_\alpha,$$

cf. the rules of §4. ¤

6.3. FACT: If each X_α is T_0, T_1, Hausdorff, regular, completely regular or zero-dimensional[7] then so are $<\kappa - \Box_\alpha X_\alpha$ (hence $\Box_\alpha X_\alpha$) and $\nabla_\alpha X_\alpha$[8].

---

[7] ≡ weakly zero-dimensional: the clopen (= closed + open) sets from a base.

[8] See Corollary 9.3 and Proposition 15.3 for a better result about $\nabla_\alpha X_\alpha$.

¤ This is an easy consequence of Fact 6.1, except perhaps for complete regularity. We use an amusing trick to handle this case. Since a space is completely regular iff it can be embedded in a power of the closed unit interval, [En_1, 2.3.23], which can be embedded in a topological group, for instance a power of the circle, and since topological groups are completely regular, [En_1, 8.1.17], we have

> a space is completely regular iff it can be embedded in a topological group.

Since $<\kappa - \square_\alpha X_\alpha$ and $\nabla_\alpha X_\alpha$ are easily seen to be topological groups if each X_α is, the result follows from Fact 6.2. [One also can base a proof on [Ke, Lemma 3, p. 114].]
⟦This fact can be found in [Kn].⟧ ¤

In an early version of [Ku_2] Kunen observed the useful fact that $G_\nu X$ can be embedded as a closed subspace in $\square^\nu X$; this will be used in §§8,15. We generalize and extend this as follows.

6.4. **PROPOSITION:** $G_\nu X$ can be embedded as a closed subspace in both $\square^\nu X$ and $\nabla^\nu X$.

¤ Denote $G_\nu X$ by G, $\square^\nu X$ by $\square$ and $\nabla^\nu X$ by ∇. It is easy to see that the diagonal map $d\colon X \to \square$, defined by

$$d(x) = \langle x, x, \ldots, x, \ldots \rangle$$

is an embedding $G \to \square$. Also, the diagonal $\Delta = d^{\to}X$ is clearly closed in $\square$ (even in ${}^\nu X$), by our tacit assumption that all spaces are (at least) Hausdorff. [This is Kunen's proof.]

Next, consider the composition $q \circ d$. It clearly is a continuous injection $G \to \nabla$. We want to show that $(q \circ d)^{\to}U$ is open in $(q \circ d)^{\to}G$ whenever U is a G_ν-set in X. Let V be a collection of at most ν open sets in X. Enumerate V as

$\langle V_\alpha \rangle_{\alpha<\nu}$ in such a way that each element of V is listed ν times. One easily verifies that

$$(q\circ d)^{\to} \cap V = (q\circ d)^{\to} G \cap \nabla_\alpha V_\alpha.$$

Finally, consider $x \in \nabla - q^{\to}\Delta$. For each $A \subseteq \nu$ with $|A| < \nu$ there are $\alpha,\beta \in \nu - A$ with $x_\alpha \neq x_\beta$. Hence we can find $\alpha_\xi, \beta_\xi < \nu$ for $\xi < \nu$ such that

$$\{\alpha_\xi,\beta_\xi\} \cap \{\alpha_\eta,\beta_\eta\} = \emptyset \text{ for } \xi < \eta < \nu, \text{ and}$$

$$x_{\alpha_\xi} \neq x_{\beta_\xi} \quad \text{for} \quad \xi < \nu.$$

We can choose a neighborhood U_α of x_α in X_α such that

$$U_{\alpha_\xi} \cap U_{\beta_\xi} = \emptyset \text{ for } \xi < \nu.$$

Then $\nabla_\alpha U_\alpha$ is a neighborhood of x that misses $q^{\to}\Delta$. Hence $(d\circ q)^{\to} G$ is closed in ∇. ¤

6.5. REMARK: More generally, if $\kappa \leq \nu^+$ then $G_{<\kappa}X$ (self-explanatory) can be embedded as a closed subset in $<\kappa - \Box^\nu X$. We have no nontrivial applications if $\kappa < \nu^+$.

The final result of this section will not be used in this paper, but is included because I like retractions.

6.6. PROPOSITION: If X is compact then $G_\omega X$ can be embedded as a retract in both $\Box^\omega X$ and $\nabla^\omega X$.

¤ We use the notation of the above proof (with $\nu = \omega$). Let $^-$ be the closure operation in X.

It suffices to find a function

$$L\colon \Box \to X$$

which is continuous as map $\Box \to G$, such that

$$L(d(x)) = x \text{ for } x \in X, \text{ and } L(x) = L(y) \text{ whenever } xEy,$$

for then we can define retractions

$$r\colon \Box \to \Delta \text{ and } \rho\colon \nabla \to q^{\to}\Delta$$

by

$$r = d\circ L, \text{ and } \rho(q(x)) = q(r(x)) \quad (x \in \Box).$$

[Note that ρ is continuous since $\rho \circ q$ is continuous and since q is a quotient map, [En$_1$, 2.4.2].]

To define L choose a free ultrafilter $\mathcal{F}$ on ω. Then for each $x \in \Box$ there is a unique $L(x) \in X$ satisfying

$$L(x) \in \bigcap_{F \in \mathcal{F}} \overline{\{x_n : n \in F\}}.$$

[$L(x)$ exists since X is compact, and is unique since X is Hausdorff. Note that $L(x)$ is the $\mathcal{F}$-limit of $\langle x_n \rangle_n$, [Be, Def. 3.1].]

We show that L is continuous. Let $x \in \Box$ and let U be a neighborhood of $L(x)$ in G. There is a sequence $\langle V_k \rangle_k$ of neighborhoods of x in X with

$$V_0 = X, \text{ and each } \overline{V}_{k+1} \subseteq V_k, \text{ and } \bigcap_k V_k \subseteq U.$$

One can choose a sequence $\langle F_k \rangle_k$ in $\mathcal{F}$ such that

$$F_0 = \omega, \text{ and each } F_{k+1} \subseteq F_k, \text{ and each } \overline{\{x_n : n \in F_k\}} \subseteq V_k, \text{ and } \bigcap_k F_k = \emptyset.$$

Define a sequence $\langle W_n \rangle_n$ of open sets in X by

$$W_n = V_k \text{ if } n \in F_k - F_{k+1}.$$

Then $\Pi_k W_k$ is a neighborhood of x in $\Box$, and if $y \in \Pi_k W_k$ then

$$L(y) \in \bigcap_k \overline{\{y_n : n \in F_k\}} \subseteq \bigcap_k \overline{V}_k = \bigcap_k V_k \subseteq U.$$

Hence L is continuous.

[I didn't try to show if G can be embedded as retract in both $\Box$ and ∇ for $\nu > \omega$.] ¤

7. NEGATIVE RESULTS ON METRIZABLE FACTORS I: TOPOLOGICAL ARGUMENTS

Throughout this section we use $\Box$ to denote $\Box^{\omega}(\omega + 1)$, and ${}^{\omega}\omega$ to denote the ordinary product ${}^{\omega}\omega$ (which is homeomorphic to the irrationals) unless we explicitly indicate that we consider it to be a subset of $\Box$.

The results of this section are a typical example of the first technique mentioned in the introduction for studying

$\square$-products: $^{\omega}\omega \times \square$ is (homeomorphic to) a $\square$-product of countably many metrizable factors, but by writing $^{\omega}\omega \times \square$ we take away its forbidding appearance, because it looks no worse than $P \times M$ where P denotes the irrationals, as subspace of $\mathbb{R}$, and M is the Michael line, i.e. $\mathbb{R}$, retopologized by making each irrational an isolated point. After retopologizing P becomes a dense set of isolated points which is not an F_σ. Similarly, if we retropologize $^{\omega}(\omega + 1)$ and make it $\square$, $^{\omega}\omega$ becomes a dense set of isolated points in $\square$ which, as we will see below, is not an F_σ in $\square$. The similarity has led to the discovery that $^{\omega}\omega \times \square$ is not normal: Michael, [M_2], has shown that $P \times M$ is not normal by proving that $\{\langle x,x\rangle : x \in P\}$ and $P \times (M - P)$ are disjoint closed sets which do not have disjoint neighborhoods. The analogous statement was proved in [vD_1]: $^{\omega}\omega \times \square$ is not normal since it has $\{\langle x,x\rangle : x \in {}^{\omega}\omega\}$ and $^{\omega}\omega \times (\square - {}^{\omega}\omega)$ as disjoint closed sets which do not have disjoint neighborhoods. We will here follow a somewhat different route to stronger results (see also §12).

7.1. THEOREM: $^{\omega}\omega \times \square$ is not paranormal.

7.2. THEOREM: $\square$ is not hereditarily paranormal.

We begin with showing that $\square$ is not perfect (= open sets are F_σ's). This result was motivated by Borges' question of whether a box product is stratifiable[9], [Bo_2, 2.5], and was my first result about $\square$-products, [vD_1].

7.3. PROPOSITION: $^{\omega}\omega$, when considered as subset of $\square$, is open but is not an F_σ.

---

[9]A space is called *stratifiable* if it is perfectly normal in a certain monotone way, which we do not have to specify here.

¤ $^{\omega}\omega$ is open in $\square$ since it is an ω_1-box. Let $\langle F_n \rangle_n$ be a sequence of subsets of $^{\omega}\omega$ which are closed in $\square$. Since $z \notin {}^{\omega}\omega$, hence $z \notin F_n$, whenever z satisfies $z_i = \omega$ for $i \geq n$, we can successively find functions $\varepsilon_n \colon \omega \to \omega$ such that if x_n is defined by

$$x_n = max_{k \leq n}\, \varepsilon_k(n),$$

then

$$F_n \cap \{y \in \square\colon y_i = x_i \text{ for } i < n, \text{ and } y_i \geq \varepsilon_n(i) \text{ for } i \geq n\} = \emptyset.$$

[We only use x_i for $i < n$ when choosing ε_n.] Clearly

$$x = \langle x_n \rangle_n \in {}^{\omega}\omega - \cup_n F_n.$$

⟦This was observed in [vD$_1$, §2].⟧ ¤

7.4. PROOF OF THEOREM 7.2: Since $(\omega + 1) \times \square$ and $\square$ are homeomorphic, it suffices to note that Theorem 5.2, which was based on [Ka, Cor. 1] and [Ze, B], and Proposition 7.3 imply that

$$Z = \omega \times \square \cup \{\omega\} \times {}^{\omega}\omega$$

is a nonparanormal subspace of $(\omega + 1) \times \square$.
⟦This is the proof in [vD$_1$, §2] that $\square$ is not hereditarily normal. Kunen has proved independently that $\square$ is not hereditarily pseudonormal, [Ku$_1$, 3]; this will be proved in §12.⟧ ¤

For the proof of Theorem 7.1 we will use the following result which will also be used in the next section. We use $\mathcal{T}$ and $\mathcal{B}$ for heuristical reasons: $\mathcal{T}$ and $\mathcal{B}$ will be the $\mathcal{T}$ychonoff and $\mathcal{B}$ox topology.

7.5. LEMMA: Let X be a set with topologies $\mathcal{T}$ and $\mathcal{B}$, with $\mathcal{T} \subseteq \mathcal{B}$. For $Y \subseteq X$ let $Y_{\mathcal{T}}$ and $Y_{\mathcal{B}}$ denote Y as subspace of $\langle X, \mathcal{T} \rangle$ and of $\langle X, \mathcal{B} \rangle$, respectively.

If $\mathcal{T}$ is a Hausdorff topology, then for each $Y \subseteq X$ the product $Y_\mathcal{T} \times X_\mathcal{B}$ has a closed subspace homeomorphic to $Y_\mathcal{B}$.

¤ For $A \subseteq X$ consider the diagonal $\Delta_A = \{\langle a,a\rangle : a \in A\}$, as subspace of $X_\mathcal{T} \times X_\mathcal{B}$, or, equivalently, as subspace of $A_\mathcal{T} \times A_\mathcal{B}$. We first observe that Δ_X is closed in $X_\mathcal{T} \times X_\mathcal{B}$ since $\mathcal{T}$ is Hausdorff and $\mathcal{T} \subseteq \mathcal{B}$, hence that Δ_Y is closed in $Y_\mathcal{T} \times X_\mathcal{B}$ since $\mathcal{T}$ is Hausdorff and $\mathcal{T} \subseteq \mathcal{B}$, hence that Δ_Y is closed in $Y_\mathcal{T} \times X_\mathcal{B}$ since $\Delta_Y = \Delta_X \cap (Y_\mathcal{T} \times X_\mathcal{B})$. We next observe that Δ_Y is homeomorphic to $Y_\mathcal{B}$ since $\mathcal{B} \supseteq \mathcal{T}$.

⟦This was observed in [vD$_6$].⟧ ¤

7.6. PROOF OF THEOREM 7.1: From 7.4 and 7.5 we see that $Z \times \square$ is not paranormal, where Z is the subspace

$$Z = \omega \times {}^{\omega}(\omega + 1) \cup \{\omega\} \times {}^{\omega}\omega \text{ of } T = (\omega + 1) \times {}^{\omega}(\omega + 1),$$

where ${}^{\omega}(\omega + 1)$ carries the Tychonoff topology. So we want to show that Z is homeomorphic to ${}^{\omega}\omega$. By a classical result of Alexandroff and Urysohn, [AfU,IV], [En$_1$, 6.2A], it suffices to check that

(1) Z is zero-dimensional and separable;

(2) Z is completely metrizable; and

(3) Z has no nonempty open compact subsets.

Indeed, (1) is trivial, (2) follows from the fact that Z is the union of two G_δ-subsets of T, [En$_1$, 4.3.23], and (3) follows from the fact that both Z and $T - Z$ are dense in T. ¤

As another application of Lemma 7.5 we mention the following result; we omit the easy proof.

7.7. THEOREM: The following are equivalent for a closed hereditary topological property $\mathbb{P}$:

(a) every countable $\square$-product of separable metrizable spaces hereditarily has $\mathbb{P}$;

(b) every countable $\square$-product of separable metrizable spaces has $\mathbb{P}$;

(c) every countable $\square$-product of separable metrizable spaces has $\mathbb{P}$ provided there is at most one noncompact factor.

⟦A variation of this was announced in [vD_4].⟧ ¤

We ask questions about ${}^{\omega}\omega \times \square$ and $\square$ in Question 13.13.

8. NEGATIVE RESULTS ABOUT COMPACT FACTORS I: TOPOLOGICAL ARGUMENTS

Kunen showed in an early version of [Ku_2] that $\square^{\omega}({}^{\underline{c}^+}2)$ is not normal. This was improved in [vD_2], where it was shown that $\square^{\omega}({}^{\omega_2}2)$ is not normal. In §13 we will further improve this result by showing that $\square^{\omega}({}^{\omega_2}2)$ is not paranormal, and $\square^{\omega}({}^{\underline{c}^+}2)$ is not metanormal. The following result is weaker (except for the statement about collectionwise Hausdorff) but has been included because it has a nice simple proof thanks to the fact that the hard work (among others in the form of information about ordinary products—another point of interest) has been done already.

8.1. THEOREM (A): $\square^{\omega}({}^{\underline{f}}2)$ is not paranormal.

(B) $\square^{\omega}({}^{\underline{c}^+}2)$ is not collectionwise Hausdorff, is not normal, and is not orthocompact.

¤ If $\kappa = \underline{c}^+$ or $\kappa = \underline{f}$, then $\square^{\omega}({}^{\kappa}2)$ has density $\underline{c}$ but has a closed discrete subset of cardinality κ, by the result below. So $\square^{\omega}({}^{\underline{f}}2)$ is not paranormal by Theorem 5.3, based on [Fl, 2.2] cf. [Jo, Thm. 1], and $\square^{\omega}({}^{\underline{c}^+}2)$ is neither collectionwise Hausdorff nor metacompact, hence is not paracompact, hence is neither normal nor orthocompact by Proposition 5.4, based on [Tm] and [Jn].

[This proof that $\square^{\omega}(\underline{c}^{+}2)$ is not normal is completely different than Kunen's proof from the early version of [Ku_2], which is published in [Ru_3, p.5].] ¤

8.2. THEOREM: (a) if $1 \leq \kappa \leq \underline{f}$ then $d(\square^{\omega}({}^{\kappa}2)) = \underline{c}$.
(b) $\square^{\omega}(\underline{f}2)$ has a closed discrete set of cardinality $\underline{f}$, and
(c) $\square^{\omega}(\underline{c}^{+}2)$ has a closed discrete set of cardinality $\underline{c}^{+}$.

¤ PROOF OF (a): The usual product ${}^{\omega}({}^{\kappa}2)$ has a dense subset D of cardinality (at most) $\underline{c}$, [En_1, 2.3.15]. Since $\underline{c}^{\omega} = \underline{c}$ there is a subset E of ${}^{\omega}({}^{\kappa}2)$ of cardinality $\underline{c}$ such that every sequence in D has a cluster point in E. Then obviously E is dense in $\square^{\omega}({}^{\kappa}2)$, or even in $G_{\omega}({}^{\omega}({}^{\kappa}2))$. But $\square^{\omega}({}^{\kappa}2)$ has a family of $\underline{c}$ pairwise disjoint nonempty open sets since ${}^{\kappa}2$ has at least 2 disjoint open sets. Hence $d(\square^{\omega}({}^{\kappa}2)) = \underline{c}$.

PROOF OF (b) AND (c): We use the following notation:

$Y \vartriangleleft X$ means that Y can be embedded as a closed subspace in X.

The nontrivial result we need for the proof is

(1) $\kappa \vartriangleleft {}^{\kappa}\omega$ (hence $\kappa \vartriangleleft G_{\omega}({}^{\kappa}\omega)$) for $\kappa = \underline{c}^{+}, \underline{f}$.

To prove this we use Juhász' result, [Jh, Thm. 1], [En_1, 3.1 H(b)], that $2^{\kappa} \vartriangleleft {}^{(2^{\kappa})}2$ successively for $\kappa = \omega$ and $\kappa = \underline{c}$ to see that

(1') $\underline{c} \vartriangleleft {}^{\underline{c}}\omega$ and $\underline{f} \vartriangleleft {}^{\underline{f}}\omega$.

[In fact, $2^{\kappa} \vartriangleleft {}^{(2^{\kappa})}\omega$ if (and only if) κ is not Ulam-measurable, [Jh, Cor.].] Also, Mycielski, [My, A], proved that

(1") $\kappa^{+} \vartriangleleft {}^{(\kappa^{+})}\kappa$.

Clearly (1') and (1") imply (1).

Let $\kappa \geq \omega$ be arbitrary. A moment's reflection shows that

(2) $G_{\omega}({}^{\kappa}\omega)$ is a closed subspace of $G_{\omega}({}^{\kappa}(\omega + 1))$.

Also, since $\omega + 1 \lhd {}^{\omega}2$ we have ${}^{\kappa}(\omega + 1) \lhd {}^{\kappa}2$, hence

(3) $G_{\omega}({}^{\kappa}(\omega + 1)) \lhd G_{\omega}({}^{\kappa}2)$.

[In fact ${}^{\kappa}(\omega + 1)$ and ${}^{\kappa}2$ are homeomorphic, hence so are $G_{\omega}({}^{\kappa}(\omega + 1))$ and $G_{\omega}({}^{\kappa}\omega)$.] Finally, using Proposition 6.4 we have

(4) $G_{\omega}({}^{\kappa}2) \lhd \square^{\omega}({}^{\kappa}2)$

Clearly (1), (2), (3) and (4) imply

$\kappa \lhd \square^{\omega}({}^{\kappa}2)$ for $\kappa = \underline{c}^{+}, \underline{f}$. ¤

8.3. REMARK: Theorem 8.2c implies that there is a compact space X with $w(G_{\omega}X) = \underline{c}^{+}$ such that $G_{\omega}X$ is not $\underline{c}$-Lindelöf[10]. This is the smallest possible weight such a space can have since for all X one has $w(G_{\omega}X) \leq w(X)^{\omega}$. An earlier example due to Isbell and Mŕowka, recorded in [WF, p. 148], is that $G_{\omega}\beta\underline{c}^{+}$ is not $\underline{c}$-Lindelöf since if κ is not Ulam-measurable then κ is closed discrete in $G_{\omega}\beta\kappa$. But $w(\beta\underline{c}^{+}) = 2^{\underline{c}^{+}}$. (We note that there are examples of compact X with $w(X)$ arbitrarily large such that $G_{\omega}X$ is $\underline{c}$-Lindelöf, see [WF] or use the fact that $G_{\omega}X$ is Lindelöf if X is Lindelöf and scattered, [Ge].)

9. COUNTABLE ∇-PRODUCTS OF COMPACT SPACES AND PARACOMPACTNESS

The following simple but useful result is credited to Rudin in [Ku_2, 2.1]. It is one of the reasons that countable ∇-products are easy to handle.

9.1. PROPOSITION: Every countable ∇-product is a *P*-space.[11]

Since we are also interested in uncountable ∇-products, we prove a more general result. In view of condition (a) we

---

[10] A space is called κ-*Lindelöf* if every open cover has a subcover of cardinality at most κ.

[11] A space is called a *P-space* if each G_{δ}-subset is open.

remind the reader that an infinite compact space is not a P-space.

9.2. PROPOSITION: The following are equivalent

(a) $cf(\nu) = \omega$ or $|\{\alpha < \nu : X_\alpha \text{ is not a } P\text{-space}\}| < \nu$.

(b) $\nabla_\alpha X_\alpha$ is a P-space.

¤ Let $A = \{\alpha < \nu : X_\alpha \text{ is not a } P\text{-space}\}$. Denote $\nabla_\alpha X_\alpha$ by ∇.

<u>(a) ⇒ (b)</u>: Let each V_k be a neighborhood of x in ∇. For each k find neighborhoods $U_{\alpha,k}$ of x_α in X_α such that $\nabla_\alpha U_{\alpha,k} \subseteq V_k$; it is important to have $x_\alpha \in U_{\alpha,k}$ for all α, not just with less than κ exceptions.

CASE 1: $cf(\nu) = \omega$.

Let $\{N_k : k < \omega\}$ be a decomposition of ν with each $|N_k| < \nu$. For $\alpha < \nu$, if $\alpha \in N_k$ let $W_\alpha = \cap_{n \le k} U_{n,k}$. Then $\nabla_\alpha W_\alpha$ is a neighborhood of x in ∇ which is included in each $\nabla_\alpha U_{k,\alpha}$, hence in each V_k.

CASE 2: $|A| < \nu$.

Left to the reader.

<u>(b) ⇒ (a)</u>: Assume $|A| = \nu$. Choose $x \in \nabla$, and a sequence $\langle U_{\alpha,k} \rangle_k$ of neighborhoods of x_α in X_α such that

$\cap_k U_{\alpha,k}$ is not a neighborhood of x_α in X_α, for $\alpha \in A$.

Since ∇ is a P-space we can find for each α a neighborhood V_α of x_α in X_α such that for all k

$\nabla_\alpha V_\alpha \subseteq \nabla_\alpha U_{\alpha,k}$, i.e. $V_\alpha \subseteq U_{\alpha,k}$ with less than ν exceptions.

Since $|A| = \nu$ and

if $\alpha \in A$ then for some k we have $V_\alpha \not\subseteq U_{\alpha,k}$,

it follows that $cf(\nu) = \omega$. ¤

9.3. COROLLARY (TO 9.1): Every countable ∇-product of regular spaces is zero-dimensional.

¤ Recall that a ∇-product of regular spaces is regular, and a regular P-space is zero-dimensional. ¤

9.4. COROLLARY (TO PROOF OF 9.2): If every X_α is first countable, and if $cf(\nu) \neq \omega$, then each point of $\nabla_\alpha X_\alpha$ is a G_δ. ¤

[But $\nabla_\alpha X_\alpha$ is nondiscrete if (and only if) $|\{\alpha < \nu: X_\alpha \text{ non-discrete}\}| = \nu$.]

We now come to two important results about ∇-products, both of which are minor improvements of results of Kunen.

9.5. THEOREM: If each X_n is σ-compact and locally compact, then the quotient map q is closed.

¤ Denote $\square_n X_n$ by $\square$ and $\nabla_n X_n$ by ∇. Recall that q is closed if (and only if)

for every $x \in \nabla$ and neighborhood U of $q^{\leftarrow}\{x\}$ in $\square$ there is a neighborhood W of x in ∇ with $q^{\leftarrow}W \subseteq U$,

[En_1, 1.4.13].

Let $x \in \square$, and let U be a neighborhood of $q^{\leftarrow}\{q(x)\}$. For each n choose a sequence $\langle X_{n,k}\rangle_k$ of compact subsets of X_n such that

$X_{n,0}$ is a neighborhood of x_n, and each $X_{n,k} \subseteq X_{n,k+1}$ and $X_n = \cup_k X_{n,k}$.

CLAIM: For each k we can choose a sequence $\langle V_{n,k}\rangle_{n\ k}$ with each $V_{n,k}$ a neighborhood of x_n such that

$$\Pi_{n\leq k} X_{n,k} \times \Pi_{n>k} V_{n,k} \subseteq U.$$

Indeed, we can identify $\square$ with $\Pi_{n\leq k} X_n \times \square_{n>k} X_n$; then U is a neighborhood of $\Pi_{n\leq k} X_{n,k} \times \{\langle x_{k+1}, x_{k+2}) \ldots\rangle\}$. As $\Pi_{n\leq k} X_n$ is compact, there is a neighborhood V of $\langle x_{k+1}, x_{k+2}, \ldots\rangle$ in $\square_{n>k} X_n$ with $\Pi_{n\leq k} X_{n,k} \times V \subseteq U$, cf. [$\text{En}_1$, 3.1.15]. We may assume that V has the form $\Pi_{n>k}V_{n,k}$.

Given the $V_{n,k}$'s, we define for each n a neighborhood W_n of x_n by

$$W_0 = X_{n,0}, \text{ and } W_n = X_{n,0} \cap \cap_{k<n}V_{n,k} \text{ if } k > 0.$$

Then $\nabla_n W_n$ is a neighborhood of $q(x)$. It remains to show that $q^{\leftarrow}\nabla_n W_n \subseteq U$. So let $y \in q^{\leftarrow}\nabla_n W_n$. There is m with

$$y_n \in W_n \text{ for } n > m.$$

Choose $k \geq m$ such that

$$y_n \in X_{n,k} \text{ for } n \leq m.$$

As $y_n \in W_n \subseteq X_{n,0} \subseteq X_{n,k}$ for $n > m$, and $y_n \in W_n \subseteq V_{n,k}$ for $n \geq k$ we see that

$$y \in \Pi_{n\leq k}X_{n,k} \times \Pi_{n>k}V_{n,k} \subseteq U,$$

as required.

[Kunen has the result for compact factors, [Ku_1, 2.2]; his proof is essentially the same.] ¤

We will see in § 14 that both σ-compactness and local compactness are essential, even if each X_n is metrizable.

The main application of the preceding theorem is the following important result, essentially due to Kunen, [Ku_2, 2.3].

9.6. THEOREM: If each X_n is σ-compact and locally compact, then $\square_n X_n$ is strongly paracompact[12] iff $\square_n X_n$ is paracompact iff $\nabla_n X_n$ paracompact.

¤ Denote $\square_n X_n$ by $\square$ and $\nabla_n X_n$ by ∇. Because of Theorem 9.5 and Michael's theorem that paracompactness is preserved by closed maps, [M_1, Cor. 1], [En_1, 5.1.33], it suffices to show that $\square$ is strongly paracompact if ∇ is paracompact.

---

[12] A space is called *strongly paracompact* if every open cover has a star-finite open refinement.

Let $\mathcal{U}$ be an open cover of $\square_n X_n$, for each $x \in \square$ the set

$$E(x) = \cup_{n<\omega}\{y \in \square : y_k = x_k \text{ for } k > n\}$$

is σ-compact. Hence for each $x \in \nabla$ we can choose a countable $\mathcal{U}_x \subseteq \mathcal{U}$ with $q^{\leftarrow}\{x\} \subseteq \cup\mathcal{U}_x$, and next choose an open V_x in ∇ such that

$$x \in V_x \text{ and } q^{\leftarrow}V_x \subseteq \cup\mathcal{U}_x.$$

Since ∇ is a paracompact P-space it is ultraparacompact[13,14]. Hence we can find an indexed open cover $\langle V_x : x \in \nabla\rangle$ of ∇ such that

$$V_x \subseteq U_x, \text{ and } V_x \cap V_y = \emptyset \text{ for distinct } x, y \in \nabla$$

(and most V_x's are empty). Then

$$\cup_{x\in\nabla}\{q^{\leftarrow}V_x \cap U : U \in \mathcal{U}_x\}$$

is a star-countable open refinement of $\mathcal{U}$. Hence $\square$ is strongly paracompact since it is regular, and a regular space is strong strongly paracompact iff every open cover has a star-countable open refinement, [En_1, 5.3.10].

[Kunen considers compact factors only [Ku_2, 2.3], (but see <u>9.8</u> below), and does not consider strong paracompactness.] ¤

9.7. COROLLARY: If each X_n is σ-compact and locally compact, then $\square_n X_n$ is ultraparacompact iff $\square_n X_n$ is paracompact and each X_n is weakly zero-dimensional.

¤ For the nontrivial implication recall that a strongly paracompact space is ultraparacompact if (and only if) it is weakly zero-dimensional, cf. [En_2, 2.4.2]. [One can also

---

[13] A space is called *ultraparacompact* if every open cover has a disjoint open refinement.

[14] For completeness sake we give a quick proof: Every open cover $\mathcal{A}$ of ∇ has a locally finite refinement $\mathcal{B}$ consisting of cozero-sets, e.g. because $\mathcal{A}$ has a partition of unity subordinate to it, [En_1, 5.1.9]. Let $<$ be a well-order on $\mathcal{B}$, then $\mathcal{C} = \{B - \cup\{A \in \mathcal{B} : A < B\} : B \in \mathcal{B}\}$ is a disjoint refinement of $\mathcal{B}$. Now $\mathcal{B}$ consists of clopen sets since ∇ is a P-space, hence $\mathcal{C}$ consists of open sets since $\mathcal{B}$ is locally finite.

prove this implication directly by slightly modifying the above proof: find for each x a countable disjoint open family V_x with $q^{\leftarrow}\{x\} \subseteq V_x$ such that each member of V_x is included in some member of U.] ⌑

We now show that the essential difficulty of studying paracompactness of □-products with σ-compact locally compact factors lies in □-products with compact factors [this is due to Kunen, [Ku$_2$, §6]], and that it is easy to get information about □-products with paracompact locally compact factors. Hence we will consider only compact factors in the sequel.

9.8. PROPOSITION: If each X_n is paracompact and locally compact, then $\Box_n X_n$ is paracompact iff $\Box_n K_n$ is paracompact whenever each K_n is a compact subspace of X_n.

⌑ ONLY IF: Obvious.

IF: We first observe that each X_n admits a disjoint open cover U_n consisting of σ-compact locally compact subspaces, cf. [En$_1$, 5.1.24 and 5.3.10]. Then $\{\Box_n U_n\colon \text{each } U_n \in U_n\}$ is a disjoint open cover of $\Box_n X_n$ every member of which is a □-product of countably many σ-compact locally compact spaces. So we may assume without loss of generality that each X_n is σ-compact.

[This argument is essentially contained in [vD$_1$, §1].]

For each n choose a sequence $\langle X_{n,k}\rangle_k$ of compact G_δ-subsets of X_n with

$$X_n = \cup_k X_{n,k}, \text{ and } X_{n,k} \subseteq Int\, X_{n,k+1}, \text{ and } X_{n,0} = \emptyset$$

and for each n,k choose a sequence $\langle X_{n,k,i}\rangle_i$ of closed subsets of X_n with

$$X_{n,k} = \cap_i X_{n,k,i} \text{ and } X_{n,k,i+1} \subseteq Int\, X_{n,k,i}, \text{ and}$$
$$X_{n,k,i} \subseteq X_{n,k+1}$$

Then for each $f \in {}^{\omega}\omega$, the set of functions $\omega \to \omega$

$$\nabla^{f} = \cap_{i}\nabla_{n}X_{n,f(k),i}$$

is a subspace of ∇ which is clopen since ∇ is P-space, and which is paracompact by Theorem 9.6 since it is a closed subspace of $\nabla_{n}X_{n,f(n)+1}$. [Note that we use Fact 6.2b here.]

CLAIM: $\cup_{f\in F}\nabla^{f}$ is closed in ∇ for each $F \subseteq {}^{\omega}\omega$.

Indeed, let $x \in \square$ with $q(x) \notin \cup_{f\in F}\nabla^{f}$ be arbitrary. Define $\phi, \psi \in {}^{\omega}\omega$ by

$$\phi(n) = max\{k\colon x_{n} \notin X_{n,k}\}$$

$$\psi(n) = min\{i\colon x_{n} \notin X_{n,k,i}\}$$

and let

$$U_{n} = X_{n} - X_{n,\phi(n),\psi(n)}, \text{ for } n \in \omega.$$

Then $\nabla_{n}U_{n}$ is a neighborhood of $q(x)$. One easily checks that $\nabla_{n}U_{n} \cap \nabla^{f} = \emptyset$ for each $f \in F$.

Now let $<$ be a well-order on ${}^{\omega}\omega$. Then

$$\{\nabla^{f} - \cup_{g<f}\nabla^{g}\colon f \in {}^{\omega}\omega\}$$

is a disjoint open cover of ∇ by paracompact subspaces. Hence ∇ is paracompact, and so is $\square$ by Theorem 9.6. ⟦This is essentially Kunen's argument, which takes place inside $\square$, [Ku_2, §6].⟧

10. POSITIVE RESULTS, I: CONSISTENCY RESULTS

The first nontrivial positive result about $\square$-products is M.E. Rudin's theorem that under CH a $\square$-product of countably many compact metrizable spaces is paracompact, [Ru_1]. The following stronger result is due to Kunen, [Ku_2, 4.1]; the proof is much easier.

10.1. THEOREM: $CH \models$ if each X_{n} is compact with $w(X_{n}) \leq \underline{c}$, then $\square_{n}X_{n}$ is paracompact.

¤ By Theorem 9.6 it suffices to show that $\nabla = \nabla_n X_n$ is paracompact. It is easy to see that $w(\nabla) \leq \mathfrak{c}$. Since ∇ is zero-dimensional, by Corollary 9.3, it follows that every open cover $\mathcal{U}$ of ∇ has a clopen refinement $\mathcal{V}$ with $|\mathcal{V}| \leq \mathfrak{c}$. Let $<$ well-order $\mathcal{V}$ in type $\leq \mathfrak{c}$. Then $\{V - \cup\{W \in \mathcal{V}: W < V\}: V \in \mathcal{V}\}$ is a refinement of $\mathcal{U}$, whose members are open provided ∇ is a $P_{\mathfrak{c}}$-space[15]. That is the case since ∇ is a P-space by 9.1, and we assume CH.

[This is Kunen's proof, [Ku_2, 3.1].] ¤

Without CH this argument breaks down even if $w(X_n) = \omega_1$; consider for example ${}^{\omega_1}2$, then $G_\omega({}^{\omega_1}2)$ is not a P_{ω_2}-space[15], hence neither is $\nabla^\omega({}^{\omega_1}2)$ by 6.4. Hence $\nabla^\omega({}^{\omega_1}2)$ is a $P_{\mathfrak{c}}$-space iff CH holds. It is unknown if $\Box^\omega({}^{\omega_1}2)$ can be paracompact and normal if CH fails. In Theorem 13.8 we will see that $\Box^\omega({}^{\omega_2}2)$ is not normal, hence CH is equivalent to the conclusion of the theorem (as stated). The crucial thing about these two examples is that they are not first countable. Indeed, we will see that a $\Box$-product of first countable compact spaces is paracompact under axioms weaker than CH.

We need some definitions. For the duration of this section we think of ${}^\omega\omega$ not as a space, but only as the set of functions $\omega \to \omega$. For $f, g \in {}^\omega\omega$ we put

$f \leq^* g$ if $f(n) \leq g(n)$ for all but finitely many n,

$f <^* g$ if $f \leq^* g$ but $g \not\leq^* f$.

A λ-*scale* is by definition an indexed subset $\langle f_\alpha \rangle_{\alpha<\lambda}$ of ${}^\omega\omega$ such that

---

[15]A space is called a P_κ-*space* if the intersection of each family of fewer than κ open sets is open. Note that P-space[11] $= P_{\omega_1}$-space.

if $\alpha < \beta$ then $f_\alpha <^* f_\beta$, and $\forall g \in {}^\omega\omega\ \exists \alpha < \lambda\ [g <^* f_\alpha]$.

The following is well known.

10.2. PROPOSITION: *CH* implies *MA* implies $\exists\underline{c}$-scale + $\underline{c}$ is regular, and none of the implications can be reversed.

Therefore the following theorem, due to Kunen, [Ku_2, 4.4], shows that *CH* is not needed in Theorem 10.1 for first countable spaces. ¤

10.3. THEOREM: [$\underline{c}$ is regular + $\exists\underline{c}$-scale] $\models$ if each X_n is compact and first countable, then $\Box_n X_n$ is paracompact.

¤ Denote $\Box_n X_n$ by $\Box$ and $\nabla_n X_n$ by ∇. Every first countable compact space has cardinality at most $\underline{c}$ by Arhangel'skiĭ's theorem, [Ar, Thm. 1], [En_1, 3.1,30], hence $w(X_n) \leq \underline{c}$, hence $w(\nabla) \leq \underline{c}$. So the proof of 10.1 shows that it suffices to prove that ∇ is a $P_{\underline{c}}$-space. To this end we only have to show that each point in ∇ has a nonincreasing neighborhood base of length $\underline{c}$. Fix $x \in \nabla$. For each n choose a nonincreasing sequence $\langle B_n(x_n,k)\rangle_k$ of neighborhoods of x_n in X_n which forms a neighborhood base. For $f \in {}^\omega\omega$ define an open set U_f by

$$U_f = \nabla_n B_n(x_n, f(n)).$$

Then $U_f \subseteq U_g$ if $g \leq^* f$ by §4, hence $\langle U_{f_\alpha}\rangle_{\alpha<\underline{c}}$ is a decreasing sequence of length $\underline{c}$ which is a neighborhood base at x. ⟦This is Kunen's proof, [Ku_2, 4.3, 4.4].⟧ ¤

10.4. REMARK: If each X_n is compact, first countable and nondiscrete, then $\nabla_n X_n$ is a $P_{\underline{c}}$-space iff $\underline{c}$ is regular and $\exists\underline{c}$-scale. We prove the sufficiency: We use the notation of the above proof. Assume x_n is nonisolated in X_n for each n, and assume $B_n(x_n, k+1) \subset (B_n(x_n,k)$[16] for all k. Then

[16] $\subset$ denotes *proper* inclusion.

$U_f \subset U_g$ iff $f <^* g$

by §4. Since $|^{\omega}\omega| = \underline{c}$ and ∇ is a $P_{\underline{c}}$-space we now see that $\underline{c}$ is regular and that it is easy to construct a $\underline{c}$-scale.

Unfortunately, there need not be a $\underline{c}$-scale, as the following result of Hechler, [He], shows.

10.5. THEOREM (a): It is consistent with *ZFC* that $\underline{c}$ be anything reasonable, and that there is a λ-scale for any prescribed regular λ with $\omega_1 \leq \lambda \leq \underline{c}$.

(b) It is consistent with *ZFC* that there be no λ-scale, no matter what λ is. ¤

This shows that the following result, announced in [vD_3] and [vD_4], is stronger than Theorem 10.3 for metrizable spaces, but still relies on an additional axiom.

10.6. THEOREM. [$\exists$ λ-scale] $\models$ If each X_n is compact and metrizable, then $\square_n X_n$ is paracompact.

⟦Earlier Williams proved the special case that if $\exists\lambda$-scale then $\square^{\omega}(\omega + 1)$ is paracompact with a more complicated totally different proof, [W_1].⟧

¤ Denote $\square_n X_n$ by $\square$ and $\nabla_n X_n$ by ∇. The proofs of Theorems 10.1 and 10.3 relied on the fact that $w(\nabla) = \underline{c}$ and ∇ is a $P_{\underline{c}}$-space. In Theorem 10.1 we simply used *CH* to assert that ∇ is a $P_{\underline{c}}$-space, and in Theorem 10.3 we used a weaker set theoretic hypothesis than *CH* combined with additional information about the X_n's to show that ∇ is a $P_{\underline{c}}$-space. Although we now have even more information about the X_n's our set theoretic hypothesis has now become so weak that we can no longer assert that ∇ is a $P_{\underline{c}}$-space, as Remark 10.4 shows. This shows that we need a different idea for the proof that

∇ is paracompact: The next proof is the only proof in this section which does not use induction of length $\underline{c}$.

We will use the concept of a κ-metrizable space, where κ is any cardinal. For our purposes the following definition is most convenient:

A space X is called *κ-metrizable* if there is a function

$$U: X \times \kappa \to \{\text{the open sets of } X\}$$

such that

(1) $\{U(x,\alpha): \alpha < \kappa\}$ is a neighborhood base at x;

(2) for all $x,y \in X$ and $\alpha,\beta < \kappa$, if $\alpha \leq \beta$ then

(a) if $y \in U(x,\alpha)$ then $U(y,\beta) \subseteq U(x,\alpha)$, and

(b) if $y \notin U(x,\alpha)$ then $U(y,\beta) \cap U(x,\alpha) = \emptyset$.

This is equivalent to the usual definition if $cf(\kappa) > \omega$ but we won't bother. Our definition allows a straightforward proof of the following two claims, the second of which is known.

CLAIM 1: ∇ is λ-metrizable.

CLAIM 2: Every κ-metrizable space is (ultra)paracompact.

PROOF OF CLAIM 1: For $n,k < \omega$ and $x \in X_n$ let

$B_n(x,k)$ denote the open 2^{-k}-ball around x

(with respect to some metric on X_n). Note that

(1') $\{B_n(x,k): k < \omega\}$ is a neighborhood base at x; and

(2') for all $x,y \in X$ and $k,l < \omega$, if $k \leq l$ then

(a) if $y \in B_n(x,k+1)$ then $B_n(y,l+1) \subseteq B_n(x,k)$; and

(b) if $y \notin B_n(x,k)$ then $B_n(y,l+1) \cap B_n(x,k+1) = \emptyset$.

For $x \in \nabla$ and $f \in {}^{\omega}\omega$ define

$$V(x,f) = \cap_k \nabla_n B_n(x_n, f(n)+k).$$ Since ∇ is a P-space, by Proposition $\underline{9.1}$, it is clear that for each $x \in \nabla$

(1'') $\{V(x,f): f \in {}^{\omega}\omega\}$ is a neighborhood base at x.

We now claim that

(2") for all $x,y \in \nabla$ and $f,g \in {}^{\omega}\omega$ with $f \leq^* g$

(a) if $y \in V(x,f)$ then $V(y,g) \subseteq V(x,f)$, and

(b) if $y \notin V(x,f)$ then $V(y,g) \cap V(x,f) = \emptyset$.

Indeed, consider $x,y \in \nabla$, and $f,g \in {}^{\omega}\omega$, and $m \in \omega$ such that

$f(n) \leq g(n)$ for $n > m$.

CASE (a): $y \in V(x,f)$.

It suffices to show that for each k

(3) $\nabla_n B_n(y_n, g(n)+k+1) \subseteq \nabla_n B_n(x_n, f(n)+k)$.

So fix k. There is $m' \geq m$ such that

$y_n \in B_n(x, f(n)+k+1)$ for $n \geq m'$;

it then follows from (2') that

$B_n(y_n, g(n)+k+1) \subseteq B_n(x_n, f(n)+k)$ for $n \geq m'$,

which in turn implies (3).

CASE (b): $y \notin V(x,f)$.

There must be a $k < \omega$ and an infinite $N \subseteq \omega$ such that

$y_n \notin B_n(x_n, f(n)+k)$ for all $n \in N$.

Then

$B_n(y_n, g(n)+k+1) \cap B_n(x_n, f(n)+k+1) = \emptyset$ for $n \in N$ with $n \geq m$

by (2'), hence $V(y,g) \cap V(x,f) = \emptyset$.

This completes the proof of (2"). If $\langle f_\alpha \rangle_{\alpha < \lambda}$ is a λ-scale we can define a function U that witnesses that ∇ is λ-metrizable by

$U(x,\alpha) = V(x, f_\alpha)$.

PROOF OF CLAIM 2: Let U witness that X is κ-metrizable. First we observe that

(4) for all $x,y \in X$ and $\alpha < \kappa$, if $y \in U(x,\alpha)$ then $U(x,\alpha) = U(y,\alpha)$,

for $x \in U(y,\alpha)$ by (2b) hence $U(y,\alpha) \subseteq U(x,\alpha) \subseteq U(y,\alpha)$ by two applications of (2a).

Now let $\mathcal{V}$ be an open cover of X. We can define $\mu: X \to \kappa$ by

$$\mu(x) = min\{\alpha < \kappa: \exists V \in \mathcal{V}[U(x,\alpha) \subseteq V]\}.$$

Clearly

$$\mathcal{U} = \{U(x,\mu(x)): x \in X\}$$

is an open refinement of $\mathcal{V}$ that covers X. We show it is disjoint: consider any

$$x,y \in X \text{ with } \mu(x) \leq \mu(y).$$

Then $U(y,\mu(y)) \cap U(x,\mu(x)) = \emptyset$ if $y \notin U(x,\mu(x))$ by (2a), so assume $y \in U(x,\mu(x))$. Then $U(y,\mu(x)) = U(x,\mu(x))$ by (4), hence the minimality of $\mu(y)$ ensures that $\mu(y) \leq \mu(x)$, so that $\mu(y) = \mu(x)$. But then $U(y,\mu(y)) = U(x,\mu(x))$.

It follows that ∇ is paracompact, hence so is $\square$ by Theorem 9.6.

⟦This proof was presented in [vD_3].⟧ ⌑

10.7. REMARK: If each X_n is infinite, compact and metrizable, then ∇ is λ-metrizable iff there is a λ-scale. For choose $x \in \square_n X_n$ with each x_n nonisolated in X_n. Then $q(x)$ has a strictly decreasing neighborhood base of length λ, hence there is a λ-scale; this is clear from the argument in Remark 10.4. ⌑

10.8. REMARK: The above proof uses metrizability in an essential way, for a T_0-space X is metrizable if (and only if) there is a function $B: X \times \omega \to$ {open sets of X} satisfying (the analogues of) (1') and (2') above. (This is the Frink metrization theorem, [Fr, Thm. 3], [En_1, 5.4.D].) It is easy to see that $\underline{c}$ is regular + $\exists\underline{c}$-scale implies that $\nabla_n X_n$ is $\underline{c}$-metrizable if each X_n is first countable and compact; I don't know if this follows from $\exists\lambda$-scale (for some $\lambda < \underline{c}$), and in fact I don't know if $\exists\lambda$-scale (for some $\lambda < \underline{c}$) implies that $\Box_n X_n$ is paracompact if each X_n is first countable and compact.

The final result of this section, due to Roitman, [Ro$_2$] is like Theorem 10.6 an improvement of Theorem 10.3, but the set theoretical hypotheses are incomparable, as shown by Theorems 10.5 and 10.9.

As usual, we say that $D \subseteq {}^\omega\omega$ *dominates* if

$$\forall f \in {}^\omega\omega \; \exists g \in D \; f \leq^* g \; .$$

Consider the statement

$$\mathbb{D}: \text{ if } D \subseteq {}^\omega\omega \text{ dominates, then } |D| = \underline{c}.$$

Clearly, if $\mathbb{D}$ and $\exists\lambda$-scale then $\lambda = \underline{c}$ and λ is regular. However, from [He] we get

10.9. THEOREM: $\mathbb{D} + \neg\exists\underline{c}$-scale is consistent with ZFC. ¤

[Note that $\mathbb{D}$ is independent from ZFC by Theorem 10.5.]

Hence the following improves Theorem 10.3.

10.10. THEOREM: $\mathbb{D} \models$ if each X_n is compact and first countable, then $\Box_n X_n$ is paracompact.

¤ Denote $\Box_n X_n$ by $\Box$ and $\nabla_n X_n$ by ∇.

From the proof of Theorem 10.3, based on [Ar, Thm. 1], we see that $w(\nabla) \leq \underline{c}$. Since we do not assume $\exists\underline{c}$-scale, it could be the case that ∇ is not a $P_{\underline{c}}$-space, by Remark 10.4. The key idea of Roitman's proof is that we did not use the full strength of ∇ being a $P_{\underline{c}}$-space in the proof of Theorem 10.3; what we used was

CLAIM 1: ∇ has a base $\mathcal{B}$ such that for all $\mathcal{A} \subseteq \mathcal{B}$, if $|\mathcal{A}| < \underline{c}$ then $\cup\mathcal{A}$ is closed.

PROOF OF CLAIM 1: We will show that

$$\mathcal{B} = \{\cap_k \nabla_n B_{n,k}: \text{each } B_{n,k} \text{ open in } X_n, \text{ each } \overline{B}_{n,k+1} \subseteq B_{n,k}\}.$$

is as required. So let $\kappa < \underline{c}$, and if $\alpha < \kappa$ let $B_{n,k,\alpha}$'s satisfy

$$\text{each } B_{n,k,\alpha} \text{ is open in } X_n, \text{ and each } \overline{B}_{n,k+1} \subseteq B_{n,k}.$$

We have to show that

$$K = \cup_{\alpha < \kappa} \cap_k \nabla_n B_{n,k,\alpha}$$

is closed. Choose any $y \in \nabla - K$. For $n \in \omega$ let $\{U_{n,k}: k < \omega\}$ be a neighborhood base for y_n in X_n, with

$$\text{each } U_{n,k+1} \subseteq U_{n,k}.$$

For each $\alpha < \kappa$ we can find $k_\alpha < \omega$ with $y_n \notin \nabla_n \overline{B}_{n,k_\alpha,\alpha}$, hence choose $f_\alpha \in {}^\omega\omega$ with

$$U_{n,f_\alpha(n)} \cap B_{n,k_\alpha,\alpha} = \emptyset \text{ for infinitely many } n.$$

We claim that there is one single $g \in {}^\omega\omega$ such that

$$U_{n,g(n)} \cap B_{n,k_\alpha,\alpha} = \emptyset \text{ for infinitely many } n;$$

if so then for each $\alpha < \kappa$ we have

$$\nabla_n B_{n,g(n)} \cap \cap_k \nabla_n B_{n,k,\alpha} \subseteq \nabla_n B_{n,g(n)} \cap \nabla_n B_{n,k_\alpha,\alpha} = \emptyset$$

hence $y \notin \overline{K}$, as required.

The existence of g follows from the nontrivial part of the following claim, also due to Roitman. Recall that $[\omega]^\omega$ denotes the family of infinite subsets of ω.

CLAIM 2: $\mathbb{D}$ holds iff for all $F \subseteq {}^\omega\omega$ and $\mathcal{I} \subseteq [\omega]^\omega$ with $|F| + |\mathcal{I}| < \underline{c}$ there is a $g \in {}^\omega\omega$ such that

$$\{n \in I: f(n) \le g(n)\} \text{ is infinite for all } f \in F \text{ and } I \in \mathcal{I}.$$

PROOF OF CLAIM 2: Without loss of generality the members of F are nondecreasing. For $f \in F$ and $I \in \mathcal{I}$ define $h_{f,I} \in {}^\omega\omega$ by

$$h_{f,I}(n) = f(min(I \cap [n,\omega))).$$

Then $\{h_{f,I}: f \in F, I \in \mathcal{I}\}$ does not dominate, hence there is $g \in {}^\omega\omega$ such that

$\{n \in \omega: h_{f,I}(n) < g(n)\}$ is infinite for all $f \in F$ and $I \in \mathcal{I}$.

Again we may assume that g is nondecreasing. Then for all $f \in F$, $I \in \omega$ and $n < \omega$, if $h_{f,I}(n) < g(n)$ then

$$g(min(I \cap [n,\omega))) \geq g(n) \geq h_{f,I}(n) = f(min(I \cap [h,\omega))),$$

hence g is as required.

⟦This is essentially Roitman's proof, [Ro$_2$, §2].⟧ ¤

10.11. REMARK: If each X_n is nondiscrete and first countable, and ∇ denotes $\nabla_n X_n$, then the following are equivalent for each κ:

(a) if $D \subseteq {}^{\omega}\omega$ dominates then $|D| \geq \kappa$;

(b) ∇ has a base $\mathcal{B}$ such that $\cup\mathcal{A}$ is closed for each $\mathcal{A} \subseteq \mathcal{B}$ with $|\mathcal{A}| < \kappa$;

(c) for $A \subseteq \nabla$, if $|A| < \kappa$ then A is closed;

(d) $\chi(\nabla) \geq \kappa$.

PROOF: Pick $x \in \nabla$ with each x_n not isolated in X_n.

(a) ⇒ (b): Same argument as before.

(b) ⇒ (c): Given $y \in \nabla - A$ find $\mathcal{A} \subseteq \mathcal{B}$ with $|\mathcal{A}| = |A|$ and $A \subseteq \cup\mathcal{A} \subseteq \nabla - \{y\}$.

(c) ⇒ (d): There is $A \subseteq \nabla - \{x\}$ with $x \in \overline{A}$ and $|A| = \chi(x,\nabla)$.

(d) ⇒ (a): It is easy to verify that $\chi(\nabla) \leq |D|$ whenever D dominates.

10.12. REMARK: The importance of Theorems 10.6 and 10.10 is that they show that neither $\exists\lambda$-scale nor $\mathbb{D}$ is necessary for $\square^{\omega}(\omega + 1)$ to be paracompact; Roitman even has models in which $\square^{\omega}(\omega + 1)$ is paracompact and yet $\exists\lambda$-scale and $\mathbb{D}$ both fail, [Ro$_1$], [Ro$_2$]. Therefore I believe that one cannot prove that it is consistent with ZFC that $\square^{\omega}(\omega + 1)$ be nonpara-compact (or nonnormal) by finding a combinatorial statement which is independent from *ZFC* and follows from $\Box^{\omega}(\omega + 1)$ being paracompact or normal.

Roitman observes in [Ro_2, Thm. 0(b)] that the following is a easy corollary to Theorems 10.6 and 10.10.

10.13. THEOREM: $\underline{c} \leq \omega_2 \models$ A countable $\Box$-product of compact metrizable spaces is paracompact.

¤ If $\mathbb{D}$ holds we use Theorem 10.10. If $\mathbb{D}$ fails then there is a dominating set of cardinality ω_1 (since we assume $\underline{c} \leq \omega_2$). One can easily show that this implies that $\exists\omega_1$-scale. Now use Theorem 10.6. ⟦This is Roitman's proof.⟧ ⟦Williams has also proved this for $\Box^{\omega}(\omega+1)$ (unpublished).⟧ ¤

10.14. REMARK: Let $\langle X_n \rangle_n$ be a sequence of compact spaces, and denote $\Box_n X_n$ and $\nabla_n X_n$ by $\Box$ and ∇. The reader will have observed that the proofs in this section show that

(1) ∇ is hereditarily paracompact

and, with exception of the proof of Theorem 10.6, involve a transfinite induction of length $\underline{c}$, made possible since

(2) ∇ is $\underline{c}$-Lindelöf

which we knew since in fact

(3) $w(\nabla) \leq \underline{c}$.

We here comment on these features of the proofs of this section: Kunen has shown that if $\Box$ is paracompact, then $\Box$ is $\underline{c}$-Lindelöf[10], [Ku_2, 3.2] (we will not include a proof), hence (2) is essential. On the other hand Kunen also has shown that (2) holds if each X_n is scattered, [Ku_2, 5.1] (we will not include a proof), hence then ∇ is paracompact under *CH* by the proof of Theorem 10.1. In particular, this is true if each $X_n = \alpha\underline{c} \times \alpha\underline{c}^+$, where $\alpha\underline{c}$ and $\alpha\underline{c}^+$ denote the one-point compactifications of the discrete spaces $\underline{c}$ and $\underline{c}^+$. But for this choice of the X_n's we have $w(\nabla) = \underline{c}^+$, and ∇ has a nonnormal subspace since $G_\omega(\alpha\underline{c} \times \alpha\underline{c}^+)$, which is easily seen to have a nonnormal subspace, can be embedded in ∇ by Proposition 6.4. Hence neither (1) nor (3) are essential.

10.15. REMARK: The proofs of this section also show that $\nabla_n X_n$ is hereditarily paracompact under the following circumstances:

(a) CH holds and each $w(X_n) \leq \underline{c}$;

(b) $\exists\lambda$-scale and each X_n is metrizable;

(c) $\mathbb{D}$ holds, and each X_n is first countable with $|X_n| \leq \underline{c}$.

10.16. QUESTIONS: The two extreme questions one would like to be answered are:

Q1: Is $\square_n X_n$ paracompact in ZFC if each X_n is compact and either first countable or has $w(X_n) \leq \omega_1$.

Q2: Is it consistent that a countable $\square$-product of infinite compact spaces never is countably orthocompact?

If the answer to both questions is no, one can look at intermediate questions: put more restrictions on the factors, or strengthen countable orthocompactness. If the answer to Q2 is yes, one can ask what happens if one strengthens countable orthocompactness, or could ask if in certain models $\square$-products of compact metrizable spaces behave better than $\square$-products of compact first countable spaces.

A question of Kunen, [Ku_2], which has many variations, is

Q3: Can a $\square$-product of compact spaces be normal without being paracompact?

Another question, also with many variations, is suggested by Q3 and Remark 10.15:

Q4: Is it consistent that $\nabla^{\omega}(\omega + 1)$ is paracompact but not hereditarily paracompact?

11. POSITIVE RESULTS II: HONEST RESULTS

Our first result does not require additional axioms, and puts no restriction on the number of factors. There also is no restriction that the factors be metrizable (or only first countable), but this, perhaps paradoxically, strikes me as a disadvantage of the result: if there were such a restriction we would know how to exploit it without at the same time assuming additional axioms, and we have to know how to use such a restriction if we want to prove that a countable box product of compact metrizable (or first countable) spaces is paracompact (if it is true in *ZFC*) since without such a restriction the result is false.

11.1. THEOREM: A box product of compact spaces, no matter how many, is pseudonormal.

¤ This is a consequence of the following claims.

CLAIM 1: A countable box product of compact spaces is pseudonormal.

CLAIM 2: A box product is pseudonormal iff each countable sub-$\square$-product is pseudonormal.

Claim 2 makes the "no matter how many factors" a lot less appealing, of course, even though the proof of Claim 2 is more complicated than that of Claim 1. We will need

CLAIM 3: The following conditions on a countable subset K of a space X are equivalent:

(a) K has arbitrarily small closed neighborhoods;

(b) every open (in X) cover of K has a discrete (in X) open (in X) refinement that covers K, and

(c) every open (in X) cover of K has a discrete (in X) open (in X) refinement that covers K.

Before we prove our three claims we point out a trivial consequence of the fact that countable ∇-products are zero-dimensional P-spaces, by Proposition 9.1 and Corollary 9.3:

TRIVIALITY: If $K \subseteq \nabla_n X_n$ is countable, then K is closed discrete and is separated[17] by a discrete clopen family in $\nabla_n X_n$.

PROOF OF CLAIM 1: Denote $\Box_n X_n$ by $\Box$ and $\nabla_n X_n$ by ∇. Let G and K be disjoint closed subsets of $\Box$ with K countable. Put

$$K^* = q^{\leftarrow} q^{\rightarrow} K,$$

then K^* is σ-compact, hence is Lindelöf, since each $q^{\leftarrow}\{x\}$ is covered by countably many homeomorphs of finite subproducts. So upon repeating the proof that regular Lindelöf spaces are normal, cf. the proof of [En_1, 1.5.14], we see that $G \cap K^*$ and K have disjoint neighborhoods U_G and U_K in $\Box$. Then $(\Box - G) \cup U_G$ is a neighborhood of K^* in $\Box$. Since $q\colon \Box \to \nabla$ is closed, by Theorem 9.5, we see from the Triviality that there is a clopen V in $\Box$ with

$$K^* \subseteq V \subseteq (\Box - G) \cup U_G.$$

Then $(\Box - V) \cup U_G$ and $V \cap U_K$ are disjoint neighborhoods of G and K.

PROOF OF CLAIM 2: Since pseudonormality is closed hereditary it suffices to prove the sufficiency. Denote $\Box_\alpha X_\alpha$ by $\Box$.

Let $K \subseteq \Box$ be countable and closed. The key to the proof is a suitable factorization of $\Box$ as a product of two factors. For $x, y \in K$ define

[17] A collection $\mathcal{C}$ is said to *separate* a set T if $\forall t \in T[\,|\{C \in \mathcal{C}\colon t \in C\}| = 1]$.

$$d(x,y) = \{\alpha < \nu: x_\alpha \neq y_\alpha\}.$$

Since K is countable, one can find a countable $N \subseteq \nu$ such that

(1) $\forall x,y \in K[d(x,y)$ finite $\Rightarrow d(x,y) \subseteq N]$; and

(2) $\forall x,y \in K[d(x,y)$ infinite $\Rightarrow N \cap d(x,y)$ infinite$]$.

We need some notation:

$\square_0$ denotes $\square_{\alpha \in N} X_\alpha$, and $\square_1$ denotes $\square_{\alpha \in \nu - N} X$;

we identify $\square$ with $\square_0 \times \square_1$;

$\pi_i: \square \to \square_i$ denotes the projection, for $i < 2$;

∇_0 denotes $\nabla_{\alpha \in N} X$; and $q: \square_0 \to \nabla_0$ denotes the quotient.

For every two $x,y \in K$ with $q(\pi_0(x)) = q(\pi_1(x))$ the set $N \cap d(x,y)$ is finite, hence $d(x,y) \subseteq N$ by (1) and (2), so that $\pi_1(x) = \pi_1(y)$. Hence if we put

$$S = q \vec{0} \vec{\pi}_0 K, \text{ and } K_s = \vec{\pi}_0 K \cap q^{\leftarrow}\{s\} \text{ for } s \in S,$$

then we can find $y_s \in \square_1$ for $s \in S$ such that

$$K = \cup_{s \in S} K_s \times \{y_s\},$$

and because of the Triviality there is a

discrete clopen family $\langle D_s : s \in S \rangle$ in $\square_0$ with

$K_s \subseteq D_s$ for $s \in S$ (and $D_s \cap D_t = \emptyset$ for distinct $s, t \in S$).

Let U be any open (in $\square$) cover of K. For each $s \in S$

$$U_s = \{U^s : U \in U\}, \text{ where } U^s = \{x \in \square_0 : \langle x, y_s \rangle \in U\},$$

is an open (in $\square$) cover of K_s; as K_s is closed, since $K_s \times \{y_s\} = K \cap \square \times \{y_s\}$, it follows from Claim 3 and from our assumption that $\square_0$ is pseudonormal, that there is

a discrete (in $\square_0$) open (in $\square_0$) family V_s that covers K_s and refines U_s;

let $c_s: V_s \to U_s$ satisfy $V \subseteq (c_s(V))^s$ for each $V \in V_s$. Then

$$\cup_{s \in S} \{\pi_0^{\leftarrow} V \cap c_s(V) \cap D_s : V \in V_s\}$$

is a discrete (in $\square$) open (in $\square$) refinement of $\mathcal{U}$ that covers K.

Because of Claim 3 this shows that $\square$ is pseudonormal.

PROOF OF CLAIM 3: (a) $\Rightarrow$ (b): We first observe that

(I) every point of K has arbitrarily small neighborhoods whose boundaries miss K.

This is obvious if X is completely regular, and can be proved with a simple inductive construction if X is merely assumed to be regular.

Now let $\mathcal{U}$ be any open (in X) cover of K. Using (I) and the fact that K is countable we can easily find a disjoint open (in X) refinement $\mathcal{V}$ of $\mathcal{U}$ that covers K. Let W be an open neighborhood of K with $\overline{W} \subseteq \cup\mathcal{V}$. Then $\{V \cap W\colon V \in \mathcal{V}\}$ is a discrete (in X) open (in X) refinement of $\mathcal{U}$ that covers K.

(b) $\Rightarrow$ (c) Obvious.

(c) $\Rightarrow$ (a) Let W be any neighborhood of K. Then

$$\mathcal{U} = \{U \subseteq X\colon U \text{ open and } \overline{U} \subseteq W\}$$

is an open (in X) cover of K. Let $\mathcal{V}$ be a locally finite (in X) open (in X) refinement of $\mathcal{U}$ that covers X. Then $\mathcal{V}$ is closure preserving (in X), hence $\cup_{V\in\mathcal{V}}\overline{V}$ is a closed neighborhood of K that is included in W.

[That a countable $\square$-product of compact space is pseudonormal and an uncountable $\square$-product of compact spaces has property D[18] was announced in [vD$_4$], and was based on an analogue of Claim 2 for property D. The current theorem was discovered when writing up this paper. Claim 3 is essentially due to

---

[18]A space is said to have *property D* if it is strongly ω-CWH[19]; this is weaker than pseudonormality, [vD$_5$,vDW].

[19]A space is called (*strongly*) κ-CWH (CWH = collectionwise Hausdorff) if every closed discrete subset of cardinality at most κ can be separated[17] by a disjoint (discrete) open family.

Kunen (who did not consider (b)), and was used by Miller, [Mi], in his proof that a countable □-product of countable spaces is pseudonormal; the restriction of the number of factors is redundant by Claim 2.⟧ ¤

Our next result is just a simple corollary to earlier results.

11.2. THEOREM: (a) A countable □-product of compact spaces is ω_1-*CWH*.[19].

(b) A countable □-product of compact metrizable spaces is strongly ω_1-*CWH*[19] and is ω_2-*CWH*.

¤ Let □ be a countable □-product of compact spaces, and let ∇ be the corresponding ∇-product. Since q is closed and has Lindelöf fibers, and since □ is pseudonormal, by Theorem 11.1, it suffices to prove the analogous statements for ∇-products.

PROOF OF (A): □ is a regular P-space by Proposition 9.1 and Corollary 9.3, hence clearly is ω_1-*CWH*.

PROOF OF (B): Define

$$\delta = min\{|D| : D \subseteq {}^{\omega}\omega \text{ dominates}\}.$$

[Recall from Theorem 10.5, due to Hechler, [He] that each of $\delta = \omega_1$, $\omega_1 < \delta < \underline{c}$ and $\delta = \underline{c}$ is consistent with *ZFC*.]

CASE 1: $\delta = \omega_1$. Then ∇ is ω_1-metrizable, hence is paracompact, as noted in the proof of Theorem 10.12, due to Roitman, [Ro_2, Thm. 0(b)], hence ∇ is strongly κ-*CWH* for all κ.

CASE 2: $\delta > \omega_1$. By the proof of Theorem 10.10, due to Roitman, [Ro_2, Thm. 0(a)], ∇ has a base $\mathcal{B}$ such that $\cup\mathcal{A}$ is closed for each $\mathcal{A} \subseteq \mathcal{B}$ with $|\mathcal{A}| < \delta$. Hence ∇ is strongly κ-*CWH* for all $\kappa < \delta$, and is κ-*CWH*.

⟦This I observed some time ago.⟧ ¤

11.3. REMARK: Let δ be as in the above proof. Then ${}^{\omega}\omega \times \square^{\omega}(\omega + 1)$ is not δ - *CWH*, by Remark 12.7.

12. NEGATIVE RESULTS ON METRIZABLE FACTORS II: COMBINATORIAL ARGUMENTS

As in §7, $\square$ denotes $\square^{\omega}(\omega + 1)$, and ${}^{\omega}\omega$ carries the Tychonoff topology unless we indicate explicitly that it is considered as subset of $\square$.

In this section we amplify the results of §7.

12.1. THEOREM: ${}^{\omega}\omega \times \square$ is not discretely pseudonormal.

12.2. THEOREM: $\square$ has a subspace that is not discretely pseudonormal.

Theorem 12.2 is essentially due to Kunen, [Ku_1, Thm. 3]. Theorem 12.1 is derived from Theorem 12.2 the same way. Theorem 7.1 was derived from Theorem 7.2. We note that Miller has shown that $\square_n X_n$ is pseudonormal if each X_n is countable, [Mi]; our example shows that the condition that each X_n is countable is essential. (We will not include a proof of Miller's result.)

We need this definition: Let $D \subseteq {}^{\omega}\omega$ and $B \subseteq \omega$, then D is called *eventually dominant* (e.d.), or *strictly dominant* (s.d.) on B, if for all $f \in {}^{\omega}\omega$ there is $d \in D$ such that $f(n) < d(n)$ for all but finitely many n, or $f(n) < d(n)$ for all $n \in B$, respectively.

[Note: D is e.d. iff D dominates, as defined in §10. We have used the term e.d. for clarity.]

12.3. LEMMA: If $D \subseteq {}^{\omega}\omega$ is e.d., then there are a finite $A \subset \omega$ and a $p: a \to \omega + 1$ such that $\{d \in D: d \supseteq p\}$ is s.d. on $\omega - A$.

¤ We first show that

(1) D is s.d. on $\omega - n$ for some n.

Indeed, if not, then for each n there is a $f_n \in {}^{\omega}\omega$ such that

$$\text{for all } d \in D \text{ there is } m \geq n \text{ with } f_n(m) \geq d(m).$$

Define $h \in {}^{\omega}\omega$ by

$$h(n) = max_{k \leq n} f_k(n).$$

Then for each $d \in D$ and $n < \omega$ there is $m \geq n$ with $h(m) \geq d(m)$, which contradicts our assumption that D is e.d.

We next show that

(2) if E is s.d. on B, and $i \in \omega - B$, then either

(a) E is s.d. on $B \cup \{i\}$, or

(b) there is $j < \omega$ such that $\{e \in E\colon e(i) = j\}$ is s.d. on B.

Indeed, suppose E is s.d. on B but not on $B \cup \{i\}$, and pick $f \in {}^{\omega}\omega$ such that

$$\text{for all } e \in E \text{ there is } k \in B \cup \{i\} \text{ with } f(k) \geq e(k).$$

Consider any $g \in {}^{\omega}\omega$. Since E is s.d. on B there is $e \in E$ such that $e(k) > max\{f(k), g(k)\}$ for all $k \in B$. Then $e(i) \leq f(i)$, so we have shown that $\cup_{j \leq g(n)} \{e \in E\colon e(i) = j\}$ is s.d. on B. Since the union of finitely many sets which are not s.d. on B is not s.d. on B, this proves (b).

Since n (as in (1)) is finite, we can construct A and p in finitely many steps using (2).

⟦This is due to Kunen, [Ku_1, 1].⟧ ¤

12.4. PROOF OF THEOREM 12.2: We will define $F, G \subseteq (\omega + 1) \times \square$ such that the subspace

$$Z = F \cup G \cup \omega \times {}^{\omega}\omega$$

of $(\omega + 1) \times \square$ is not discretely pseudonormal. This proves our theorem since $(\omega + 1) \times \square$ and $\square$ are homeomorphic.

Enumerate

$$P = \cup\{{}^{A}\omega : A \subseteq \omega \text{ is finite}\}$$

as $\langle p_n \rangle_n$ in such a way that each member of P is listed ω times. For $n < \omega$ define $f_n \in \Box$ by

$$f_n = p_n \cup (\omega - dom(p_n)) \times \{\omega\}.$$

Then we can define F and G by

$$F = \{\langle n, f_n \rangle : n < \omega\}.$$

$$G = \{\omega\} \times {}^{\omega}\omega.$$

We begin with observing that

(1) $\{[n,\omega] \times \{g\} : n < \omega\}$ is a neighborhood base in Z for $\langle n, g \rangle$.

We next point out that

(2) F and G are disjoint closed discrete subsets of Z.

Indeed, since the points of $Z - (F \cup G) = \omega \times {}^{\omega}\omega$ are isolated in Z, and since $F \cap G = \emptyset$, it suffices to show that $F \cup G$ is relatively discrete: for each $n < \omega$ we have

$$(F \cup G) \cap (\{n\} \times \Box) = \{\langle n, f_n \rangle\},$$

and for each $g \in {}^{\omega}\omega$ we have

$$(F \cup G) \cap ((\omega + 1) \times \{g\}) = \{\langle \omega, g \rangle\}.$$

Since F is countable, it remains to show that F and G do not have disjoint neighborhoods. Let U_F and U_G be any neighborhoods of F and G. For $m < \omega$ define

$$D_m = \{g \in {}^{\omega}\omega : [m,\omega] \times \{g\} \subseteq U_G\}.$$

Then $\cup_m D_m = {}^{\omega}\omega$ by (1). Since ${}^{\omega}\omega$ is trivially s.d. on ${}^{\omega}\omega$, and since the union of countably many non-e.d. sets is non-e.d. (cf. the proof of (1) in the proof of Lemma 12.3), some D_m is e.d. By Lemma 12.3 there are a finite $A \subseteq \omega$ and a $p: A \to \omega$ such that

(3) $\{d \in D_m : d \supseteq p\}$ is s.d. on $\omega - A$.

There is $n \geq m$ with $p_n = p$. Since U_F is a neighborhood of $\langle n,f_n\rangle$, and $f_n(k) = \omega$ for $k \notin dom(p)$, there is $h \in {}^{\omega}\omega$ such that the neighborhood

$$V = \{\langle n,g\rangle : g \in {}^{\omega}\omega \text{ with } g \supseteq p \text{ and } g(k) > h(k) \text{ for } k \notin dom(p)\}$$

of $\langle n,f_n\rangle$ is included in U_F. By (3) there is $g \in D_m$ with

$$g \supseteq p, \text{ and } g(k) > h(k) \text{ for } k \notin dom(p).$$

As $p = p_n$ and $n \geq m$, we now see from the definition of D_m that

$$\langle n,g\rangle \in V \cap (\omega - m)\times\{g\} \subseteq U_F \cap U_G.$$

Hence $U_F \cap U_G \neq \emptyset$, as required.

[Kunen gives the same argument to show that

$$F' = \{\langle n,f\rangle : f(k) = \omega \text{ for all but finitely many } k\}$$

and G do not have disjoint neighborhoods in $F' \cup G \cup \omega \times {}^{\omega}\omega$, [Ku$_1$, §3].] ¤

12.5. PROOF OF THEOREM 12.1: The set Z of the above proof is the union of three G_δ-sets, when considered as subset of $(\omega + 1) \times {}^{\omega}(\omega + 1)$ since F is relatively discrete as subset of $(\omega + 1) \times {}^{\omega}(\omega + 1)$. The result now follows from the argument in 7.6, based on [AfU, IV].

[This result was announced in [vD$_4$].] ¤

12.6. REMARK: In the notation of 12.4, if

$$Z' = F' \cup G \cup \omega \times {}^{\omega}\omega$$

then Z', as subspace of $(\omega + 1) \times {}^{\omega}(\omega + 1)$, is not homeomorphic to ${}^{\omega}\omega$, since Z' is not completely metrizable, for

$$\{\langle 0,f\rangle \in Z' : \text{each } f(n) \in \{0,\omega\}\}$$

is a dense in itself countable closed subspace of Z'. So we have to thin out F' and get F if we want the noncompact factor to be ${}^{\omega}\omega$, even if one does not care about *discrete* pseudonormality.

12.7. REMARK: Let $D \subseteq {}^{\omega}\omega$ be e.d., of minimal cardinality δ. Then the argument in 12.4 shows that F and $\{\omega\} \times D$ do not have disjoint neighborhoods, hence ${}^{\omega}\omega \times \square$ is not δ-CWH[16], (Recall from Theorem 10.5, due to Hechler, [He], that it is consistent with ZFC that $\delta < \underline{c}$.) By contrast we have seen in the proof of Theorem 11.2 that a $\square$-product of countably many compact metrizable spaces is $max\{\omega_2, \delta\}$-CWH.

We ask questions about ${}^{\omega}\omega \times \square$ and $\square$ in Question 13.13.

13. NEGATIVE RESULTS ABOUT COMPACT FACTORS II: COMBINATORIAL ARGUMENTS

The key observation for handling compact factors is the following

13.1. LEMMA: Let κ, λ, ν be cardinals with κ regular and $\kappa \leq \nu$. If there is $\mu < \kappa$ with $\lambda \leq 2^{\mu}$, then $<\kappa - \square^{\nu}2$ and $<\kappa - \square^{\nu}\lambda$ are homeomorphic.

¤ For each cardinal α with $2 \leq \alpha \leq 2^{\mu}$ we have that

$$<\kappa - \square^{\nu}\alpha \text{ and } <\kappa - \square^{\nu}(<\kappa - \square^{\mu}\alpha) \text{ are homeomorphic}$$

by the associative law 6.1c, since $\mu < \nu$. As $<\kappa - \square^{\mu}\alpha$ and α^{μ} are clearly homeomorphic, and $\alpha^{\mu} = 2^{\mu}$ since $2 \leq \alpha \leq 2^{\mu}$, we see that

$$<\kappa - \square^{\nu}\alpha \text{ is homeomorphic to } <\kappa - \square^{\nu}(2^{\mu}).$$

The result now follows from the cases $\alpha = 2$ and $\alpha = \lambda$. [This is a straightforward generalization of the case $\kappa = \lambda = \omega_1$ and $\nu = \omega_2$ which occurs in [vD_2, 2].] ¤

13.2. COROLLARY: If $\kappa \leq \nu$ then $\square^{\kappa}({}^{\nu}2)$ and $\nabla^{\kappa}({}^{\nu}2)$ have a closed subspace homeomorphic to $\underline{<}\kappa - \square^{\nu}\lambda$ for $2 \leq \lambda \leq 2^{\kappa}$.

¤ Since $G_{\kappa}({}^{\nu}2)$ and $\underline{<}\kappa - \square^{\nu}2$ are homeomorphic, this follows from Proposition 6.3. ¤

So we are interested in spaces of the form $<\kappa - \Box^{\nu}\lambda$, at least for $\kappa = \omega_1$ (or, more generally, for successor ω). For $\kappa = \lambda = \omega$ and $\nu = \omega_1$ a well known result of Stone, [St, Thm. 3] is that

(1) ${}^{\omega_1}\omega$ is not normal.

Borges, in [Bo_1], points out that a straightforward generalization of Stone's proof yields

(2) if κ is regular, then $<\kappa - \Box^{\kappa^+}\kappa$ is not normal.

Scott, in [Sc_2, 3.1.0], see also [Sc_3, 2.4], showed that the central idea also leads to

(3) if κ is regular, then $<\kappa - \Box^{\kappa^+}\kappa$ is not orthocompact.

Nagami, in [Na, Lemma 2.6], shows that Stone's proof can be modified so as to prove the stronger, by Proposition 5.5, result that ${}^{\omega_1}\omega$ is not countably paracompact; and little change in his proof, or Remark 5.6, shows that in fact

(4) ${}^{\omega_1}\omega$ is not paranormal.

It is easy to generalize this, and prove (2) with "paranormal" instead of "normal". Borges has asked if the condition that κ be regular is essential in (2). Our first result shows that it is not; although a large part of the proof of (3) is identical with that of (2), I am unable to avoid the assumption that κ be regular in the proof of (3).

13.3. THEOREM: $<\kappa - \Box^{\kappa^+}\kappa$ is not paranormal.

13.4. THEOREM: If κ is regular than $<\kappa - \Box^{\kappa^+}\kappa$ is not orthocompact.

For suitable κ we can improve Theorem 13.3.

13.5. THEOREM: If $\kappa^{\underset{\smile}{\lambda}} = \kappa$ [20], then $<\kappa - \Box^{\kappa^+}\kappa$ is not metanormal.

---

[20] $\kappa^{\underset{\smile}{\lambda}} = sup\{\kappa^{\mu}: 1 \leq \mu < \lambda\}$.

13.6. COROLLARY: If $\kappa^{<\kappa} = \kappa$, then $<\kappa$ - $\square^{\kappa^+}\kappa$ is not metanormal. ¤

I do not know if the hypothesis that $\kappa^{<\kappa} = \kappa$ (which implies that κ is regular) is essential in the Corollary, this would lead to some obvious improvements in the results below.

[In this context it should be pointed out that the *GCH* implies that $\kappa^{<\kappa} = \kappa$ iff κ is regular, but that it is consistent with *ZFC* that $\kappa^{<\kappa} = \kappa$ iff $\kappa = \omega$: it is consistent with *ZFC* that $2^\lambda = \lambda^{++}$ for all regular $\lambda \geq \omega$, and $2^\lambda = \lambda^+$ for all singular λ, and that there are no strongly inaccessible cardinals, [Ea], and then $\kappa^{<\kappa} = \kappa^+$ for all $\kappa > \omega$.]

Before we proceed to the proofs of Theorems 13.3, 13.4 and 13.5 we present the corollaries.

13.7. THEOREM: ${}^{\omega_1}\omega$ is not metanormal and is not countably orthocompact.

¤ ${}^{\omega_1}\omega$ is not metanormal by 13.5, hence is not countably orthocompact by 5.5b, which is based on [Sc_1, 1.3], cf. [Do, Thm. 4].

[I find the following alternative proof that ${}^{\omega_1}\omega$ is not countably orthocompact of interest: ${}^{\omega_1}\omega$ is separable, [En_1, 2.3.15], and it is easy to show that a separable countably orthocompact space is countably metacompact.]

⟦This I knew when I announced that ${}^{\omega_1}\omega$ is not countably metacompact, [vD_4]. After being informed that ${}^{\omega_1}\omega$ is not metanormal, Roman Pol informed me that this can also be proved from the results of [PP-P], which were obtained independently.⟧

⟦The corollary that ${}^{\omega_1}\omega$ is not subparacompact is due to Alster and Engelking, [AE, Thm. 2]. The corollary that ${}^{\omega_1}\omega$ is neither countably subparacompact nor submetacompact is due, independently, to Pol and Puzio-Pol, [PP-P, p. 63].⟧ ¤

13.8. THEOREM: $\square^{\omega}({}^{\omega_2}2)$ is neither paranormal nor orthocompact.

¤ Use 13.2 and 13.3.

⟦That (2) implies that $\square^{\omega}({}^{\omega_2}2)$ is nonnormal was observed in [vD_2]; Scott subsequently proved (3), and hence that $\square^{\omega}({}^{\omega_2}2)$ is not orthocompact, [Sc_2, 7.0.1]. That $\square^{\omega}({}^{\omega_2}2)$ is not countably paracompact was announced in [vD_4].⟧ ¤

13.9. THEOREM: $\square^{\omega}({}^{\underline{c}^+}2)$ is neither metanormal nor countably orthocompact.

¤ $\square^{\omega}({}^{\underline{c}^+}2)$ is not metanormal by 13.5 and 13.2, hence is not countably orthocompact by 5.5b, which is based on [Sc_1, 1.3], cf. [Do, Thm. 4].

⟦This I realized when writing up this paper.⟧ ¤

Our final result shows that *CH* is essential in Theorem 10.1.

13.10. THEOREM: The following statements are equivalent:

(a) *CH*;

(b) if each X_n is compact, with $w(X_n) \leq \underline{c}$, then $\square_n X_n$ is paracompact;

(c) $\square^{\omega}({}^{\underline{c}}2)$ is paracompact;

(d) $\square^{\omega}({}^{\underline{c}}2)$ is normal;

(e) $\square^{\omega}({}^{\underline{c}}2)$ is paranormal; and

(f) $\square^{\omega}({}^{\underline{c}}2)$ is orthocompact.

¤ (a) ⇒ (b): This is Theorem 10.1.

(e) ⇒ (a) and (f) ⇒ (a): $\square^{\omega}({}^{\omega_2}2)$ cannot be embedded as a closed subset in $\square^{\omega}({}^{\underline{c}}2)$ by 13.8, hence ${}^{\underline{c}}2$ cannot be embedded in ${}^{\omega_2}2$.

⟦(a) ⇒ (b) is due to Kunen, [Ku_2, 4.1]; (d) ⇒ (a) was proved in [vD_2], and inspired Scott's (f) ⇒ (a), [Sc_2, p. 79].⟧

13.11. JOINT PROOF OF THEOREMS 13.3, 13.4 AND 13.5: Throughout the proof we keep κ and λ, with $\lambda \le \kappa$ fixed, and denote $<\lambda - \Box^{\kappa^+}\kappa$ by $\Box$.

We will have $\kappa = \lambda$ in the proofs of Theorems 13.3 and 13.4, but will always distinguish between κ and λ since they play different roles.

We emphasize that the underlying set of $\Box$ is the set of functions $\kappa^+ \to \kappa$. For the duration of the proof we introduce the following terminology and define $E(p)$ as follows.

A *partial function* is any $p \subseteq \kappa^+ \times \kappa$ which is a function.

A *partial injection* is a partial function that is injective.

For $p \subseteq \kappa^+ \times \kappa$ the set of functions that *E*xtend p is denoted by

$$E(p) = \{x \in \Box: x \supseteq p\}.$$

Note that $E(p) \neq \emptyset$ iff p is a partial function. Also note that

$\{E(x \restriction D): D \in [\kappa^+]^{<\lambda}\}$[21] is a neighborhood base in $\Box$ for x, for each $x \in \Box$.

Following Stone we define for each $\alpha < \kappa$

$$F_\alpha = \{x \in \Box: \forall\beta \in \kappa - \{\alpha\}\ [|x^{\leftarrow}\{\beta\}| \le 1]\}.$$

CLAIM 1: $\langle F_\alpha: \alpha < \kappa\rangle$ is a discrete closed family in $\Box$, with $F_\alpha \neq F_\beta$ if $\alpha \neq \beta$.

PROOF OF CLAIM: Consider any $x \in \Box$. Since $\kappa < \kappa^+$ there are distinct $\xi,\eta < \kappa^+$ and $\alpha < \kappa$ such that $x(\xi) = x(\eta) = \alpha$. Then

---

[21] $[X]^{<\lambda} = \{A \subseteq X: |A| < \lambda\}$.

$E(\{\xi,\nu\} \times \{\alpha\})$ is a neighborhood of x (even in ${}^{\kappa^+}\kappa$) that misses F_β for $\beta \neq \alpha$. If $x \notin F_\alpha$ there are $\xi',\eta' < \kappa^+$ and $\beta \neq \alpha$ with $x(\xi') = x(\eta') = \beta$, and then $E(\{\xi',\eta'\} \times \{\beta\})$ is a neighborhood of x that misses F_α.

Because it does not require additional work, we will prove more than that □ is not paranormal, or not even metanormal if $\kappa^\lambda = \kappa$: Throughout the remainder of the proof we let

$U_{\alpha,\beta}$ be an arbitrary open set with $U_{\alpha,\beta} \supseteq F_\alpha$, for $\alpha,\beta < \kappa$.

Our goal will be to prove the following two statements.

(I) $\bigcap_{\alpha,\beta}\overline{U}_{\alpha,\beta} \neq \emptyset$.

(II) If $\kappa^\lambda = \kappa$ then even $\bigcap_{\alpha,\beta}U_{\alpha,\beta} \neq \emptyset$.

For proof technical reasons we reenumerate the $U_{\alpha,\beta}$'s as $\langle U_\xi: \xi < \kappa\rangle$ in such a way that

each $U_{\alpha,\beta}$ is listed κ times as U_ξ[22].

Let $a: \kappa \to \kappa$ satisfy

$F_{a(\xi)} \subseteq U_\xi$, for $\xi < \kappa$.

Also, for $\xi < \kappa$ and $x \in F_{a(\xi)}$ define

$W_{x,\xi} = U_\xi$.

[The reason for doing so will become clear.]

For the proof that □ is not orthocompact we consider an arbitrary

open family $\mathcal{W}$ in □ which covers F_0 and refines

$\{E(\{\langle\alpha,0\rangle\}): \alpha < \kappa^+\}$,

and for each $x \in F_0$ and $\xi < \kappa$ we define

$W_x = W_{x,\xi} = \bigcap\{W \in \mathcal{W}: x \in W\}$.

[The reason for having the subscript ξ will become clear.]

We will prove that

---

[22]This is implicit in [St].

(III) if κ is regular then some W_x is not open.

This shows that the open cover

$\{E(\{\langle\alpha,0\rangle\}): \alpha < \kappa^+\} \cup \{\square - F_0\}$

of $\square$ witnesses that $\square$ is not orthocompact.

The proofs of (I), (II) and (III) have much in common, in order to make this clear we proceed jointly when possible, and split our proof when needed. Also, when proving (III) we assume $\neg$(III), and redefine a to be the constant function $\kappa \times \{0\}$; this was the advantage that part of the proofs can be done simultaneously, since we now have

$W_{x,\xi}$ is a neighborhood of x, for each $x \in F_{a(\xi)}$.

We plan to construct for each $\xi < \kappa$

a partial injection p_ξ, and either

a $D_\xi \in [\kappa^+]^{<\lambda}$ (IF WE PROVE (I) OR (III)), or

a set T_ξ of partial functions (IF WE PROVE (II))

such that

(1) $p_\xi \subseteq p_\eta$ if $\xi < \eta < \kappa$,

(2) IF WE PROVE (I) OR (III) THEN

(a) $D_\xi = dom(p_{\xi+1}) - dom(p_\xi)$;

(b) if $x_\xi \in \square$ is defined to be

$$x_\xi = p_\xi \cup (\kappa^+ - dom(p_\xi)) \times \{a_\xi\}$$

then $E(x_\xi \restriction dom(p_{\xi+1})) \subseteq W_{x_\xi,\xi}$;

(c) p_ξ does not take the value 0.

(3) IF WE PROVE (II) THEN

(a) $T_\xi = \kappa^+$, and $|t| < \lambda$ for $t \in T_\xi$;

(b) distinct members of $\{p_\xi\} \cup T_\xi$ have disjoint domains;

(c) $E(p_\xi \cup t) \subseteq U_\xi$.

THE CONSTRUCTION: For $\gamma < \kappa$ a limit, including $\gamma = 0$, if we know p_ξ for $\xi < \gamma$ then we put $p_\gamma = \cup_{\xi<\gamma} p_\xi$, so $p_0 = \emptyset$. Then clearly p_ξ is a partial injection because of (1). It remains to show how to construct $p_{\gamma+1}$ and D_γ or T_γ if we know p_γ.

Since it is possible that $E(p_\xi) \not\subseteq U_\xi$ (if we prove (I)), we need an inductive hypothesis which prevents $range(p_\xi) = \kappa$, so that we can properly extend p_ξ. For regular κ one can follow Stone, [St], and require that $|p_\xi| < \kappa$ for all ξ. For singular κ this only assures that we can perform $cf(\kappa)$ steps, but, as will become clear, it is essential to have κ steps (even if we only want to show that $\overline{U}_{0,0} \cap F_1 \neq \emptyset$, i.e. that $\square$ is not normal). Fortunately we need no more than

$$|\kappa - range(p_\xi)| = \kappa, \text{ for } \xi < \kappa,$$

and this is easily built in: Before we start the construction we choose

$K_\xi \in [\kappa]^\kappa$ for $\xi < \kappa$ such that $K_\xi \cap K_\eta = \emptyset$ whenever $\xi < \eta < \kappa$

and our inductive hypothesis will be that

$$range(p_\xi) \subseteq \cup_{\eta<\xi} K_\xi.$$

CASE 1: We construct D_γ and $p_{\gamma+1}$ from p_γ.

Let x_γ be as in (2b). Then $x_\gamma \in F_{a(\gamma)}$, so $W_{x_\gamma,\gamma}$ is well defined. There is a $D \in [\kappa^+]^{<\lambda}$ with $E(x \restriction D) \subseteq W_{x_\gamma,\gamma}$. Put $D_\gamma = D - dom(p_\gamma)$, let f be any injection $D_\gamma \to K_\gamma - \{0\}$, and let $p_{\gamma+1} = p_\gamma \cup f$.

[Note that if we do not assume $|p_\gamma| < \kappa$, then $E(x \restriction dom(p_\gamma))$ is not open, but that there is no need for it to be open.]

CASE 2: We construct T_γ and $p_{\gamma+1}$ from p_γ (assuming $\kappa^\lambda = \kappa$).

Let $\sigma = sup(dom(p_\gamma)) + 1$. For $\xi \in \kappa^+ - \sigma$ choose $y_\xi \in \square$ such that

(a) $p_\gamma \subseteq y_\xi$;

(b) $y_\xi \restriction \xi$ is an injection;

(c) $y_\xi(\eta) = a(\xi)$ for $\xi \leq \eta < \kappa$;

(d) $y_\xi^{\rightarrow}(\xi - dom(p_\xi)) \subseteq K_\xi$.

[Since κ is regular if $\kappa^{\underline{\kappa}} = \kappa$ we do not really need the K_ξ's: If $\lambda < \kappa$ we assume $|p_\xi| \leq |\xi| \cdot \lambda$ for all $\xi < \kappa$, and if $\lambda = \kappa$ we assume $|p_\xi| < \kappa$ for all $\xi < \kappa$.]

Then each $y_\xi \in F_{a(\xi)}$, hence there is $B_\xi \in [\kappa^+]^{<\lambda}$ such that

$E(y_\xi \restriction B_\xi) \subseteq U_\xi$, and $\xi \in B_\xi$.

Since κ^+ is regular and $\kappa^\mu < \kappa^+$ for $\mu < \lambda$, we see from the Δ-system Lemma that

there are $L \in [\kappa^+ - \sigma]^{\kappa^+}$ and $B \subseteq \kappa^+$ such that

$B_\xi \cap B_\eta = B$ for any two distinct $\xi, \eta \in L$.

[Use either the indexed version of the Δ-system Lemma, [CN, 3.2], or observe that $|\{B_\xi : \xi \in \kappa^+ - \sigma\}| = \kappa^+$.]

Clearly $|B| < \lambda$, hence $|{}^B\kappa| \leq \kappa^\lambda = \kappa < \kappa^+$. It follows that there are $L' \in [L]^{\kappa^+}$ and $b \in {}^B\kappa$ such that

$y_\xi \restriction B = b$ for $\xi \in L'$.

Now define $p_{\gamma+1}$ and T_γ by

$p_{\gamma+1} = p_\gamma \cup b$;

[Note that $p_{\gamma+1}$ is a partial injection since $p_\gamma \cup b \subseteq y_\xi$ for each $\xi \in L'$, and since there is $\xi \in L'$ with $sup(p_\gamma \in b) < \xi$.]

$T_\gamma = \{(y_\xi \restriction B_\xi) - p_\gamma : \xi \in L'\}$.

[Note that $|T_\gamma| = \kappa^+$ since $\xi \in B_\xi$ for each $\xi \in L'$, and $|p_\gamma| \leq \kappa < \kappa^+ = |L'|$.]

This completes the construction.

Because of (1) we can define a partial function p by

$p = \cup_{\xi<\kappa} p_\xi$.

PROOF OF (I): Consider any $f \in E(p)$. Let $\alpha, \beta < \kappa$ be arbitrary. Let $D \in [\kappa^+]^{<\lambda}$. Since the D_ξ's clearly are pairwise disjoint, and since $U_{\alpha,\beta}$ is listed κ times as U_ξ, and since we have κ many D_ξ's, there is $\gamma < \kappa$ with

$$D_\gamma \cap D = \emptyset, \text{ and } U_\gamma = U_{\alpha,\beta}.$$

If x_γ is as in (2b), then $x_\gamma \restriction dom(p_\gamma) = p_\gamma \subseteq p \subseteq f$. As $(dom(p_{\gamma+1}) - dom(p_\gamma)) \cap D = D_\gamma \cap D = \emptyset$, it follows that $E(x_\gamma \restriction dom(p_{\gamma+1})) \cap E(f \restriction D) \neq \emptyset$. Hence $E(f \restriction D) \cap U_{\alpha,\beta} \neq \emptyset$ since $W_{x_\gamma,\gamma} = U_\gamma$. Consequently $f \in \cap_{\alpha,\beta} \overline{U}_{\alpha,\beta}$.

PROOF OF (II): Since $|p| \leq \kappa$, for p is an injection, we see from (a) and (b) that we can successively choose $t_\xi \in T_\xi$ for $\xi < \kappa$ such that

$$dom(p) \cap dom(t_\xi) = \emptyset = dom(t_\xi) \cap dom(t_\eta) \text{ whenever } \xi < \eta < \kappa.$$

This ensures that $s = p \cup \cup_{\xi<\kappa} t_\xi$ is a partial function, hence that $E(s) \neq \emptyset$. But clearly

$$E(s) = \cap_{\xi<\kappa} E(p_\xi \cup t_\xi) \subseteq \cap_{\xi<\kappa} U_\xi.$$

PROOF OF (III): Since p is a partial injection, we can define $f \in F_0$ by

$$f = p \cup (\kappa - dom(p)) \times \{0\}.$$

We still assume that each W_x is open, hence there is $D \in [\kappa^+]^{<\kappa}$ with $E(f \restriction D) \subseteq W_f$. Since κ is regular, we see from (1) that there is $\gamma < \kappa$ such that $D \cap dom(p) \subseteq dom(p_\gamma)$. Since $x_\gamma \restriction (\kappa - dom(p)) = p \restriction (\kappa - dom(p))$, and since $p_\gamma \subset x_\gamma$ it follows that

$$x_\gamma \in E(f \restriction D), \text{ hence } x_\gamma \in W_f, \text{ hence } W_{x_\gamma} \subseteq W_f,$$

because of the definition of the W_x's.

The definition of the W_x's also tells that there is an $\alpha < \kappa^+$ with

$W_f \subseteq E(\{\langle\alpha,0\rangle\})$, hence $W_{x_\gamma} \subseteq E(\{\langle\alpha,0\rangle\})$.

Since $W_{x_\gamma} = W_{x_\gamma,\gamma}$, we see from (2b) that $\alpha \in dom(p_{\xi+1}) \subseteq dom(p)$. As $p(\alpha) \neq 0$ by (2c), and $p \subseteq f$, it follows that $f(\alpha) \neq 0$. This contradicts $f \in W_f$.

REMARK: The difference in the proofs of (I) and (III) is that in the proof of (I) we only need γ with $E(x_\gamma \restriction dom(p_{\gamma+1})) \cap E(f\restriction D) \neq \emptyset$, while the proof of (III) we need γ with $x_\gamma \subseteq E(f\restriction D)$: if $y \in E(f\restriction D)$ we have no control over the sets $A \in [\kappa^+]^{<\kappa}$ such that $E(y\restriction A) \subseteq W_y$ unless y is some x_γ. But (if each $D_\xi \neq \emptyset$ then) there is a $C \in [\kappa^+]^{cf(\kappa)}$ such that $C \cap dom(p) \nsubseteq dom(p_\xi)$ for each $\xi < \kappa$, and then $x_\xi \notin E(f\restriction C)$ for each $\xi \in \kappa$. So while we can find f even if κ is singular, we need to know that κ is regular for the proof that W_f is not open.

⟦The only real difference in the proof of 13.3 and the proof of the special case $\kappa = \omega$, [St, Thm. 3], is that we avoid the assumption that κ is regular. The proof of 13.4 is Scott's, [Sc_2, 3.1.0], [Sc_3, 2.4]. The proof of 13.5 is new.⟧ ¤

13.12. REMARK AND QUESTION: For $\kappa \geq \omega$ let

$$I(\kappa) = \cup_{\alpha<\kappa^+}\{f \in {}^{\alpha}\kappa\colon f \text{ is an injection}\},$$

and let $\Delta(\kappa)$ be the following statement:

$\Delta(\kappa)$: for each $D\colon I(\kappa) \to P(\kappa)$ such that each $D(f) \in [dom(f)]^{<\kappa}$ there is a κ^+-sequence $\langle f_\alpha\colon \alpha < \kappa^+\rangle$ in $I(\kappa)$ such that

a) $dom(f_\alpha) \subset dom(f_\beta)$[16] if $\alpha < \beta < \kappa^+$

b) $\cup_\alpha f_\alpha \restriction D(f_\alpha)$ is a partial function $\kappa^+ \to \kappa$.

The above proof makes clear that $\Delta(\kappa)$ is a weak form of the Δ-system Lemma, and implies that $<\kappa - \square^{\kappa^+}\kappa$ is not metanormal. Now the Δ-system lemma for κ^+ elements of $[\kappa^+]^{<\kappa}$ holds (if and) only if $\kappa^{\underline{\kappa}} = \kappa$, [CN, 3.2]. I do not know if

$\kappa^{\underline{\kappa}} = \kappa$ is necessary for $\Delta(\kappa)$ to be true; perhaps $\Delta(\kappa)$ is true in *ZFC* for all κ, or at least for all regular κ. If $\Delta(\omega_2)$ is true in *ZFC* one can strengthen Theorem 13.10.

13.13. QUESTION: We have seen that $\square^{\omega}(\underline{c}^{+}2)$ is not countably orthocompact, but is pseudonormal, and that ${}^{\omega}\omega \times \square^{\omega}(\omega + 1)$ is not (discretely) pseudonormal, and that $\square^{\omega}(\omega + 1)$ is not hereditarily (discretely) pseudonormal. Is it true that ${}^{\omega}\omega \times \square^{\omega}(\omega + 1)$ is not countably orthocompact, and that $\square^{\omega}(\omega + 1)$ is not hereditarily countably orthocompact? Or is a (countable) $\square$-product of metrizable spaces always metacompact (or subparacompact)? See Theorem 7.7.

14. WHEN THE QUOTIENT MAP IS NOT CLOSED

In Theorem 9.5 we saw that $q\colon \square_n X_n \to \nabla_n X_n$ is closed if each X_n is σ-compact and locally compact. We now show that both local compactness and σ-compactness are essential, and that the result fails for uncountable products.

14.1. EXAMPLE: Let each X_n be metrizable but not locally compact. Then q is not closed.

¤ Let $X = \{\infty\} \cup \omega \times \omega$, with $\infty \notin \omega \times \omega$, and topologize X as follows: points of $\omega \times \omega$ are isolated, and basic neighborhoods of ∞ have the form

$$B_n = \{\infty\} \cup [n,\omega) \times \omega.$$

Then each X_n has a closed subspace homeomorphic to X. It follows from Fact 6.2b that we may assume without loss of generality that each $X_n = X$. Let $\square$ denote $\square^{\omega}X$.

Define $p \in \square$ and a neighborhood U_p of p by

$$p = \langle\infty,\infty, \ldots\rangle, \text{ and } U_p = B_1 \times X \times X \times \ldots .$$

For $x \in E(p) - U_p$ we can define $m(x) \geq 1$ and $a(x), b(x) \in \omega$ by

$$m(x) = min\{n: x_n = \infty\}, \text{ and } \langle a(x), b(x)\rangle = x_{m(x)-1},$$

and we can define a neighborhood U_x of x by

$$U_x = \{y \in \Box: y_n = x_n \text{ for } n < m(x), \text{ and } y_{m(x)} \in B_{b(x)+1}\}.$$

Then

$$U = U_p \cup \cup\{U_x: x \in E(p) - U_p\}$$

is a neighborhood of $E(p)$ in $\Box$ (even in ${}^{\omega}X$). Suppose q is closed. Then there is a neighborhood V of $q(p)$ with $q^{\leftarrow}V \subseteq U$, [En$_1$, 1.4.13], or equivalently, there is a sequence $\langle k_n\rangle_n$ in ω such that $W \subseteq U$, where

$$W = \{x \in \Box: x_n \in B_{k_n} \text{ for all but finitely many } n\}.$$

Define $y \in \Box$ as follows

$$y_0 = \langle 0,0\rangle, \text{ and } y_n = \langle k_n, k_{n+1}\rangle \text{ for } n > 0.$$

Then clearly $y \in W$, and $y \notin U_p$. Consider any $x \in E(p) - U_p$. If $x_{m(x)-1} \neq y_{m(x)-1}$ then obviously $y \notin U_x$. If $y_{m(x)-1} = x_{m(x)-1}$ then $y \notin U_x$, too, since then $b(x) = k_{m(x)}$, so that

$$y_{m(x)} = \langle k_{m(x)}, k_{m(x)+1}\rangle \notin B_{b(x)+1}.$$

Therefore $y \notin U$. Consequently q is not closed.

⟦This I discovered December 1977.⟧ ¤

14.2. COROLLARY: If X is separable metrizable, then $q: \Box^{\omega}X \to \nabla^{\omega}X$ is closed iff X is locally compact. ¤

So if $\Box^{\omega}Q$ (Q = rationals) is (consistently) paracompact, the proof will require entirely different tools; all that is known now is that $\Box^{\omega}Q$ (and more generally $\Box_n X_n$ with each X_n countable) is pseudonormal in *ZFC*, [Mi].

14.3. EXAMPLE: There are a locally compact non-σ-compact (but paracompact) X_0, and compact X_n for $n > 0$, such that $q: \Box_n X_n \to \nabla_n X_n$ is not closed.

¤ This example can be obtained from our next example by making the first factor discrete. ¤

Even having one Lindelöf (but not locally compact) factor and all other factors compact is not good enough, as we now show.

14.4. EXAMPLE: There are a separable metrizable X_0 and compact metrizable X_n's, $n > 0$, such that $q\colon \square_n X_n \to \nabla_n X_n$ is not closed.

¤ Let $X_0 = {}^{\omega}\omega$, and $X_n = \omega + 1$ for $n > 0$. Use $\square$ to denote $\square_n X_n$, and ∇ to denote $\nabla_n X_n$.

First note that q is not closed if λ-scale or if $\mathbb{D}$. For in either case ∇ is paracompact by Remark 10.15, so if q were closed then $\square$ is paracompact by the proof of Theorem 9.6, but $\square$ is not even normal, by Theorem 7.1 or 12.1.

We now prove that q is not closed without using additional axioms. Upon identifying $\square$ with ${}^{\omega}\omega \times \square_{n>0} X_n$ we see that

$$\Delta = \{\langle x, x\rangle : x \in {}^{\omega}\omega\}$$

is closed in $\square$. We claim that $q^{\to}\Delta$ is not closed, or, equivalently, that $q^{\leftarrow}q^{\to}\Delta$ is not closed in $\square$. A moment's reflection shows that

$$q^{\leftarrow}q^{\to}\Delta = \{x \in \square : x_n \neq \omega \text{ for all but finitely many } n > 0\},$$

hence $q^{\leftarrow}q^{\to}\Delta$ is dense in $\square$ since another moment's reflection shows that even $\{x \in \square : x_n \neq \omega \text{ for all } n > 0\}$ is dense in $\square$. Since $q^{\leftarrow}q^{\to}\Delta \neq \square$, as our formula for $q^{\leftarrow}q^{\to}\Delta$ shows, it follows that $q^{\leftarrow}q^{\to}\Delta$ is not closed.

⟦This I discovered December 1977.⟧ ¤

Our next example takes care of uncountable $\square$-products.

14.5. THEOREM: If each X_α is infinite and compact, and $\nu = \omega_1$, then $q\colon \square_\alpha X_\alpha \to \nabla_\alpha X_\alpha$ is not closed.

¤ Each X_α has a subspace which is homeomorphic to some compactification $b_\alpha\omega$ of ω. Because of Fact 6.2b we may assume without loss of generality that each $X_\alpha = b_\alpha\omega$. Denote $\square_\alpha b_\alpha\omega$ by $\square$, and define $m\colon \square \to \omega + 1$ by

$$m(x) = min(\{\omega\} \cup \omega\cap\{x_\alpha\colon \alpha < \omega_1\}).$$

For each $x \in \square$ define a neighborhood U_x as follows:

$$U_x = \begin{cases} \{y \in \square\colon y_n \not\leq n \text{ for } n < \omega\} & \text{if } m(x) = \omega; \\ \{y \in \square\colon y_\alpha = x_\alpha \text{ if } x_\alpha \in \omega, \text{ else } y_\alpha \not\leq m(x)\} & \\ & \text{if } m(x) < \omega. \end{cases}$$

Pick any $p \in \square$ with $m(p) = \omega$, i.e. $p \in \Pi_\alpha(b_\alpha\omega - \omega)$. Then

$$V = \cup_{x\in E(p)} U_x$$

is a neighborhood of $E(p)$. Now suppose q is closed. Then there is a neighborhood W of $q(p)$ such that $q^{\leftarrow}W \subseteq V$, or, equivalently, there is an indexed family $\langle W_\alpha\colon \alpha < \omega_1\rangle$ such that

(1) each W_α is a neighborhood of x_α in $b_\alpha\omega$; and

(2) for all $y \in \square$, if $y_\alpha \in W_\alpha$ for all but countably many α, then $y \in V$.

Since ω is dense in each $b_\alpha\omega$, and each $x_\alpha \not\in \omega$, we see from (1) that

(3) each $\omega \cap W_\alpha$ is infinite.

Use (3) first to see that one can find a $k \in \omega$ such that

$$\{\alpha < \omega_1\colon min(\omega\cap W_\alpha) = k\} \text{ is uncountable.}$$

Use (3) again to see that we can define $z \in \square$ by

$$z_\alpha = \begin{cases} k & \text{if } \alpha < \omega; \\ min([k,\omega) \cap W_\alpha) & \text{if } \alpha \geq \omega. \end{cases}$$

Now consider any $x \in E(p)$. If $m(x) = \omega$ then $z \notin U_x$ since $z_k = k \leq k$, so assume $m(x) < \omega$. If $z_\alpha \neq x_\alpha \in \omega$ for some α then

$z \notin U_x$, so assume $z_\alpha = x_\alpha$ whenever $x_\alpha \in \omega$. Then $m(x) \geq m(z)$, hence again $z \notin U_x$ since

$$|\{\alpha < \omega_1 \colon z_\alpha = m(x)(=k)\}| = \omega_1$$

but if $y \in U_x$ then

$$|\{\alpha < \omega_1 \colon y_\alpha \leq m(x)\}| = \omega,$$

since $x_\alpha \notin \omega$ for all but countably many α's, for $p_\alpha \notin \omega$ for all α, and xEp. Consequently $z \notin U$. This contradicts (2). ⟦This I discovered when writing up this paper.⟧ ⌑

15. THE ∇-PRODUCT OF ARBITRARILY MANY FACTORS

We have seen that ∇-products are useful for the study of countable $\Box$-products of compact factors since

(a) countable ∇-products are easier to work with than $\Box$-products

(b) if each X_n is compact, then $\Box_n X_n$ is paracompact iff $\nabla_n X_n$ is.

In this section we see that uncountable ∇-products also are easy to work with, but that this is of no use for the study of $\Box$-products since the analogue of (b) is not true in *ZFC*. This is hardly suprising, since the proof of (b) depends on the fact that the quotient map of a countable $\Box$-product of compact spaces is closed and has Lindelöf fibers, but the quotient map of a $\Box$-product of ω_1 infinite compact spaces is not closed, by Theorem 14.5, and does not have Lindelöf fibers. [For each fiber has a closed discrete set of cardinality $\underline{c}$.]

We begin with an easy positive result.

15.1. THEOREM: $2^\nu = \nu^+ \models$ If each X_α is completely regular and $w(X_\alpha) \leq 2^\nu$, then $\nabla_\alpha X_\alpha$ is ultraparacompact.

For $\nu = \omega$ this is essentially Theorem 10.1. The natural approach to proving Theorem 15.1 would be to imitate the proof of Theorem 10.1, hence to show that $\nabla_\alpha X_\alpha$ is a P_{ν^+}-space. But we have seen that $\nabla_\alpha X_\alpha$ need not even be a P-space ($= P_{\omega_1}$-space), Proposition 9.2. Only while writing up this paper did I realize that the proof to be imitated is not that of Theorem 10.1, but that of 10.10: Since clearly $w(\nabla_\alpha X_\alpha) \leq 2^\nu$ if each $w(X_\alpha) \leq 2^\nu$, and we assume $2^\nu = \nu^+$, it suffices to generalize the key idea of Roitman's proof, [Ro$_2$, Thm. 5]: we only need a nice base. This is furnished by the following simple result, which is the promised improvement of 6.3. ¤

15.2. PROPOSITION: If each X_α is completely regular, then $\nabla_\alpha X_\alpha$ has a base $\mathcal{B}$ such that $\cup\mathcal{A}$ is closed for each $\mathcal{A} \subseteq \mathcal{B}$ with $|\mathcal{A}| \leq \nu$.

¤ Denote $\nabla_\alpha X_\alpha$ by ∇. Since each X_α is completely regular, it is easy to see that

$$\mathcal{B} = \{\cup_k \nabla_\alpha B_{\alpha,k}\colon \text{ each } B_{\alpha,k} \text{ open in } X_\alpha, \text{ and } \overline{B}_{\alpha,k} \subseteq B_{\alpha,k+1}\}^{23}$$

is a base for ∇. To show that $\mathcal{B}$ is as required it suffices to prove the following simple

CLAIM: If $\mathcal{F} = \{\nabla_\alpha F_\alpha\colon$ each F_α is closed in $X_\alpha\}$, then $\mathcal{G}$ is closed for each $\mathcal{G} \subseteq \mathcal{F}$ with $|\mathcal{G}| \leq \nu$.

PROOF OF CLAIM: Let $F_{\alpha,\beta}$ be closed in X_α, for $\alpha,\beta < \nu$, and consider an arbitrary $x \in \nabla - \cup_\beta \nabla_\alpha F_{\alpha,\beta}$. For each β we have

$$|A_\beta| = \nu, \text{ where } A_\beta = \{\alpha < \nu\colon x_\alpha \notin F_{\alpha,\beta}\}.$$

---

[23] There are no misprints here: we did NOT want $\cap_k \nabla_\alpha B_{\alpha,k}$ or $B_{\alpha,k} \supseteq \overline{B}_{\alpha,k+1}$, as in the proof of Theorems 10.6 or 10.10.

Construct $\langle D_\beta : \beta < \nu \rangle$ such that

each $D_\beta \subseteq A_\beta$, and $|D_\beta| = \nu$; and $D_\beta \cap D_\gamma = \emptyset$ whenever $\beta \neq \gamma$.

Define $\langle U_\alpha : \alpha < \nu \rangle$ by

$$U_\alpha = \begin{cases} X_\alpha - F_{\alpha,\beta} & \text{if } \alpha \in D_\beta \text{ for a (unique) } \beta < \nu; \\ X_\alpha & \text{otherwise.} \end{cases}$$

Then $\nabla_\alpha U_\alpha$ is a neighborhood of x which misses $\cup_\beta \nabla_\alpha F_{\alpha,\beta}$.
⟦This I discovered when writing up this paper.⟧ ¤

This leads to the following analogue of Theorem 13.10.

15.3. THEOREM: The following conditions on ν are equivalent:

(a) $2^\nu = \nu^+$;

(b) if each X_α is completely regular, with $w(X_\alpha) \leq 2^\nu$, then $\nabla_\alpha X_\alpha$ is paracompact;

(c) $\nabla^\nu({}^{(2^\nu)}2)$ is paracompact;

(d) $\nabla^\nu({}^{(2^\nu)}2)$ is paranormal;

(e) $\nabla_\nu({}^{(2^\nu)}2)$ is orthocompact;

(f) $G_\nu({}^{(2^\nu)}2)$ is paracompact;

(g) $G_\nu({}^{(2^\nu)}2)$ is paranormal; and

(h) $G_\nu({}^{(2^\nu)}2)$ is orthocompact.

¤ (a) ⇒ (b): This is Theorem 15.1.

(b) ⇒ (c) ⇒ (d) and (c) ⇒ (e): Obvious.

(c) ⇒ (f), (d) ⇒ (g) and (e) ⇒ (h): $G_\nu({}^{(2^\nu)}2)$ can be embedded as a closed subspace in $\nabla_\nu({}^{(2^\nu)}2)$ by Proposition 6.4.

(g) ⇒ (a) and (h) ⇒ (a): $G_\nu({}^{(2^\nu)}2)$ is homeomorphic to $\leq\nu - \Box^{2^\nu}2$, which is homeomorphic to $\leq\nu - \Box^{2^\nu}\nu^+$ by Lemma 13.1. Hence if $2^\nu \geq \nu^{++}$ then $G_\nu({}^{(2^\nu)}2)$ has a closed subspace homeomorphic to $<\nu^+ - \Box^{\nu^{++}}\nu^+$, which is neither paranormal nor orthocompact, by Theorems 13.3 and 13.4.

[This I discovered when writing up this paper.] ¤

Theorems 15.1 and 13.8 lead to the following result, which shows that the analogue of (b), i.e. of Theorem 9.6 fails for uncountably many factors.

15.4. PROPOSITION: $2^{\omega_1} = \omega_2 \models \nabla^{\omega_1}({}^{\omega_2}2)$ is paracompact but $\square^{\omega_1}({}^{\omega_2}2)$ is not. ¤

15.5. REMARK: Consider the following statements about ${}^{\nu}\omega$:

$\mathbb{D}(\nu,\kappa)$: if $F \subseteq {}^{\nu}\omega$ has $|F| < \kappa$, then there is a $g \in {}^{\nu}\omega$ such that $|\{\alpha < \nu: f(\alpha) < g(\alpha)\}| = \nu$ for each $f \in F$; and

$\mathbb{E}(\nu,\kappa)$: if $F \subseteq {}^{\nu}\omega$ and $\mathsf{A} \subseteq [\nu]^{\nu}$ have $|F|, |\mathsf{A}| < \kappa$, then there is a $g \in {}^{\nu}\omega$ such that $|\{\alpha \in A: f(\alpha) < g(\alpha)\}| = \nu$ for each $f \in F$ and $A \in \mathsf{A}$.

So $\mathbb{D}(\omega,\underline{c})$ is what we called $\mathbb{D}$ in §10. We have seen that $\mathbb{D}(\omega,\kappa)$ and $\mathbb{E}(\omega,\kappa)$ are equivalent, [Ro_2], in the proof of Theorem 10.10, and that $\mathbb{D}(\omega,\underline{c})$ is independent from *ZFC*, but properly weaker than *CH*, [He], Theorem 10.5. For $\nu > \omega$ our knowledge is incomplete. It is clear from the argument in 15.2 that $\mathbb{E}(\nu,\nu^+)$ holds in *ZFC*, hence that $\mathbb{E}(\nu,2^{\nu})$ holds under $2^{\nu} = \nu^+$. But one can have $\mathbb{E}(\nu,2^{\nu})$ while $2^{\nu} \neq \nu^+$. For example, if $\omega \leq \nu < \underline{c}$ then $2^{\nu} = \underline{c}$ under *MA*, and it is easy to prove that $\mathbb{E}(\nu,\underline{c})$ under *MA* if $\omega \leq \nu < \underline{c}$. It is apparently unknown if $\mathbb{E}(\nu,\kappa)$ and $\mathbb{E}(\nu,\kappa)$ are equivalent in general, and also if $\mathbb{E}(\nu,2^{\nu})$ is independent from *ZFC* for $\nu > \omega$, (we are especially interested in these questions if $\nu = \omega_1$). However, Kunen has shown that if $\nu^{\omega} = \nu$ then $\mathbb{D}(\nu,2^{\nu})$ holds in *ZFC* (unpublished); the proof does not show that $\mathbb{E}(\nu,2^{\nu})$ holds if $\nu^{\omega} = \nu$. Also, Galvin has shown that $\mathbb{D}(\omega_1,2^{\omega_1})$ holds if $\underline{c} = \omega_2$.

The relevance of $\mathbb{E}(\nu,\kappa)$ for ∇-products is this: the argument of 15.2 shows that if each X_α is first countable, then

$\nabla_\alpha X_\alpha$ has a base $\mathcal{B}$ such that $\cup\mathcal{A}$ is closed for each $\mathcal{A} \in [\mathcal{B}]^{<\kappa}$ if $\mathbb{E}(\nu,\kappa)$. (The argument in Remark 10.11 shows that the converse holds if each X_α is nondiscrete.) Hence if $\mathbb{E}(\nu,2^\nu)$ then $\nabla_\alpha X_\alpha$ is paracompact if each X_α is first countable with $w(X_\alpha) \le 2^\nu$. So $\nabla^{\omega_1}(\omega + 1)$ is paracompact under less than $2^{\omega_1} = \omega_2$, and could be paracompact in *ZFC*.

REFERENCES

[AfU] P. Alexandroff and P. Urysohn, *Über nulldimensionale Punktmengen*, Math. Ann. 98 (1928), 89-106 (7.6).

[AE] K. Alster and R. Engelking, *Subparacompactness and product spaces*, Bull. Acad. Pol. Math. Astron. Phys. 20 (1976), 763-767 (13.7).

[Ar] A.V. Arhangel'skiĭ, *The cardinality of first countable bicompacta*, DAN SSSR 187 (1969), 967-970 ≡ Sov. Math. Dokl. 10 (1969), 951-955 (10.3,10.10).

[Be] A.R. Bernstein, *A new kind of compactness for topological spaces*, Fund. Math. 66 (1970), 185-193 (6.5).

[Bo_1] C.J.R. Borges, *On a counter-example of A.H. Stone*, Quart. J. Math. (2) 20 (1969), 91-95 (13.2^+).

[Bo_2] ________, *A survey of M_i-spaces: open questions and partial results*, GTA 1 (1971), 79-84 (7.2^+).

[Bu] D.K. Burke, *On p-spaces and wΔ-spaces*, Pacific J. Math. 55 (1970), 285-296 (5.1^+).

[CN] W.W. Comfort and S. Negrepontis, *The theory of ultrafilters*, Springer-Verlag, Berlin 1974.

[vD_1] E.K. van Douwen, *The box product of countably many metrizable spaces need not be normal*, Fund. Math. 88 (1975), 127-132 (1,7.2,7.3,7.4,9.8).

[vD_2] ________, *Another nonnormal box product*, GTA 7 (1977), 71-76 (8,13.1,13.8).

[vD_3] ________, *κ-scales and the box product of compact metrizable spaces*, talk, Ohio Univ. Top. Conf., May 1976 (1.4,3,10.6).

[vD_4] ________, *Separation and covering properties of box products and products*, BAMS 24 (1977) A-145 (1,7.7, 10.6,12.5,13.8)

[vD_5] E.K. van Douwen, *Functions from the integers to the integers and topology*, in preparation, (5.1, ftnote 18).

[vD_6] ________, *A technique for constructing honest locally compact submetrizable spaces*, (7.5).

[vDW] E.K. van Douwen and M.L. Wage, *Small subsets of first countable spaces*, Fund. Math. 103 (1979), 103-110 (ftnote 18).

[Do] C.H. Dowker, *On countably paracompact spaces*, Canad. J. Math. 3 (1951), 219-224 (5.5, 13.7, 13.9).

[Ea] W.B. Easton, *Powers of the regular cardinals*, Ann. Math. Logic 1 (1970), 139-170 (13.6^+).

[En_1] R. Engelking, *General topology*, PWN-Polish Scientific Publishers, Warszawa, 1977.

[En_2] ________, *Dimension theory*, PWN-Polish Scientific Publishers, Warszawa, 1978.

[ER] P. Erdös and M.E. Rudin, *A non-normal box product*, Coll. Math. Soc. János, Bolyai, Vol. 10, Infinite and finite sets, Keszthely (Hungary) (1973), 629-631 (footnote 1).

[Fl] W.G. Fleissner, *Separation properties in Moore spaces*, Fund. Math. 98 (1978), 279-286 (5.3, 8.1).

[Fr] A.H. Frink, *Distance functions and the metrization problem*, BAMS 43 (1937), 133-142 (10.8).

[Ge] M. Gewand, *Lindelöf degree of the G_δ-topology*, NAMS 24 (1977), A-143 (8.3).

[Gi] R.F. Gittings, *Some results on weak covering conditions*, Canad. J. Math. 26 (1974), 1152-1156 (5.1^+, 5.8).

[He] S.H. Hechler, *On the existence of certain cofinal subsets of $^\omega\omega$*, Proc. Symp. Pure Math., Vol. 13, Part II, AMS, Providence (1974), 155-174 (10.5, 10.9, 11.2, 12.7, 15.5).

[Jo] F.B. Jones, *Concerning normal and completely normal spaces*, BAMS 43 (1937), 671-677 (5.3, 8.1).

[Jh] I. Juhász, *On closed discrete subspaces of product spaces*, Bull. Acad. Pol. Math. Astron. Phys. 17 (1969), 217-223 (8.2).

[Jn] H.J.K. Junnila, *Metacompactness, paracompactness and interior-preserving open covers*, preprint (5.4, 8.1).

[Ka] M. Katětov, *Complete normality of cartesian products*, Fund. Math. 35 (1948), 271-274 (5.2, 7.4).

[Ke] J. Kelley, *General topology*, Van Nostrand, Princeton, 1955.

[Kn] C.J. Knight, *Box topologies*, Quart. J. Math. Oxford (2) 15 (1964), 41-54 (1,3.1,6.1,6.3).

[Ku_1] K. Kunen, *Some comments on box products*, Coll. Math. Soc. Janas Bolyai, Vol. 10, Infinite and finite sets, Keszthely (Hungary) (1973), 1011-1016 (7.4,12.2$^+$, 12.3,12.4).

[Ku_2] ________, *On paracompactness of box products of compact spaces*, TAMS 240 (1978), 307-316 (1, footnote 1, 6.3$^-$,8.1$^-$,8.2,9.1$^-$,9.5,9.6,9.8,10.1,10.3,10.14).

[M_1] E. Michael, *Another note on paracompact spaces*, PAMS 8 (1957), 822-828 (9.6).

[M_2] ________, *The product of a normal space and a metric space need not be normal*, BAMS 69 (1963), 375-376 (7.1$^-$).

[Mi] A.W. Miller, *On the pseudonormality of box products*, 1977 manuscript, (11.1,12.2$^+$,14.2$^+$).

[My] J. Mycielsky, *α-incompactness of N^α*, Bull. Acad. Pol. Math. Astron. Phys. 12 (1964), 437-438 (8.2).

[Na] K. Nagami, *Countable paracompactness of inverse limits and products*, Fund. Math. 73 (1972), 261-270 (13.2$^+$).

[PP-P] R. Pol and E. Puzio-Pol, *Remarks on Cartesian products*, Fund. Math. 93 (1976), 57-69 (13.7).

[Ro_1] J. Roitman, *Paracompact box products in forcing extensions*, to appear (10.12).

[Ro_2] ________, *More paracompact box products*, PAMS 74 (1979), 71-176 (1,10.10,10.12,10.13,11.2,15.1,15.5).

[Ru_1] M.E. Rudin, *The box product of countably many compact metric spaces*, GTA 2 (1972), 293-298 (1,10.1$^-$).

[Ru_2] ________, *Countable box products of ordinals*, TAMS 192 (1974), 121-128 (footnote 1).

[Ru_3] ________, *Lectures on set-theoretic topology*, Regional Conf. Series in Math. #23, AMS, Providence, 1975.

[Sc_1] B.M. Scott, *Toward a product theory for orthocompactness*, in K.R.Allen and N.M. Stavrakas, eds., *Studies in topology*, Academic Press, New York (1975), 517-537 (5.5,13.7,13.9).

[Sc_2] ________, *Orthocompact: metacompact:: normal: paracompact (a.e.)*, University of Wisconsin, Madison, 1975, Doctoral Dissertation (5.4,5.5,13.2$^+$,13.8).

[Sc_3] ________, *More on orthocompactness*, preprint (5.5, 13.2$^+$).

[So] R.H. Sorgenfrey, *On the topological product of paracompact spaces*, BAMS 53 (1947), 631-632 (1).

[St] A.H. Stone, *Paracompactness and product spaces*, BAMS 54 (1948), 977-982 (13.2^+,13.11).

[Tl] F.D. Tall, *P-points in βN-N, normal nonmetrizable Moore spaces and other problems of Hausdorff*, TOPO'72, Second Pittsburgh Top. Conf., Springer Lecture notes in Math #378 (1972), 501-512 (5.1^+).

[Tm] H. Tamano, *On compactifications*, J. Math. Kyoto Univ. 1 (1962), 162-193 (5.4,8.1).

[W_1] S.W. Williams, *Is $\square^{\omega}(\omega+1)$ paracompact?*, Top. Proc. 1 (1976), 141-146 (10.6).

[W_2] ________, *Boxes of compact ordinals*, Top. Proc. 3 (1977), 631-642 (footnote 1).

[WF] S.W. Williams and W. Fleischman, *The G_δ-topology on compact spaces*, Fund. Math. 83 (1974), 143-149 (8.3).

[Z] P. Zenor, *Countable paracompactness in product spaces*, PAMS 30 (1971), 199-201 (5.2,7.4).

TRANSFINITE DIMENSION

R. Engelking
Warsaw University

The terminology and notations in this survey follow my Dimension Theorey book, [E1], to which this text is a kind of supplement. I am grateful to E. Pol, R. Pol and T. Przymusiński for their valuable comments on an earlier version of this survey. □

1. Definitions. We start with the definition of the small transfinite dimension and the large transfinite dimension.

1.1. Definition ([Hu]; cf. [U1], p. 66). Let X denote a regular space and α be either an ordinal number ≥ 0 or the integer -1 (which we shall assume to be less than each ordinal number). The following conditions:

(ti1) $\operatorname{trind} X = -1$ if and only if $X = \emptyset$;

(ti2) $\operatorname{trind} X \leq \alpha$ if for every point $x \in X$ and each neighborhood $V \subset X$ of the point x there exists an open set $U \subset X$ such that

$$x \in U \subset V \text{ and } \operatorname{trind} \operatorname{Fr} U < \alpha;$$

(ti3) $\operatorname{trind} X = \alpha$ if $\operatorname{trind} X \leq \alpha$ and the inequality $\operatorname{trind} X < \alpha$ does not hold;

define inductively the *small transfinite dimension* trind on a subclass of the class of regular spaces. If trind X is defined, we say that X has small transfinite dimension;

ISBN: 0-12-584960-5

otherwise, we say that X has no small transfinite dimension.

One proves that the Hilbert cube $I^{\aleph_0}$ has no small transfinite dimension (see Example 2.2 or Corollary 4.4); hence, the class of spaces for which trind is defined is a proper subclass of the class of all regular spaces.

Applying transfinite induction, one can easily verify that the small transfinite dimension is a topological invariant.

1.2. Definition ([Sm1]). Let X denote a normal space and let α be either an ordinal number ≥ 0 or the integer -1. The following conditions:

(tI1) trInd X = -1 if and only if $X = \emptyset$;

(tI2) trInd X $\leq \alpha$ if for every closed set $A \subset X$ and each open set $V \subset X$ which contains the set A there exists an open set $U \subset X$ such that

$$A \subset U \subset V \text{ and trInd Fr}U < \alpha;$$

(tI3) trInd X = α if trInd X $\leq \alpha$ and the inequality trInd X $< \alpha$ does not hold;

define inductively the *large transfinite dimension* trInd on a subclass of the class of normal spaces. If trInd X is defined, we say that X has large transfinite dimension; otherwise, we say that X has no large transfinite dimension.

Since the Hilbert cube $I^{\aleph_0}$ has no trind, it follows from Proposition 3.2 that it has no trInd either; hence, the class of spaces for which trInd is defined is a proper subclass of the class of all normal spaces.

One easily verifies that the large transfinite dimension is a topological invariant.

Obviously, conditions (ti2) and (tI2) can also be stated in terms of partitions, i.e., we have

(ti2') trind $X \leq \alpha$ if and only if for every point $x \in X$ and each closed set $B \subset X$ such that $x \notin B$ there exists a partition L between x and B such that trind $L < \alpha$;

(tI2') trInd $X \leq \alpha$ if and only if for every pair A, B of disjoint closed subsets of X there exists a partition L between A and B such that trInd $L < \alpha$.

We shall now define three classes of spaces which are only "slightly" larger than the class of finite-dimensional spaces. For the sake of simplicity, we restrict ourselves to metric spaces.

1.3. Definition ([Hu]; cf. [U2], p. 351). A metric space X is *countable-dimensional (strongly countable-dimensional)* if X can be represented as the union of a sequence $X_1, X_2, \ldots$ of subspaces (closed subspaces) such that Ind $X_i < \infty$ for $i = 1, 2, \ldots$. (The reader should be warned here, that strongly countable-dimensional spaces are sometimes called weakly countable-dimensional.)

By virtue of the second decomposition theorem ([E1], Theorem 4.1.17), a metric space X is countable-dimensional if and only if X can be represented as the union of a sequence $X_1, X_2, \ldots$ of subspaces such that Ind $X_i \leq 0$ for $i = 1, 2, \ldots$.

1.4. Remark. Let us note that there exist countable-dimensional compact metric spaces which are not strongly countable-dimensional; every countable-dimensional compact metric space all of whose non-empty open subsets are infinite-dimensional is such a space. The existence of spaces with the latter property was announced in [Hu]; the first example was given in [Sm1] (a simpler example can be obtained by defining an appropriate upper semicontinous decomposition of the Cantor set C1, cf. [AP], p. 518; the decomposition defined in [AP] can be somewhat simplified: represent C as C^2, arrange all vertical Cantor sets sticking out at rational points of the horizontal copy of C into a sequence, for n = 1,2,... decompose the nth term of the sequence in 2^{n-1} "portions" of diameter $\leq 1/n$ homeomorphic to the Cantor set and replace each "portion" by its decomposition determined by a continuous mapping of C onto I^n).

1.5. Definition. A metric space X is *locally finite-dimensional* if for every point $x \in X$ there exists a neighborhood U such that Ind $U < \infty$.

Clearly, for every locally finite-dimensional metric space X we have trind $X \leq \omega$; on the other hand, trInd X does not necessarily exist for such a space (cf. Example 2.1). From the paracompactness of metric spaces and the locally finite sum theorem ([E1], Theorem 4.1.10) it follows (cf. [W1]), that every locally finite-dimensional metric space is strongly countable-dimensional.

2. *Examples.* We shall describe now some examples to illustrate the notions introduced in the preceding section.

2.1. Example ([Sm1]). Set S denote the class of all normal spaces of the form $X = \oplus_{n=0}^{\infty} X_n$, where Ind $X_n \leq n$ and Ind $X = \infty$. No space in S has large transfinite dimension. Suppose the contrary, and consider a space $X = \oplus_{n=1}^{\infty} X_n$ in S which has the smallest possible trInd. Choose for every n such that $X_n \neq \emptyset$ a pair A_n, B_n of disjoint closed subsets of

X_n with the property that each partition $L_n \subset X_n$ between A_n and B_n satisfies the equality $\text{Ind } L_n \geq \text{Ind } X_n - 1$. By the definition of trInd, there exists a partition $L \subset X$ between the union of all A_n's and the union of all B_n's such that $\text{trInd } L < \text{trInd } X$; since $L \in S$, we have a contradiction.

In particular, the space $S = \oplus_{n=0}^{\infty} I^n$ has no trInd; on the other hand, one easily checks that $\text{trind } S = \omega_0$. It is equally easy to verify that for ωS, the one-point compactification of S, we have $\text{trind } \omega S = \text{trInd } \omega S = \omega_0$.

2.2. *Example: Smirnov's spaces* ([Sm1]). *Smirnov's spaces* $S_0, S_1, \ldots, S_\alpha, \ldots, \alpha<\omega_1$ are defined by transfinite induction:

$$S_0 = I^0 = \{0\} \text{ is a one-point space,}$$

and for $\alpha > 0$

$$S_\alpha = \begin{cases} S_\beta \times I & \text{if } \alpha = \beta + 1 \\ \omega \bigoplus_{\beta<\alpha} S_\beta & \text{if } \alpha \text{ is a limit number,} \end{cases}$$

where ωX denotes the one-point compactification of X. Let us note, that S_ω is the space ωS considered in Example 2.1. One easily checks that the spaces S_α are strongly countable-dimensional compact metric spaces. One proves that

(i) $\text{trInd } S_\alpha = \alpha$ for every $\alpha < \omega_1$ ([Sm1]);

(ii) $\sup\{\text{trind } S_\alpha : \alpha<\omega_1\} = \omega_1$ ([Lv2], cf. Remark 4.8).

Since all Smirnov's spaces are embeddable in the Hilbert cube $I^{\aleph_0}$, from Corollary 3.9, (i), (ii) and Proposition 3.3 it follows that $I^{\aleph_0}$ has neither trind nor trInd.

In [Lj3] it is announced that $\text{trind } S_{\omega_0+3} < \omega_0+3$; the following two problems arise in this context.

2.3. *Problem.* Compute $\text{trind } S_\alpha$ for $\alpha < \omega_1$.

Partial results are announced in [Lj4] and [Lj3]: for every limit ordinal $\lambda < \omega_1$ and $n = 3, 4, \ldots$ we have $\operatorname{trind} S_{\lambda+n} \leq \lambda + [(n+2)/2]$, and for every limit ordinal $\lambda < \omega_1$ such that $\lambda = \omega_0 + \beta$, where $\omega_0 \cdot \beta = \beta$, and $n = 0,1,2$ we have $\operatorname{trind} S_{\lambda+n} = \lambda + n$.

2.4. *Problem*. Define a transfinite sequence $L_0, L_1, \ldots, L_\alpha, \ldots, \alpha < \omega_1$ of "nice" compact metric spaces such that $\operatorname{trind} L_\alpha = \alpha$ for $\alpha < \omega_1$.

2.5. *Remark*. By a modification of Smirnov's spaces one can obtain compact metric AR-spaces with similar properties (see [He]).

2.6. *Example* ([Hu]). Let K_ω be the subspace of the Hilbert cube $I^{\aleph_0}$ consisting of all points in $I^{\aleph_0}$ which have only finitely many non-zero coordinates. Clearly, K_ω is a strongly countable-dimensional space. One prove that

(i) K_ω is a universal space for the class of all strongly countable-dimensional separable metric spaces ([Sm1] and [N]);

(ii) K_ω has no countable-dimensional metrizable compactification ([Sk]; announcement [Hu]);

(iii) K_ω has no small transfinite dimension.

The last property follows either from (i), Corollary 3.9, Example 2.2 and Proposition 3.3 or from (ii) and Theorem 4.15 below.

For the sake of historical accuracy, one should note that in [Hu] not the space K_ω is being considered, but the subspace K'_ω of the Hilbert space l^2 consisting of all points which have only finitely many non-zero coordinates. Since, as one can easily see, K'_ω contains topologically K_ω and vice versa, properties (i) - (iii) are equivalent to similar properties of K'_ω; moreover, the spaces K_ω and K'_ω are homeomorphic (see [BP], Secs. 5 and 6).

2.7. Example ([N]). Let N_ω be the subspace of the Hilbert cube $I^{\aleph_0}$ consisting of all points in $I^{\aleph_0}$ which have only finitely many rational coordinates. Clearly, N_ω is a countable-dimensional space and K_ω is embeddable in N_ω. It is proved in [N] that N_ω is a universal space for the class of all countable-dimensional separable metric spaces. It follows from (ii), Example 2.6, that N_ω has no countable-dimensional compactification. Arguing as in Example 2.6, one shows that N_ω has no small transfinite dimension.

2.8. Problem. Prove in a direct way that there exists a countable-dimensional separable metric space which has no trind.

2.9. Remark. Let us add that, as proved in [W2], the subspace J^* of $I^{\aleph_0}$ consisting of all points $(x_1, x_2, \ldots, x_n, 0, 0, \ldots) \in I^{\aleph_0}$, where $n = 1, 2, \ldots$, such that $x_i \leq 1/n$ for $i = 1, 2, \ldots, n$ and at least one x_i does not vanish, is universal for the class of all locally finite-dimensional separable metric spaces. Clearly, trind $J^* = \omega$ and trInd J^* does not exist. The one-point compactification $J = J^* \cup \{(0, 0, \ldots, 0, \ldots)\}$ of J^* is a strongly countable-dimensional space (discussed earlier in [Hu]) such that trind J = trInd $J = \omega_0$.

3. *Basic properties of transfinite dimensions.* The following four propositions follow directly from the definitions.

3.1. Proposition. For every regular space X, if ind $X < \infty$, then trind X = ind X and, if trind $X < \omega$, then ind X = trind X. For every normal space X, if Ind X < ∞, then trInd X = Ind X and, if trInd X < ω, then Ind X = trInd X.

3.2. Proposition. For every normal space X, if trInd X exists, then trind X also exists and trind X $\leq$ trInd X.

3.3. Proposition. For every regular space X and a subspace A of X, if trind X exists, then trind A also exists and trind A $\leq$ trind X. For every normal space X and a closed subspace A of X, if trInd X exists, then trInd A also exists and trInd A $\leq$ trInd X.

In the second part of the above proposition, the assumption that A is a closed subspace is essential (see Example 2.1).

3.4. Proposition. If trind X = α, then for every ordinal number $\beta < \alpha$ there exists a closed subspace A of X such that trind A = β. If trInd X = α, then for every ordinal number $\beta < \alpha$ there exists a closed subspace A of X such that trInd A = β.

3.5. Theorem ([T]; for $\alpha = 0$, [Hu]). If a regular space X of weight $w(X) \leq \aleph_\alpha$ has small transfinite dimension, then it satisfies the inequality trind X < $\omega_{\alpha+1}$.

Proof. Suppose that there exists a regular space X such that $w(X) \leq \aleph_\alpha$ and trind X $\geq \omega_{\alpha+1}$. By virtue of Proposition 3.4, we can assume that trind X = $\omega_{\alpha+1}$; so, X has a base $\mathcal{B}$ such that trind FrU < $\omega_{\alpha+1}$ for every $U \in \mathcal{B}$. The inequality $w(X) \leq \aleph_\alpha$ implies (cf. [E2], Theorem 1.1.15) that $\mathcal{B}$ contains a subfamily $\mathcal{B}_0$ which also is a base for X such that $|\mathcal{B}_0| \leq \aleph_\alpha$. From the regularity of $\aleph_{\alpha+1}$ it follows that trind X < $\omega_{\alpha+1}$, and we have a contradiction.

3.6. Definition ([Sa]). A family $\mathcal{B}$ of open subsets of a topological space X is called a *large base* for the space X if for every closed set $A \subset X$ and each open set $V \subset X$ which contains the set A there exists a $U \in \mathcal{B}$ such that $A \subset U \subset V$. The *large weight* of a space X is defined as the smallest cardinal number of the form $|\mathcal{B}|$, where $\mathcal{B}$ is a large base for X; this cardinal number is denoted by W(X).

Clearly, if X is a T_1-space, then every large base for X is a base for X and $w(X) \leq W(X)$. One easily checks that for each infinite compact space X we have $W(X) = w(X)$.

3.7. Lemma. If $W(X) \leq m \geq \aleph_0$, then for every large base $\mathcal{B}$ for X there exists a large base $\mathcal{B}_0$ such that $|\mathcal{B}_0| \leq m$ and $\mathcal{B}_0 \subset \mathcal{B}$.

Proof. Let $\mathcal{B}$ be an arbitrary large base for X and $\mathcal{D}$ a large base of cardinality $\leq m$. Consider the family of all pairs $(V,W) \in \mathcal{D} \times \mathcal{D}$ such that there exists a set $U \in \mathcal{B}$ satisfying $V \subset U \subset W$ and for each such pair pick up a set $U \in \mathcal{B}$ with the above property. The family $\mathcal{B}_0 \subset \mathcal{B}$ obtained in this way has cardinality $\leq m$ and is a large base for X.

Arguing as in the proof of Theorem 3.5 and applying the above lemma we obtain

3.8. Theorem ([Sm1]). If a normal space X of large weight $W(X) = \aleph_\alpha$ has large transfinite dimension, then it satisfies the inequality $\text{trInd}\ X < \omega_{\alpha+1}$.

Theorems 3.5 and 3.8 yield

3.9. Corollary. If a separable metric space X has small transfinite dimension, then it satisfies the inequality $\text{trind}\ X < \omega_1$. If a compact metric space X has large transfinite dimension, then it satisfies the inequality trInd $X < \omega_1$.

3.10. Remark. We shall show later (see Theorem 3.19) that if an arbitrary metric space X has large transfinite dimension, then it satisfies the inequality trInd $X < \omega_1$. On the other hand, the (non-separable) metric space X defined as the sum $\oplus_{\alpha<\omega_1} S_\alpha$ of Smirnov spaces S_α (see Example 2.2) satisfies the equality trind $X = \omega_1$. Since, as announced by Ljuksemburg in [Lj3], even for a compact metric space X, it may happen that trind $(X\times I)$ = trind X, the following problems arise:

3.11. Problem. Do there exist metric spaces of arbitrarily high small transfinite dimension?

3.12. Problem. Does there exist a (preferably compact) metric space X which has trind such that trind $(X\times I^n)$ = trind X for $n = 1,2,\ldots$?

The final part of this section is devoted to metric spaces which have large transfinite dimension. Except for Theorem 3.21, announced in [Lj2], this is a variation on [Sk] and [Sm2]. Our presentation of the material is a result of conversations with E. Pol.

For every metric space X let

$$S(X) = X\setminus\bigcup\{U: U \text{ is an open subspace of } X \text{ and } \operatorname{Ind} U < \infty\}.$$

Obviously, $X\setminus S(X)$ is a locally finite-dimensional space. We shall show that if X has large transfinite dimension, then S(X) is a compact space and X is in a sense "condensend" around S(X). We start with two lemmas. Let us recall that the class $\mathcal{S}$ is defined in Example 2.1.

3.13. Lemma. If a locally finite-dimensional metric space contains no closed subspace which belongs to the class S, then it is finite-dimensional.

Proof. Consider a locally finite-dimensional metric space X such that Ind X = ∞. Applying the locally finite sum theorem ([E1], Theorem 4.1.10), one easily defines a sequence of points $x_1, x_2, \ldots$ in X and a sequence $V_1, V_2, \ldots$ of open sets, where $x_i \in V_i$ for $i = 1,2,\ldots$, such that $\text{Ind } V_1 < \text{Ind } V_2 < \ldots$ and x_i has no neighborhood V such that $\text{Ind } V < \text{Ind } V_i$. Since X is locally finite-dimensional the family of all one-point sets $\{x_i\}$ is discrete, and thus (cf. [E2], Theorem 2.1.14) there exist a family $\{U_i\}_{i=1}^{\infty}$ of open subsets of X such that $x_i \in U_i \subset \overline{U}_i \subset V_i \cap B(x_i, 1/i)$ for $i = 1,2,\ldots$ and $\overline{U}_i \cap \overline{U}_j = \emptyset$ for $i \neq j$. The union of all $\overline{U}_i$'s is a closed subspace of X which belongs to the class S.

By a similar argument one obtains the following lemma.

3.14. Lemma. If a metric space X has large transfinite dimension, then the subspace S(X) is compact.

3.15. Remark. In the same way one can prove that if a separable metric space X (or, more generally, a metric space X on all subspaces of which ind and Ind coincide) has small transfinite dimension and contains no closed subspace which belongs to the class S, then the subspace S(X) is compact.

3.16. Theorem. If a metric space X has large transfinite dimension, then the subspace S(X) is compact and for every open set $U \subset X$ which contains S(X) we have $\text{Ind } (X \setminus U) < \infty$.

If in a metric space X there exists a closed subspace M which has large transfinite dimension and is such that for every open set $U \subset X$ which contains M we have $\operatorname{Ind}(X\setminus U) < \infty$, then X has large transfinite dimension; more exactly, $\operatorname{trInd} X \leq \omega_0 + \operatorname{trInd} M+1$.

Proof. The first part follows from Lemmas 3.14 and 3.13. The second part is proved by transfinite induction with respect to trInd M. If $\operatorname{trInd} M = -1$, then $M = \emptyset$; taking $U = \emptyset$ we see that $\operatorname{Ind} X < \infty$, so that $\operatorname{trInd} X \leq \omega_0 = \omega_0 + (\operatorname{trInd} M+1)$. Assume now that the theorem is proved if the large transfinite dimension of the subspace under consideration is less than $\alpha \geq 0$ and take a space X which contains a subspace M satisfying the assumptions and such that $\operatorname{trInd} M = \alpha$. For any pair A, B of disjoint closed subspaces of X there exists a partition L' in the space M between $M \cap A$ and $M \cap B$ such that $\operatorname{trInd} L' < \alpha$. Applying the standard device of extending partitions (see [E1], Lemma 1.2.9) we obtain a partition L in the space X between A and B which satisfies the inclusion $M \cap L \subset L'$. Now, the space L and its subspace $M \cap L$ satisfy the assumptions of the second part of our theorem and $\operatorname{trInd}(M \cap L) \leq \operatorname{trInd} L' < \alpha$, so that by the inductive hypothesis $\operatorname{trInd} L \leq \omega_0 + \operatorname{trInd}(M \cap L) + 1 < \omega_0 + \alpha + 1$; thus, $\operatorname{trInd} X \leq \omega_0 + \alpha + 1$.

3.17. Remark. The assumption that M is a closed subspace can be omitted in the second part of the above theorem. In the case of an arbitrary subspace we take two open sets $V_1, V_2 \subset X$ such that $A \subset V_1$, $B \subset V_2$ and $\overline{V}_1 \cap \overline{V}_2 = \emptyset$, consider a partition L' in M between $M \cap \overline{V}_1$ and $M \cap \overline{V}_2$ and extend it to L in conformity with the first part of Lemma 1.2.9 in [E1]; to obtain the inequality $\operatorname{trInd}(M \cap L) \leq \operatorname{trInd} L'$ we apply Theorem 5.1 below.

From Theorem 3.16 we obtain

3.18. Corollary. If a metric space X has large transfinite dimension, then trInd X $\leq \omega_0$ + trInd S(X) + 1.

Corollaries 3.9, 3.18 and the first part of Theorem 3.16 yield

3.19. Theorem. If a metric space X has large transfinite dimension then it satisfies the inequality trInd X $< \omega_1$.

The following lemma is a special case of Theorem 5.15; its proof parallels the proof of the second part of Theorem 3.16.

3.20. Lemma. If a metric space X can be represented as the union A ∪ B, where trInd A $\leq \alpha \geq \omega_0$ and Ind B $< \infty$, then trInd X $\leq \alpha$.

Applying Lemma 3.20 and the first part of Theorem 3.16 we obtain

3.21. Theorem. Let X be a metric space which has large transfinite dimension. For $\alpha \geq \omega_0$t the space X satisfies the inequality trInd X $\leq \alpha$ if and only if for every pair A, B of disjoint closed subsets of S(X) there exists a partition L in X between A and B such that trInd L $< \alpha$.

Since the class of compact metric spaces which have large transfinite dimension coincides with the class of compact countable-dimensional spaces (see Corollary 4.7), Theorem 3.16 yields an internal topological characterization of metric spaces which have large transfinite dimension in terms of the dimension function Ind.

3.22. Problem. Find an internal topological characterization of metric spaces (or separable metric spaces) which have small transfinite dimension in terms of the dimension function ind.

Let us observe that the class of separable metric spaces which have small transfinite dimension coincides with the class of spaces which have countably dimensional metrizable compactifications (see Theorem 4.15), but this is not an internal characterization.

4. Relations between trind, trInd and countable dimensionality. We start with four theorems which exhibit the relations between the existence of transfinite dimensions and countable dimensionality.

4.1. Theorem ([Hu]). Every countable-dimensional completely metrizable space has small transfinite dimension.

Proof. Assume that a countable-dimensional complete metric space X has no trind; let $X = \bigcup_{i=1}^{\infty} X_i$, where $\mathrm{Ind}\, X_i \leq 0$. By our assumption, there exists a point $x \in X$ and a neighborhood $V \subset X$ of the point x such that for every open set U satisfying $x \in U \subset V$ the boundary $F = \mathrm{Fr}U$ has no trind. Consider an open set U_1 such that $x \in U_1 \subset V \cap B(x,1)$ and $X_1 \cap \mathrm{Fr}U_1 = \emptyset$ (see [E1], Theorem 4.1.13), and let $F_1 = \mathrm{Fr}U_1$. Since F_1 has no trind, we can repeat the argument and define inductively a decreasing sequence $F_1 \supset F_2 \supset \ldots$ of closed subsets of X such that $X_i \cap F_i = \emptyset$, $\delta(F_i) \leq 1/i$ and F_i has no trind (in particular $F_i \neq \emptyset$). The space X being complete, $\bigcap_{i=1}^{\infty} F_i \neq \emptyset$; on the other hand, $\bigcap_{i=1}^{\infty} F_i = (\bigcup_{i=1}^{\infty} X_i) \cap (\bigcap_{i=1}^{\infty} F_i) \subset \bigcup_{i=1}^{\infty} (X_i \cap F_i) = \emptyset$, a contradiction.

In a similar way one obtains the following theorem.

4.2. Theorem ([Sm1]). Every countable-dimensional compact metric space has large transfinite dimension.

4.3. Theorem ([Hu]). Every separable metric space which has small transfinite dimension is countable-dimensional.

Proof. We apply transfinite induction with respect to the small transfinite dimension of the space under consideration. If trind X = -1, the theorem holds. Assume that the theorem is established for all separable metric spaces with trind less than $\alpha \geq 0$ and consider a space X such that trind X = α. The space X has a base $\mathcal{B}$ such that trind FrU < α for every U $\in \mathcal{B}$; without loss of generality one can assume that $|\mathcal{B}| \leq \aleph_0$. The space Y = $\bigcup$\{FrU: U $\in \mathcal{B}$\} is countable-dimensional by the inductive assumption, and the space Z = X\Y satisfies the equality Ind Z = ind Z = 0, so X is countable-dimensional.

Since the Hilbert cube $I^{\aleph_0}$ is not countable-dimensional ([Hu]; cf. [E1], Theorem 1.8.20), we have

4.4. Corollary ([Hu]). The Hilbert cube $I^{\aleph_0}$ has no small transfinite dimension and, a fortiori, no large transfinite dimension.

Let us note that, as shown in [Sm2], the last theorem can be generalized to completely paracompact metric spaces (i.e. metric spaces that have a base which can be represented as the union of countably many star-finite covers); in the proof methods developed in [Z] to show that ind and Ind coincide in such spaces are applied.

4.5. Problem. Give an example of a metric space which has small transfinite dimension and is not countable-dimensional.

One might also ask if there exists a metric space X such that ind X = 0 which is not countable-dimensional. Let us add that when, finally, metric spaces X_n such that ind $X_n = 0$ and Ind $X_n = n$ are defined for all n, their sum $X = \oplus_{n=1}^{\infty} X_n$ will be an example of a metric space X such that ind X = 0 and X has no large transfinite dimension (for the time being only X_1 is defined; it is the famous Roy's space, cf. [E1], p. 197).

For every metric space X we have $X = S(X) \cup (X \setminus S(X))$; the locally finite-dimensional space $X \setminus S(X)$ is countable-dimensional, and if X has large transfinite dimension, the space (X) is compact by Theorem 3.16, so that from Theorem 4.3 we obtain

4.6. Theorem ([Sm2]). Every metric space which has large transfinite dimension is countable-dimensional.

Since there exist countable-dimensional compact metric spaces which are not strongly countable-dimensional (see Remark 1.4), it follows from Theorem 4.2 that in Theorems 4.3 and 4.6 "countable-dimensional" cannot be replaced by "strongly countable-dimensional".

Theorems 4.1, 4.2 and 4.3 yield

4.7. Corollary. For every compact metric space X the following conditions are equivalent:

(i) X has small transfinite dimension.

(ii) X has large transfinite dimension.

(iii) X is countable-dimensional.

4.8. Remark. Let us note that in [Lv2] the equivalence of (i) and (ii) above was established for all hereditarily normal compact spaces; more exactly, it was proved that if a hereditarily normal compact space X has small transfinite dimension, then X has also large transfinite dimension and $\text{trInd}\, X \leq \omega_0 \cdot \text{trind}\, X$. It was announced recently in [F] that (i) and (ii) are equivalent for all compact spaces and that every hereditarily normal compact space X which can be represented as the union of countably many subspaces each of which has small transfinite dimension has itself small transfinite dimension. In connection with these results Fedorčuk states the following two problems:

4.9. Problem. Does the inequality $\text{trInd}\, X \leq \omega_0 \cdot \text{trind}\, X$ hold for every compact space X which has transfinite dimensions?

4.10. Problem. Is it true that every compact space X which can be represented as the union of countably many subspaces each of which has small transfinite dimension has itself small transfinite dimension?

As already noted, in [Lj3] it is announced that for Smirnov's space S_{ω_0+3} we have $\text{trind}\, S_{\omega_0+3} < \text{trInd}\, S_{\omega_0+3}$. Thus, although a compact metric space has small transfinite dimension if and only if it has large transfinite dimension, it may happen that the transfinite dimensions are distinct. In fact, it is shown in [Lj3] that for a closed subspace L of S_{ω_0+3} we have $\text{trind}\, L = \omega_0 + 1 < \omega_0 + 2 = \text{trInd}\, L$. In a sense this is the best possible example; indeed, the following proposition is easily established.

4.11. Proposition ([W1]). For every compact metric space X such that $\text{trind}\, X = \omega_0$ the equality $\text{trind}\, X = \text{trInd}\, X$ holds.

Further examples, showing that trind and trInd can widely diverge in the realm of compact metric spaces, as well as some positive results on the coincidence of both transfinite dimensions, are announced in [Lj3] and [Lj4].

Now, we are going to discuss the relations between the existence of small and large transfinite dimensions in separable metric spaces. We start with a partial generalization of Corollary 4.7.

4.12. Theorem. A separable metric space has large transfinite dimension if and only if it has small transfinite dimension and contains no closed subspace which belongs to the class S.

Proof. It suffices to prove that if a separable metric space X has small transfinite dimension and contains no closed subspace which belongs to the class S, then trInd X is defined. By virtue of Remark 3.15, the subspace S(X) is compact; since S(X) has small transfinite dimension, it follows from Corollary 4.7 that S(X) has large transfinite dimension, too. To conclude the proof is suffices to apply Lemma 3.13 and the second part of Theorem 3.16.

It follows from Remark 3.15 that Theorem 4.12 remains valid in the class of all completely paracompact metric spaces; we have however, the following problem:

4.13. Problem. Give an example of a metric space which has small transfinite dimension, contains no closed subspace which belongs to the class S, and yet has no large transfinite dimension.

One can also ask if Theorem 4.12 can be improved to become a generalization of Corollary 4.7, i.e. if every countable-dimensional separable metric space that contains no closed subspace which belongs to the class S has small transfinite dimension. One easily checks that this is equivalent to the following

4.14. Problem. Does every countable-dimensional separable metric space X satisfying the equality Ind $X = \infty$ contain for every natural number n a closed subspace A such that Ind $A = n$?

Similarly, Problem 4.13 is equivalent to exhibiting a metric space X which has small transfinite dimension, satisfies the equality Ind $X = \infty$, and yet for some natural number n has no closed subspace A such that Ind $A = n$.

4.15. Theorem (announcement [Hu]). A separable metric space X has small transfinite dimension if and only if X has a countable-dimensional metrizable compactification.

Proof. It suffices to show that every separable metric space X such that trind X exists has a countable-dimensional metrizable compactification. By virtue of the enlargement theorem (see Theorem 5.5 below), X is dense in a completely metrizable space X^* which satisfies the equality trind X^* = trind X. Now, it is shown in [Le] that every completely metrizable separable space has a metrizable compactification with countable-dimensional remainder; such a compactification $\tilde{X}$ of the space X^* is the required compactification of X.

It seems that no proof of the above theorem was ever published; one might, however, suspect that Hurewicz had a similar argument in mind, because he notes in [Hu] that every completely metrizable separable countable-dimensional space has a countable-dimensional metrizable compactification.

Let us note that a characterization of separable metric spaces which have strongly countable-dimensional metrizable compactification was given in [Sh1].

4.16. Theorem. A separable metric space X has large transfinite dimension if and only if X has a countable-dimensional metrizable compactification $\tilde{X}$ such that every point of the remainder $\tilde{X}\setminus X$ has a neighborhood $U \subset \tilde{X}$ satisfying the inequality $\operatorname{Ind} U < \infty$.

Proof. Consider a separable metric space X which has a countable-dimensional metrizable compactification $\tilde{X}$ with the above property, i.e., such that $S(\tilde{X}) \subset X$. From the second part of Theorem 3.16 applied to the subspace $M = S(\tilde{X})$ of X it follows that X has large transfinite dimension.

Now, consider a separable metric space X which has large transfinite dimension. Let $F_i = X\setminus B(S(X),1/i)$ for $i = 1,2,\ldots$; clearly, $\operatorname{Ind} F_i < \infty$. By virtue of Theorem 2 in [K], X is dense in a compact metrizable space $\tilde{X}$ such that $\operatorname{Ind} \overline{F}_i = \operatorname{Ind} F_i$ for $i = 1,2,\ldots$, where $\overline{F}_i$ is the closure of F_i in $\tilde{X}$. One easily shows, using compactness of S(X), that every point of the remainder $\tilde{X}\setminus X$ has a neighborhood $U \subset \tilde{X}$ contained in one of the sets $\overline{F}_i$ and thus such that $\operatorname{Ind} U < \infty$. To conclude the proof it suffices to note that $\tilde{X} = X \cup (\tilde{X}\setminus X) \subset \tilde{X} \cup \bigcup_{i=1}^{\infty}\overline{F}_i$, i.e. X is countable-dimensional.

The last theorem was communicated to me by T. Przymusiński, who in fact proved it under the weaker assumption that X is a completely paracompact metric space (obviously, the compactification $\tilde{X}$ is then no longer metrizable; the notion of countable dimensionality is defined as in 1.3 above). The assumption is used in the first part of the proof; the required compactification can be obtained for any metric space which has large transfinite dimension by an application of Theorem 1' in [Pa].

4.17. Problem. Give an example of a metric space X which has a countable-dimensional compactification $\tilde{X}$ such that every point of the remainder $\tilde{X}\setminus X$ has a neighborhood $U \subset \tilde{X}$ satisfying the inequality Ind U < ∞, and yet X has no large transfinite dimension.

Let us note that the hypothetical space $X = \oplus_{n=1}^{\infty} X_n$, mentioned when commenting on Problem 4.5, would be such an example.

5. *Status of main dimension theory theorems in the case of transfinite dimensions.*

A. Subspace theorems. In addition to Proposition 3.3 and the final part of Example 2.1 we have the following result:

5.1. Theorem ([Lj2]). For every metric space X and a subspace A of X, if trInd X and trInd A exist, then trInd A $\leq$ trInd X.

Proof. This follows immediately from Theorem 3.21, because-- by virtue of Lemma 3.14 -- S(A) is a closed subspace of X.

The last theorem can be supplemented by the following proposition which is a consequence of Lemma 3.14, Theorem 4.12 and Remark 3.17.

5.2. Proposition. For a subspace A of a metric space X such that trInd X exists, trInd A exists if and only if A has no closed subspace which belongs to the class S.

Let us note, in connection with Proposition 5.2, that by Theorem 3.19 the space $X = \oplus_{\alpha<\omega_1} S_\alpha$, satisfying trind X = ω_1, considered in Remark 3.10, is not contained in any metric space which has large transfinite dimension.

Dimension Ind not being monotone in normal spaces (cf. [E1], Example 2.2.11), the assumption that X is metrizable cannot be omitted in Theorem 5.1. We have, however, the following problem (cf. [E1], Theorem 2.3.6):

5.3. Problem. Does Theorem 5.1 hold under the assumption that X is a strongly hereditarily normal space?

B. Separation theorems. Since there exists a compact metric space L such that trind L $\neq$ trInd L, the counterpart for trind of the first separation theorem for ind (cf. [E1], Theorem 1.5.12), and, a fortiori, of the second separation theorem for ind (cf. [E1], Theorem 1.5.13) does not hold even in compact metric spaces. On the other hand, Lemma 1.2.9 in [E1] implies the following separation theorem for trInd (cf. [E1], Theorem 2.2.4):

5.4. Theorem. If X is a hereditarily normal space and M is a subspace of X such that trInd M $\leq \alpha >$ 0, then for every pair A, B of disjoint closed subsets of X there exists a partition L between A and B such that trInd $(L \cap M) < \alpha$.

C. Enlargement theorems. The following two theorems were announced in [Lj2]; we include a proof of the first one to make the proof of Theorem 4.15 complete.

5.5. Theorem. For every subspace M of a separable metric space X satisfying the inequality trind M $\leq \alpha$ there exists a G_δ-set M* in X such that $M \subset M^*$ and trind M* $\leq \alpha$.

Proof. We apply transfinite induction with respect to α. The theorem holds for $\alpha = -1$. Assume that the theorem is established for all subspaces with trind less than $\alpha \geq 0$ and consider a separable metric space and its subspace M such that trind $M \leq \alpha$.

Let i be a fixed natural number. For every $x \in M$ consider an open set $U \subset X$ such that $x \in U$, $\delta(U) < 1/i$ and trind $(M \cap \mathrm{Fr}U) < \alpha$ (cf. [E1], Problem 2.2.A(a)). By the inductive assumption applied to the subspace $M \cap \mathrm{Fr}U$ of the space FrU, there exists a G_δ-set U^* in FrU such that $M \cap \mathrm{Fr}U \subset U^*$ and trind $U^* < \alpha$; the set $F(U) = (\mathrm{Fr}U)\setminus U^*$ is an F_σ-set in X and $M \cap F(U) = \emptyset$. The family of all U's contains a countable subfamily $\mathcal{U}_i$ with the same union.

The intersection $G = \bigcap_{i=1}^{\infty}(\bigcup\mathcal{U}_i)$ is a G_δ-set in X and contains M; similarly, $H = X\setminus\bigcup_{i=1}^{\infty}(\bigcup\{F(U): U \in \mathcal{U}_i)$ is a G_δ-set in X and contains M. One easily checks that the subspace $M^* = H \cap G$ of X has the required properties.

5.6. Theorem. For every subspace M of a metric space X satisfying the inequality trInd $M \leq \alpha$ there exists a G_δ-set M^* in X such that $M \subset M^*$ and trInd $M^* \leq \alpha$.

A weaker version of Theorem 5.6, viz. that trInd M^* exists, was established in [Sh2].

From Theorem 5.5 it follows that if the space X in the last theorem is separable one can assume, moreover, that trind M^* = trind M.

As shown by E. Pol, Theorem 5.5 holds for completely paracompact metric spaces.

5.7. Problem. Does Theorem 5.5 hold for all metric spaces?

The problem seems open even for finite α's, i.e. for the dimension ind.

D. Compactification theorems. The compactification theorem for trind does not hold: a separable metric space X is described in [Lj2] such that trind $X = \omega_0$ and X is not contained in any compact metric space $\tilde{X}$ such that trind $\tilde{X} = \omega_0$.

On the other hand, the following result is announced in [Lj2].

5.8. Theorem. For every separable metric space X which has large transfinite dimension there exists a compact metric space $\tilde{X}$ which contains a dense subspace homeomorphic to X and satisfies the equalities trInd $\tilde{X}$ = trInd X and trind $\tilde{X}$ = trind X.

We have also the following compactification theorem proved in [AP] (the theorem was announced in [Pa]; the special case of a metric space X was obtained independently by Ljuksemburg and announced in [Lj2] (if, in this special case, trind $X \geq \omega_0^2$, one can moreover assume that trind $\tilde{X}$ = trind X)):

5.9. Theorem. For every normal space X which has large transfinite dimension there exists a compact space $\tilde{X}$ which contains a dense subspace homeomorphic to X and satisfies the equalities trInd $\tilde{X}$ = trInd X and $w(\tilde{X}) = w(X)$.

Let us add that a normal space X has large transfinite dimension if and only if its Čech-Stone compactification βX has large transfinite dimension and, moreover, trInd X = trInd βX; the standard proof of the corresponding theorem for Ind (cf. [E1], Theorem 2.2.9) remains valid for trInd.

Since every separable metric space which has small transfinite dimension has a compactification $\tilde{X}$ which also has small transfinite dimension (see Theorems 4.15 and 4.1), the following problem arises:

5.10. Problem. Evaluate the increase of small transfinite dimension in the process of compactification of a separable metric space.

E. Universal space theorems. From Example 2.2 it follows that there is no universal space either in the class of all separable metric spaces which have small transfinite dimension or in the class of all separable metric spaces which have large transfinite dimension.

5.11. Problem. Does there exist for a fixed $\alpha \geq \omega_0$ a universal space in the class of all separable metric spaces whose small (large) transfinite dimension is at most α?

Partial negative results are announced in [Lj1]; for example, there is no universal space either in the class of all strongly countable-dimensional completely metrizable separable spaces whose small transfinite dimension is at most $\alpha \geq \omega_0$ or in the class of all strongly countable-dimensional compact metric spaces whose large transfinite dimension is at most $\alpha \geq \omega_0$. On the other hand, as shown in [Po], for every $\alpha < \omega_1$ there exists a complete separable metric space X such that trind X = α which topologically contains each compact metric space whose small transfinite dimension is at most α.

F. Decomposition theorems. Theorems 4.3 and 4.6 are decomposition theorems for small and large transfinite dimension, respectively.

The following problem is stronger than Problem 4.5.

5.12. Problem. Give an example of a metric space which has small transfinite dimension and cannot be represented as the union of a sequence $X_1, X_2, \ldots$ of subspaces such that $\operatorname{ind} X_i \leq 0$ for $i = 1,2,\ldots$.

G. Addition theorems. To state addition theorems and sum theorems for transfinite dimensions one introduces two kinds of addition for ordinal numbers.

5.13. Definition ([T]). Let α and β be arbitrary ordinal numbers and let A and B be any disjoint well-ordered sets with order types α and β, respectively. Consider all possible well-orders in the union $A \cup B$ which restricted to A and B coincide with the original well-orders in these sets; the set of all corresponding ordinal numbers has a smallest element: this is the *lower sum* $\alpha \pm \beta$ of α and β, and a least upper bound: this is the *upper sum* $\alpha \mp \beta$ of α and β.

One easily checks that if $\alpha, \beta < \omega$, then $\alpha \pm \beta = \alpha \mp \beta = \alpha + \beta$ and that, in contradistinction to the usual sum of ordinal numbers, both lower sum and upper sum are commutative. As shown in [T], if $\alpha = \alpha' + p$ and $\beta = \beta' + q$, where α', β' are limit ordinal numbers and p, q are finite ordinal numbers, then

$$\alpha \pm \beta = \begin{cases} \alpha, & \text{if } \alpha' > \beta', \\ \alpha + q = \beta + p, & \text{if } \alpha' = \beta', \\ \beta, & \text{if } \alpha' < \beta'. \end{cases}$$

It is also shown in [T] that $\alpha \mp \beta$ coincides with the so called natural sum of α and β (see [KM], p. 253).

5.14. Theorem ([T]). Let X be a hereditarily normal space which is the union of two subspaces A and B. If trind A and trind B exist, then trind X also exists; moreover

$$1 + \operatorname{trind} X \leq (1 + \operatorname{trind} A) \mp (1 + \operatorname{trind} B).$$

5.15. Theorem ([Lv2]; cf. [Pe]). Let X be a hereditarily normal space which is the union of two non-empty subspaces A and B. If trInd A and trInd B exist, then trInd X also exists; moreover

$$\text{trInd } X \leq \max(\text{trInd } A, \text{trInd } B) \pm (\min(\text{trInd } A, \text{trInd } B) + 1).$$

Let us note that in [La] the counterpart of Theorem 5.14 for trInd was established, but this result is weaker than Theorem 5.15. On the other hand the counterpart of Theorem 5.15 for trind does not hold, even under the assumption that A is closed (such a counterpart under the assumption that both A and B are closed follows from Theorem 5.16 and was stated in [Lv2]). Indeed, $S_{\omega_0+1} = S(S_{\omega_0+1}) \cup (S_{\omega_0+1} \setminus S(S_{\omega_0+1})) = I \cup \oplus_{n=1}^{\infty} I^n$, and from Proposition 4.11 it follows that $\text{trind } S_{\omega_0+1} = \omega_0 + 1$.

H. Sum theorems. We have the following two sum theorems.

5.16. Theorem ([T]). If a hereditarily normal space X can be represented as the union $F_1 \cup F_2$ of its closed subspaces which have small transfinite dimension, then

$$\text{trind } X \leq \max(\text{trind } F_1, \text{trind } F_2) \pm (\text{trind}(F_1 \cap F_2) + 1).$$

5.17. Theorem ([La] and [Pe]). If a hereditarily normal space X can be represented as the union $F_1 \cup F_2$ of its closed subspaces which have large transfinite dimension, then

$$\text{trInd } X \leq \max(\text{trInd } F_1, \text{trInd } F_2) \pm (\text{trInd}(F_1 \cap F_2) + 1).$$

From sum theorems two useful corollaries follow.

5.18. Corollary. If a hereditarily normal space X can be represented as the union $F_1 \cup F_2$ of its closed subspaces such that $\text{trind } F_i \leq \alpha \geq \omega_0$ for $i = 1,2$ and $\text{trind } (F_1 \cap F_2) < \omega_0$, then $\text{trind } X \leq \alpha$.

5.19. Corollary. If a hereditarily normal space X can be represented as the union $F_1 \cup F_2$ of its closed subspaces such that trInd $F_i \leq \alpha \geq \omega_0$ for i = 1,2 and trInd $(F_1 \cap F_2) < \omega_0$, then trInd X $\leq \alpha$.

Applying induction, and using Theorems 5.14 and 5.15, one generalizes the above corollaries to finite unions $F_1 \cup F_2 \cup \ldots \cup F_k$, where the corresponding dimension of all the intersections $F_i \cap F_j$ with $i \neq j$ is finite.

Let us add that, as shown in [Lv1], the compact space S_{ω_0+1} can be represented as the union $F_1 \cup F_2$, where F_1, F_2 are closed and trind F_i = trInd $F_i = \omega_0$ for i = 1,2; an example of a separable metric space with similar property with respect to trind was given earlier in [T].

Finally, let us observe that since all Smirnov's spaces S_α are strongly countable-dimensional, there is no evaluation of transfinite dimensions of countable unions of closed subspaces by transfinite dimensions of the subspaces.

I. Cartesian products theorems. In [T] the following theorem is established.

5.20. Theorem. If X and Y are hereditarily normal spaces such that trind X and trind Y exist, then trind $(X\times Y)$ also exists; moreover, trind $(X\times Y) \leq$ (trInd X ± trind Y) + n, where n is a finite ordinal number depending on X and Y.

An evaluation of n can be found in [T]. It is announced in [F] that the first part of the above theorem holds for arbitrary compact spaces.

We have also

5.21. Theorem ([T]). If X is a metric space such that X exists, then trind$(X\times I) \leq$ trind X + 1.

It is announced in [Lj3], that the last theorem holds for regular spaces. As already noted in Remark 3.10, even for a compact metric space X it may happen that trind(X×I) < trind X + 1.

5.22. Problem. Are there any evaluations of trInd (X×Y) or of trInd (X×I)?

References

[AP] P.S. Aleksandrov and B.A. Pasynkov, *Vvedenie v teoriju razmernosti*, Moskva 1973. (Introduction to Dimension Theory)

[BP] C. Bessaga and A. Pełczyński, *The estimated extension theorem, homogeneous collections and skeletons and their applications to the topological classification of linear metric spaces and convex sets*, Fund. Math. 69 (1970), 153-190.

[E1] R. Engelking, *Dimension Theory*, Warszawa 1978.

[E2] R. Engelking, *General Topology*, Warszawa 1977.

[F] V.V. Fedorčuk, *Infinite-dimensional compact spaces*, talk at the Colloquium on Topology, Budapest, August 7-12, 1978.

[He] D.W. Henderson, *A lower bound for transfinite dimension*, Fund. Math. 63 (1968), 167-173.

[Hu] W. Hurewicz, *Ueber unendlich-dimensionale Punktmengen*, Proc. Akad. Amsterdam 31 (1928), 916-922.

[K] K. Kuratowski, *Sur les théorèmes du "plongement" dans la théorie de la dimension*, Fund. Math. 28 (1937), 336-342.

[KM] K. Kuratowski and A. Mostowski, *Set Theory*, Amsterdam, 1976.

[La] M. Landau, *Strong transfinite ordinal dimension*, Bull. Amer. Math. Soc. 21 (1969), 591-596.

[Le] A. Lelek, *O razmernosti narostov pri kompaktnyh rasširenijah*, Dokl. Akad. Nauk SSSR 160 (1965), 534-537. (On the dimension of remainders in compact extensions, Soviet Math. Dokl. 6 (1965), 136-140)

[Lj1] L.A. Ljuksemburg, *O beskonečnomernyh prostranstvah, imejuščih transfinitnuju razmernost'*, Dokl. Akad. Nauk SSSR 199 (1971), 1243-1246. (On infinite-dimensional spaces with transfinite dimension, Soviet Math. Dokl. 12 (1971), 1272-1276)

[Lj2] L.A. Ljuksemburg, *O transfinitnyh induktivnyh razmernostjah*, Dokl. Akad. Nauk SSSR 209 (1973), 295-298. (On transfinite inductive dimensions, Soviet Math. Dokl. 14 (1973), 388-393)

[Lj3] L.A. Ljuksemburg, *O kompaktah s nesovpadajuščimi transfinitnymi razmernostjami*, Dokl. Akad. Nauk SSSR 212 (1973), 1297-1300. (On compact spaces with non-coinciding transfinite dimensions, Soviet Math. Dokl. 14 (1973), 1593-1597)

[Lj4] L.A. Ljuksemburg, *O transfinitnoĭ razmernosti metričeskih prostranstv*, Uspehi Mat. Nauk 34 (1979), vyp. 1, 233-234.

[Lv1] B.T. Levšenko, *O beskonečnomernyh prostranstvah*, Dokl. Akad. Nauk SSSR 139 (1961), 286-289. (On infinite-dimensional spaces, Soviet Math. Dokl. 2 (1961), 915-918)

[Lv2] B.T. Levšenko, *Prostranstva transfinitnoĭ razmernosti*, Mat. Sb. 67 (1965), 255-266. (Spaces of transfinite dimensionality, Amer. Math. Soc. Transl. Ser 2, 73 (1968), 135-148)

[NR] K. Nagami and J.H. Roberts, *A note on countable-dimensional metric spaces*, Proc. Japan Acad. 41 (1965), 155-158.

[N] J. Nagata, *On the countable sum of zero-dimensional spaces*, Fund. Math. 48 (1960), 1-14.

[Pa] B.A. Pasynkov, *O razmernosti normal'nyh prostranstv*, Dokl. Akad. Nauk SSSR 201 (1971), 1049-1052. (On the dimension of normal spaces, Soviet Math. Dokl. 12 (1971), 1784-1787)

[Pe] A.R. Pears, *A note on transfinite dimension*, Fund. Math. 71 (1971), 215-221.

[Po] R. Pol, *Note on classification and universality in the class of weakly-infinite dimensional compacta*, preprint, 1979.

[Sn] N.A. Shanin, *On the theory of bicompact extensions of topological spaces*, C.R. (Doklady) Acad. Sci. URSS 38 (1943), 154-156.

[Sh1] Z. Shmuely, *On strongly countable-dimensional sets*, Duke Math. Journ. 38 (1971), 169-173.

[Sh2] Z. Shmuely, *Some embeddings of infinite-dimensional spaces*, Israel Journal of Math. 12 (1972), 5-10.

[Sk] E.G. Skljarenko, *O razmernostnyh svoǐstvah beskonečnomernyh prostranstv*, Izv. Akad. Nauk SSSR, Ser. Math. 23 (1959), 197-212. (On dimensional properties of infinite dimensional spaces, Amer. Math. Soc. Transl. Ser. 2, 21 (1962), 35-50)

[Sm1] Ju. M. Smirnov, *Ob universal'nyh prostranstvah dlja nekotoryh klassov beskonečnomernyh prostrastv*, Izv. Akad. Nauk SSSR, Ser. Mat. 23 (1959), 185-196. (On universal spaces for certain classes of infinite dimensional spaces, Amer. Math. Soc. Transl. Ser. 2, 21 (1962), 21-33)

[Sm2] Ju. M. Smirnov, *Neskol'ko zamečaniǐ o transfinitnoǐ razmernosti*, Dokl. Akad. Nauk SSSR 141 (1961), 814-817. (Some remarks on transfinite dimension, Soviet Math. Dokl. 2 (1961), 1572-1575)

[T] G.H. Toulmin, *Shuffling ordinals and transfinite dimension*, Proc. London Math. Soc. 4 (1954), 177-195.

[U1] P. Urysohn, *Mémoire sur les multiplicités Cantoriennes*, Fund. Math. 7 (1925), 30-137.

[U2] P. Urysohn, *Mémoire sur les multiplicités Cantoriennes (suite)*, Fund. Math. 8 (1926), 225-359.

[W1] B.R. Wenner, *Finite-dimensional properties of infinite-dimensional spaces*, Pacific Journ. of Math. 42 (1972), 267-276.

[W2] B.R. Wenner, *A universal separable metric locally finite-dimensional space*, Fund. Math. 80 (1973), 283-286.

[Z] A.V. Zarelua, *O teoreme Gureviča*, Mat. Sb. 60 (1963), 17-28. (On a theorem of Hurewicz, Amer. Math. Soc. Transl. Ser. 2, 55 (1966), 141-152).

APPLICATIONS OF STATIONARY SETS IN TOPOLOGY

William G. Fleissner[1]
University of Pittsburgh

In the past fifteen years, the application of techniques from set theory to general topology has grown into a discipline in its own right. There are many useful techniques and notions of set theoretic topology--ramification arguments, partition relations, cardinal functions, large cardinals, forcing, Martin's Axiom, Jensen's combinatorial principles, and more. This paper concentrates on the notion of stationary set and its associated combinatorics.

The paper is organized as follows. After we establish our notations and conventions, we show how stationary sets arise in point set topology. Then we develop the combinatorics of stationary sets. Having established the existence of disjoint stationary sets, we give some examples of their use in topology. After two short combinatorial proofs, the nature of the paper changes. Instead of proofs, we present some directions in which the study of stationary sets might lead. We conclude with a sketch of the diamond constellation.

While we hope that this paper will be useful to people with different backgrounds, we have aimed at at persons with some knowledge of topology and a little knowledge of set theory.

---

[1]Partially supported by NSF Grant MCS 78-09484.

ISBN: 0-12-584960-5

For expository reasons we have placed the historical and bibliographic notes in a section at the end.

0. *Notation and Conventions*

An ordinal is the set of its predecessors. Cardinals are initial ordinals. The cardinality of a set A is denoted by $|A|$. The cofinality of an ordinal α by cf α. Exponentiation is cardinal exponentiation. The least cardinal greater than κ is denoted by κ^+.

Every ordinal α can be uniquely written as a sum $\lambda(\alpha) + n(\alpha)$ where $\lambda(\alpha)$ is a limit ordinal and $n(\alpha)$ is a natural number. The first infinite ordinal is denoted as ω. Several classes of ordinals are given names. LIM is the class of limit ordinals, REG is the class of regular cardinals; CFω is the class of ordinals of cofinality ω.

We think of a sequence as a set indexed by an ordinal. When we want to emphasize that the order is important, we write $(x_\alpha: \alpha < \beta)$ instead of $\{x_\alpha: \alpha < \beta\}$. A partition of a set A is a disjoint family $\{A_i: i \in I\}$ such that $\cup_{i \in I} A_i = A$. If f is a function and A is a set, let $f^{\rightarrow}A = \{f(x): x \in A\}$ and let $f^{\leftarrow}A = \{x \in \text{dom } f: f(x) \in A\}$.

A topological space X is paracompact iff whenever $\mathcal{U}$ is a family of open subsets of X such that $\cup\mathcal{U} = X$, then there is a family $\mathcal{V}$ of open subsets of X satisfying $\cup\mathcal{V} = X$ and

i) for all $V \in \mathcal{V}$ there is $U \in \mathcal{U}$ such that $V \subset U$

ii) for all $x \in X$ there is an open set W such that $x \in W$ and W meets only finitely many elements of $\mathcal{V}$.

Let $(L,<)$ be a linear (or total) order. We make L into a linearly ordered topological space by declaring the following types of sets to be open: $(a,b) = \{x \in L: a < x < b\}$,

$(a,\infty) = \{x \in L: a < x\}$, and $(-\infty,a) = \{x \in L: x < a\}$. If we additionally declare some closed or half closed intervals to be open we obtain a generalized ordered space.

1. *Examples of a Stationary Set in Point Set Topology*

Let us consider ω_1, the set of countable ordinals, as a linearly ordered topological spcae. Why is it not metrizable? (If you use the Pressing Down Lemma to show that it is not paracompact, you already know that stationary sets are useful). We will examine an older proof and see how it leads to the ideas of stationary sets and the Pressing Down Lemma.

The idea of the proof is to show that a countably compact metric space is separable. This will be sufficient, because ω_1 is clearly countably compact and not separable. Suppose that X is countably compact and has a metric, d. For each $n \in X$, let D_n be a subset of X maximal with respect to the following property

$$x,y \in D_n \text{ implies } d(x,y) \geq \frac{1}{n}$$

Note that each D_n is closed and discrete. That is, every $x \in X$ has a neighborhood N_x which contains at most one element of D_n. In this case, we can let $N_x = \{y \in X: d(x,y) < 1/2n\}$. Set $D = \cup_{n\in\omega} D_n$. We claim that D is dense. For if z has a neighborhood disjoint from D, then for some n, $\{y \in X: d(z,y) < \frac{1}{n}\}$ is disjoint from D. But then D_n was not maximal. So D is dense.

Returning to D_n, we note that D_n must be finite because X is countably compact. So D is countable, and X is separable.

A space is called $\aleph_1$-compact if every closed discrete subset is countable--equivalently, if every uncountable subset has a limit point. Examining the above proof, we see that D_n

being finite could be weakened to D_n countable. Thus the above proof shows that $\aleph_1$-compact metric spaces are separable.

Now let us ask, which subspaces of ω_1 are metrizable? Obviously, the countable subspaces are. The set of successor ordinals, being discrete, is an uncountable metrizable subspace.

If X is an uncountable metrizable subspace of ω_1, then X must be not $\aleph_1$-compact. Let $(\beta_\nu: \nu < \omega_1)$ be an uncountable closed (in X), discrete subset of X, indexed in the order preserving way. For each λ a limit ordinal less than ω_1, set $\gamma_\lambda = \sup_{\nu<\lambda}\beta_\nu$. Then $C = \{\gamma_\lambda: \lambda \in \mathrm{LIM} \cap \omega_1\}$ is closed and unbounded in ω_1 and $C \cap X = \emptyset$. The notion of closed and unbounded is fundamental to this paper, and we abbreviate it cub. A necessary condition for $X \subset \omega_1$ to be metrizable is that X be disjoint from a cub set.

This condition is also sufficient. Let $C = (\gamma_\nu: \nu < \omega_1)$ be a cub set, indexed in the order preserving way. Let X be a subset of ω_1 disjoint from C. For each $\nu < \omega_1$, $X_\nu = \{\delta \in X: \gamma_\nu < \delta < \gamma_{\nu+1}\}$ is a closed and open subset of X. Each X is countable, hence metrizable. So X itself is metrizable. We have established

1.1. *A subspace* X *of* ω_1 *is nonmetrizable iff* X *meets every cub set.*

Let us now examine a proof of the well known fact that every continuous real valued function from ω_1 is eventually constant. Let f be such a function. We begin by noting

1.2. *For every* $r \in \mathbb{R}$, $h > 0$, *either* $f^{\leftarrow}(r+h,\infty)$ *or* $f^{\leftarrow}(-\infty,r-h)$ *is countable.*

If not, we can choose α_n, $n \in \omega$ by induction so that

i) $\alpha_{n+1} > \alpha_n$

ii) if n is even, $\alpha_n \in f^{\leftarrow}(r+h,\infty)$

iii) if n is odd, $\alpha_n \in f^{\leftarrow}(-\infty,r-h)$

Set $\alpha = \sup_{n\in\omega}\alpha_n$. Because $\alpha = \{\sup \alpha_n : n \in \omega,\ n \text{ even}\}$, $f(\alpha) \geq r+h$. Because $\alpha = \{\sup \alpha_n : n \in \omega,\ n \text{ odd}\}$, $f(\alpha) \leq r-h$. Contradiction.

By applying 1.2 for each $n \in \omega$ with $h = \frac{1}{n}$ we see

1.3. *For every* $r \in \mathbb{R}$ *either* $f^{\leftarrow}(r,\infty)$ *or* $f^{\leftarrow}(-\infty,r)$ *is countable.*

For each $q \in \mathbb{Q}$, the rationals, choose α_q so that either $f^{\leftarrow}(q,\infty)$ or $f^{\leftarrow}(-\infty,q)$ is bounded by α_q. Let $Q' = \{q: f^{\leftarrow}(-\infty,q)$ is bounded by $\alpha_q\}$. Let $\alpha = \{\sup_{q\in\mathbb{Q}}\alpha_q\}$. For each $\beta > \alpha$ $f(\beta) = \sup_{q\in Q'}\alpha_q$. Hence f is constant above β.

Now let us ask, which subspaces share with ω_1 the property that every continuous real valued function is eventually constant?

Let X be an uncountable subset of ω_1, f a continuous real valued function from X. If we try to establish 1.2, it may be that $\alpha = \sup_{n\in\omega}\alpha_n$ is not in X. We continue defining α_ν, $\nu < \omega_1$ by induction so that i), ii), and iii) are satisfied. If for any $\lambda \in \omega_1 \cap \text{LIM}$, $\sup_{\nu<\lambda}\alpha_\nu \in X$, we may conclude 1.2 (for the particular r and h used in defining α_ν). Note that $\{\sup_{\nu<\lambda} \alpha_\nu : \lambda \in \omega_1 \cap \text{LIM}\}$ is cub. So a sufficient condition on X is that X meets every cub set.

This condition is also necessary. Suppose that $C = (\gamma_\nu : \nu < \omega_1)$ is a cub set disjoint from X, enumerated in the natural order. Define f from X to $\mathbb{R}$ by $f(\alpha) = 0$ if there is $\nu < \omega_1$ such that $\gamma_\nu < \alpha < \gamma_{\nu+1}$ and ν is even; $f(\alpha) = 1$ otherwise. Then f is continuous but not eventually constant.

We have established

1.4. Let X *be an uncountable subspace of* ω_1. *Every continuous real valued function from* X *is eventually constant iff* X *meets every cub set.*

The results 1.1 and 1.4 suggest that the concept of meeting every cub set is important. Indeed it is the subject of this entire paper. Let us now give the basic definitions.

1.5. Definitions. An uncountable cardinal κ is *regular* if the sup of less than κ ordinals less than κ is less than κ. A subset C of an ordinal δ (usually δ is an uncountable regular cardinal) is *cub in* δ if C is closed and unbounded in δ. A subset A of δ (again, usually an uncountable regular cardinal) is *stationary in* δ if A meets every set C which is cub in δ.

1.6. Theorem. A linearly ordered topological space X is paracompact if it does not contain a closed subset homeomorphic to a stationary subset of an uncountable regular cardinal.

Proof. Let $\mathcal{U}$ be an open cover of X. Define $\mathcal{V}$ to be the family of open sets contained in some element of $\mathcal{U}$. For $x,y \in X$, say that $x \sim y$ iff there is a locally finite (in X) subfamily $\mathcal{S}$ of $\mathcal{V}$ such that $\cup\mathcal{S}$ is an open interval containing x and y. It is easy to verify that $\sim$ is an equivalence relation, and that each equivalence class is open. It will suffice, therefore, to cover each equivalence class with a locally finite subfamily of $\mathcal{V}$.

Let E be an equivalence class; let x be an element of E. We will find a locally finite subfamily $\mathcal{F}$ covering

$E^+ = \{y \in E: x \leq y\}$--a parallel argument will take care of $\{y \in E: y \leq x\}$.

Choose $V^* \in \mathcal{V}$ so that $x \in V^*$. Let $Z = (z_\alpha: \alpha < \kappa)$ be a cofinal subset of E of minimal cardinality, listed in increasing order, with $x = z_0$. There are three cases.

CASE 1: $\kappa = 2$ (or 1)--that is E has a last element z_1.

Let $\mathcal{F}$ witness $x \sim z$.

CASE 2: $\kappa = \omega$. For each $n \in \omega$, let $\mathcal{W}_n$ witness $z_n \sim z_{n+2}$. For each $n \in \omega$, define $\mathcal{W}'_n$ to be

$$\mathcal{W}'_n = \{(z_n, z_{n+2}) \cap W: W \in \mathcal{W}_n\}.$$
$$\mathcal{F} = \{V^*\} \cup \cup_{n\in\omega}\mathcal{W}'_n.$$

CASE 3: κ is an uncountable regular cardinal.

We will use the hypothesis about stationary sets to make this case analogous to case 2.

Define A to be the set of $\lambda < \kappa$ such that $\{z_\alpha: \alpha < \lambda\}$ has a least upper bound in X. If $\lambda \in A$, let $g(\lambda)$ be that least upper bound. It is easy to verify that g is a homeomorphism from A to a closed subset of X. By hypothesis A is not stationary, so there is a cub set $C = (\gamma(\nu): \nu < \kappa)$ listed in increasing order disjoint from A. For each $\nu < \kappa$, let $\mathcal{W}_\nu$ witness that $x \sim z_{\gamma(\nu)}$. For W an open interval and $\nu < \kappa$, define

$$I(W,\nu) = \{x \in W: \forall\alpha < \gamma(\nu)\ z_\alpha < x \text{ and } \forall\alpha > \gamma(\nu+1)\, z_\alpha > x\}.$$
$$\mathcal{W}'_\nu = \{I(W,\nu): W \in \mathcal{W}_\nu\}$$
$$\mathcal{F} = \{V^*\} \cup \cup_{\nu<\kappa}\mathcal{W}'_\nu.$$

1.7. *Corollary.* A linearly ordered topological space is hereditarily paracompact if it does not contain a subset homeomorphic to a stationary subset of an uncountable regular cardinal.

Theorem 1.6 and Corollary 1.7 are true for generalized ordered spaces. The above proof works with the following modification. Define A to be the set of $\lambda > \kappa$ such that $\{z_\alpha : \alpha < \kappa\}$ has a least upper bound and that upper bound is in the closure of $\{z_\alpha : \alpha < \lambda\}$.

The converse of Theorem 1.6 and Corollary 1.7 are easy consequences of the Pressing Down Lemma, which we prove in the next section.

2. *The Combinatorics of Stationary Sets*

2.1. *The Sup Function Lemma.* Let κ be a regular cardinal. Let $g: \kappa \to \kappa$ satisfy for all $\alpha < \kappa, g(\alpha) \geq \alpha$ and for all $\lambda \in \kappa \cap \mathrm{LIM}$, $g(\lambda) = \sup_{\delta<\lambda} g(\delta)$. Then there is a cub set C such that for all $\gamma \in C$, $f(\gamma) = \gamma$. Loosely speaking, a sup function has a cub set of fixed points.

Proof. A moment's thought verifies that C is closed. Towards showing that C is unbounded, let $\alpha_0 < \kappa$ be arbitrary. For $n \in \omega$ define $\alpha_{n+1} = g(\alpha_n)$. Then $\sup_{n\in\omega}\alpha_n \in C$.

2.2. *The Pressing Down Lemma.* (Weak form, Neumer). Let S be a stationary subset of an uncountable regular cardinal κ. Let $f: S \to \kappa$ press down, i.e. $f(\alpha) < \alpha$ for all $\alpha \in S$. Then there is $\beta < \kappa$ such that $|f^{\leftarrow}\{\beta\}| = \kappa$.

Proof. Suppose not, that for all $\beta \in \kappa$, $|f^{\leftarrow}\{\beta\}| < \kappa$. Define, for all $\delta < \kappa$, $g(\delta) = \sup \cup_{\beta<\delta} (f^{\leftarrow}\{\beta\} \cup \{\beta\})$. By the previous lemma g has a cub set of fixed points. So there is $\gamma \in S$ with $g(\gamma) = \gamma$. Let $\beta = f(\gamma)$. Then $\beta < \gamma$, so that $\gamma \in \cup_{\beta<\gamma} f^{\leftarrow}\{\beta\}$, and $g(\gamma) \geq \sup f^{\leftarrow}\{\beta\} > \gamma$. Contradiction.

2.3. *Corollary*. Let S be a stationary subset of a regular cardinal κ. Consider as a linearly ordered topological space, S is not paracompact.

Proof. Because if $\lambda \in S \cap LIM$, an open set containing λ also contains a point $f(\lambda) < \lambda$.

Of course, paracompact can be considerably weakened; e.g. to metacompact, metalindelöf, weakly $\delta\theta$-refinable, or even very weakly refinable.

For some purposes it is useful to have a strengthened form of the Pressing Down Lemma. We first prove some closure properties of the family of cub sets.

2.4. *Lemma*. Let $\rho < \kappa$, κ a regular cardinal, and let $\{C_\alpha: \alpha < \rho\}$ be a family of cub subsets of κ. Then $C = \cap\{C_\alpha: \alpha < \rho\}$ is cub in κ.

Proof. Obviously C is closed. Towards showing that C is unbounded, let $\beta < \kappa$ be arbitrary. Set $\beta_0^\alpha = \beta$ for all $\alpha < \rho$. For $n \in \omega$ let β_{n+1}^α be an element of C_α greater than $\sup_{\gamma<\rho} \beta_n^\gamma < \kappa$. Then for all $\alpha < \rho$

$$\sup_{n\in\omega} \beta_n^\alpha = \sup_{n\in\omega} \beta_n^0 = \beta^*$$

so $\beta^* \in \cap\{C_\alpha: \alpha < \rho\}$.

2.5. *Corollary*. Let κ be an uncountable regular cardinal. Then the union of less than κ sets nonstationary in κ is nonstationary in κ.

Proof. The complement of such a union contains the intersection of less than κ cub sets.

Lemma 2.4 cannot be strengthened to allow $\rho = \kappa$. However, by changing slightly the notion of intersections, we obtain an important result. Let $(A_\alpha: \alpha < \kappa)$ be a sequence of subsets of κ. Define the *diagonal intersection*, $\Delta(A_\alpha: \alpha < \kappa)$ by

$$\Delta(A_\alpha: \alpha < \kappa) = \{\beta < \kappa: \beta \in \cap_{\alpha<\beta} A_\alpha\}.$$

2.6. Lemma. If $(C_\alpha: \alpha < \kappa)$ is a sequence of cub subsets of κ then $\Delta = \Delta(C_\alpha: \alpha < \kappa)$ is cub in κ.

Proof. Towards showing that Δ is closed, suppose that $\gamma = \sup B$, where $B = (\beta_\eta: \eta < \rho)$ listed in increasing order, $B \subset \Delta$, and $\rho < \kappa$. For each $\delta = \beta_\eta < \rho$, $\{\beta \in B: \beta > \delta\} \subset C_\delta$. Since C_δ is closed, $\gamma \in C_\delta$. Hence $\gamma \in \cap_{\delta<\gamma} C_\delta$ and $\gamma \in \Delta$.

Towards showing that Δ is unbounded, let $\gamma_0 < \kappa$ be arbitrary. By Lemma 2.4, $\cap_{\alpha<\gamma} C_\alpha$ is cub in κ for each $\gamma < \kappa$, so we may define γ_n, $n \in \omega$ by induction to satisfy

$$\gamma_{n+1} > \gamma_n \text{ and } \gamma_{n+1} \in \cap_{\alpha<\gamma_n} C_\alpha.$$

Set $\gamma^* = \sup_{n\in\omega} \gamma_n$. Then $\gamma^* \in \cap_{\alpha<\gamma^*} C_\alpha$, so $\gamma_0 < \gamma^* \in \Delta$.

The statement of the contrapositive of Lemma 2.6 in terms of complements can have a surprising form.

2.7. The Pressing Down Lemma. (Strong form, Fodor). Let S be a stationary subset of a regular uncountable cardinal κ. Let $f: S \to \kappa$ press down, i.e. $f(\alpha) < \alpha$ for all $\alpha \in S$. Then there is $\beta < \kappa$ such that $f^{\leftarrow}\{\beta\}$ is stationary in κ.

Proof. Suppose not, that for all $\beta < \kappa$, $f^{\leftarrow}\{\beta\}$ is not stationary. Let C_β be cub in κ, $C_\beta \cap f^{\leftarrow}\{\beta\} = \emptyset$. Choose $\gamma \in \Delta(C_\beta: \beta < \kappa) \cap S$. Let $\beta = f(\gamma) < \gamma$. Then $\gamma \in f^{\leftarrow}\{\beta\} \cap C_\beta$. Contradiction.

Our next topic is partitioning stationary sets into stationary sets. It is not obvious that the notion "stationary" differs from the notion "contains a cub set". An easy example showing the difference is $\{\delta < \omega_2: \text{cf } \delta = \omega\}$, which is stationary in ω_2 but does not contain a set cub in ω_2. There is no such easy example for ω_1, and we can disallow that easy example by changing "contains a cub set" to "contains a ρ-closed unbounded set for some $\rho < \kappa$". The next lemma deals with ω_1 but of course can be generalized to other cardinals. Let us remark that the use of the axiom of choice is necessary and not merely a convenience.

2.8. *Lemma.* (Banach). There are two disjoint stationary subsets of ω_1.

Proof. Let $(x_\alpha : \alpha < \omega_1)$ be a set of irrationals. For each rational q, let $A_q = \{\alpha : x_\alpha < q\}$ and $B_q = \{\alpha : x_\alpha > q\}$. If for all q either A_q or B_q contains a cub set then

$$C = (\cap\{A_q : \forall \text{ a cub } C_q,\ A_q \supset C_q\}) \cap (\cap\{B_q : \text{ a cub } C_q,\ B_q \supset C_q\})$$

is a cub set with at most one point. So for some q, A_q and B_q are disjoint subsets of ω_1.

The Pressing Down Lemma can be used to find ω_1 disjoint subsets of κ.

2.9. *Lemma.* (Ulam). Every uncountable regular cardinal κ can be partitioned into κ disjoint stationary subsets of κ.

Proof. It is sufficient to partition $A = \{\delta < \kappa : \text{cf } \delta = \omega\}$. For each $\delta \in A$, let $(\eta_\delta(n) : n \in \omega)$ enumerate a cofinal subset of δ. For each $n \in \omega$ define a pressing down function $h_n : A \to \kappa$ by $h_n(\delta) = \eta_\delta(n)$. For $n \in \omega$, let $H_n = \{\beta < \kappa : h_n^{\leftarrow}\{\beta\} \text{ is stationary}\}$. If for some n, $|H_n| = \kappa$, $\{h_n^{\leftarrow}\{\beta\} : \beta \in H_n\}$ is a family of κ disjoint stationary sets.

We will show by contradiction that for some n, $|H_n| = \kappa$. If not then $\nu = \sup(\cup_{n\in\omega} H_n) < \kappa$, because κ is regular. The set $A' = A - (\nu+1)$ is stationary, and for each $\delta \in A'$ there is

$n \in \omega$ such that $z_\delta(n) > \nu$. So for some $n \in \omega$, $A_n = \{\delta \in A': h_n(\delta) > \nu\}$ is stationary. Applying the Pressing Down Lemma to A_n and h_n we get $\beta > \kappa$ such that $h_n^{\leftarrow}\{\beta\}$ is stationary. Thus $\beta \in H_n$. However, A_n was chosen so that $\beta > \nu$, hence $\beta \notin H_n$. This contradiction shows that for some $n \in \omega$, $|H_n| = \kappa$.

What about partitioning arbitrary subsets of κ? The argument above works in "most" cases. We introduce some definitions to make that statement precise.

2.10. Definition. A regular limit cardinal is called *weakly inaccessible*. A weakly inaccessible cardinal is called *weakly Mahlo* if REG $\cap\ \kappa$ is stationary in κ. The first weakly inaccessible cardinal θ is not weakly Mahlo because the singular cardinals are cub in θ.

2.11. Lemma. Let κ be an uncountable regular cardinal, and let B be a stationary subset of κ disjoint from $\kappa \cap$ REG. Then B can be partitioned into κ disjoint stationary subsets. Hence if κ is not weakly Mahlo, every stationary subset of κ can be partitioned into κ disjoint stationary subsets.

Proof. Define $f: B \to \kappa$ by $f(\alpha) = \mathrm{cf}\ \alpha$. By hypothesis, f presses down; so f is constantly ρ on a stationary set A. Now for each $\delta \in A$ let $(\eta_\delta(\nu): \nu \in \rho)$ be a cofinal subset of δ and argue as in Lemma 2.9.

2.12. Theorem. (Solovay). Let κ be an uncountable regular cardinal. Every stationary subset of κ can be partitioned into κ disjoint stationary subsets.

Proof. By Lemma 2.10, we may assume that $A \subset$ REG. Let $A' = \{\alpha \in A: A \cap \alpha$ is <u>not</u> stationary in $\alpha\}$. We claim that A' is stationary in κ. Let C be any cub subset of κ. C', the set of limit points of C is also cub in κ. Since A is stationary, $C' \cap A$ is not empty, and so has a least element α. Now because $\alpha \in C'$, $C \cap \alpha$ is cub in α, and because α is the first element of $C' \cap A$, $(A \cap \alpha) \cap (C \cap \alpha) = \emptyset$, i.e. $A \cap \alpha$ is not stationary in α. So $\alpha \in A' \cap C$, showing that A' is stationary in κ.

For each $\delta \in A'$, let $(\eta_\delta(\beta): \beta < \delta)$ enumerate in increasing order a cub subset of δ disjoint from A. Note that for all β, $\eta_\delta(\beta) \geq \beta$; and if $\beta \in A$ then $z_\delta(\beta) > \beta$. For each $\alpha < \kappa$ define a pressing down function $h_\alpha: A' - (\alpha+1) \to \kappa$ by $h_\alpha(\delta) = \eta_\delta(\alpha)$. Let $H_\alpha = \{\beta: h_\alpha^{\leftarrow}\{\beta\}$ is stationary$\}$. Note that because h_α is pressing down, H_α contains at least one element notless than α. If some H_α has cardinality κ, then $\{h^{\leftarrow}\{\beta\}: \beta \in H_\alpha\}$ is a family of κ disjoint stationary subsets of A', and we are done. We will show by contradiction that there is such an H_α.

Towards a contradiction, assume that for all $\alpha < \kappa$, $|H_\alpha| < \kappa$. With this assumption, for all $\nu < \kappa$, $\sup(\cup_{\alpha<\nu} H_\alpha) < \kappa$. By the Sup Function Lemma (2.1) there is $\nu \in A'$ such that $\sup(\cup_{\alpha<\nu} H_\alpha) = \nu$. For each $\gamma \in A'$, $\gamma > \nu$, we have $\sup_{\beta<\nu} z_\gamma(\beta) = z_\gamma(\nu) > \nu$ so there is a function $a: (A' - (\nu+1)) \to \nu$ such that $z_\gamma(\alpha(\gamma)) = h_{a(\gamma)}(\gamma) > \gamma$. Then because $\nu < \kappa$, $B = a^{\leftarrow}\{\alpha\}$ is stationary for some $\alpha < \nu$; and h_α presses down on B, so there is β such that $h_\alpha{}^{\leftarrow}\{\beta\}$ is a stationary subset of B. We chose α and B so that $\beta > \nu$, but $\beta \in H_\alpha$, so $\beta < \sup H_\alpha \leq \nu$. Contradiction.

3. Applications of disjoint stationary sets

The classical descriptive set theory of separable metric spaces is a beautiful and well developed theory. One of the several directions in which it has been generalized is to nonseparable metrizable spaces. In this extension, $B(\kappa)$ plays the role of the irrationals and Cantor set, and the notion σ-LW($<\kappa$) corresponds to countable. The following theorem is an example of this analogy.

3.1. Theorem. (A. Stone). A Borel subset of $B(\kappa)$ which is not σ-LW($<\kappa$) contains a homeomorph of $B(\kappa)$.

3.2. Definition. $B(\kappa)$, *the generalized Baire zero dimensional space of weight* κ, is the space of functions from ω to κ with the metric given by

$$d(f,g) = 2^{-n}, \text{ where } n \text{ is least such that } f|n \neq g|n.$$

A space X is σ-LW($<\kappa$) (σ-*local weight less than* κ) if X can be partitioned into $\{X_n: n \in \omega\}$ such that for each $n \in \omega$, each $x \in X_n$ has a neighborhood N_x (in X_n) of weight less than κ. (Equivalently, N_x has a dense subset of cardinality less than κ.) Thus, when $\kappa = \omega$, Theorem 3.1 says that every uncountable Borel subset of the irrationals contains a copy of the irrationals.

Stationary sets were introduced to the theory of nonseparable metrizable spaces by the following lemma.

3.3. Lemma. (Pol). Let X be a metrizable space of weight κ, where κ is an uncountable regular cardinal. Let $(d_\alpha: \alpha < \kappa)$ be a dense subset of X. For $\delta < \kappa$, set

$$\Gamma(\delta) = \text{closure}\,\{d_\alpha: \alpha < \delta\} - \cup_{\beta<\delta}\text{closure}\{d_\alpha: \alpha < \beta\}.$$

Then, X is not σ-LW($<\kappa$) iff $\{\delta: \Gamma(\delta) \neq \emptyset\}$ is stationary.

Sometimes the theory of metrizable spaces of weight κ has the same theorems as the separable theory, but the same proofs don't work. The construction of Bernstein sets illustrates this. Recall that the Cantor set contains a copy of $B(\omega)$, the irrationals, and that the subsets of $B(\omega)$ homeomorphic to $B(\omega)$ are exactly the uncountable G_δ subsets.

3.4. Theorem. (Bernstein). There is a subset Z of $B(\omega)$ such that both Z and $B(\omega) - Z$ meet every subset of $B(\omega)$ homeomorphic to $B(\omega)$.

Proof. The key is to note that there are 2^ω G_δ subsets of $B(\omega)$ and that each uncountable G_δ has 2^ω points. We simply well order the uncountable G_δ subsets and inductively choose two new points from each--one to be in Z, the other to be in $B(\omega) - Z$. The induction continues because at each step there are less than 2^ω points already chosen, so there are plenty of possible new points.

We would like to repeat the above construction with $B(\kappa)$ in place of $B(\omega)$. However, when $\kappa^\omega = \kappa$--for example $\kappa = 2^\omega$ or $\kappa = (2^\omega)^+$--there is a problem. There are 2^κ open sets, but only κ points in the space. An inductive definition of Z simply won't work. The existence of disjoint stationary subsets is exactly what is needed.

3.5. Theorem. (Pol). Let κ be an uncountable regular cardinal. There is a subset Z of $B(\kappa)$ such that both Z and $B(\kappa) - Z$ meet every subset of $B(\kappa)$ homeomorphic to $B(\kappa)$.

Proof. Let us introduce some notation useful in working with $B(\kappa)$. For $\alpha < \kappa$, let

$$\Sigma_\alpha = \{\alpha\colon \alpha \text{ is a function, dom } \alpha \in \omega, \text{ ran } \sigma \subset \alpha\}$$

For $\sigma \in \Sigma_\kappa$, define

$$[\sigma] = \{f \in B(\kappa)\colon \sigma \subset f\}.$$

Each $[\sigma]$ is a closed and open subset of $B(\kappa)$, and

$\{[\sigma]: \sigma \in \Sigma_\kappa\}$ is a base for $B(\kappa)$. For $f \in B(\kappa)$, define $f^* = \sup \operatorname{ran} f$; for $\sigma \in \Sigma_\kappa$, define $\sigma^* = \sup \operatorname{ran} \sigma$. Note that for each $\alpha < \kappa$, $|\{\sigma: \sigma^* < \alpha\}| < \kappa$.

By Lemma 2.10, there is $A \subset CF\ \omega \cap \kappa$ such that both A and $CF\ \omega \cap \kappa$ are stationary in κ. Set $z = \{f \in B(\kappa): f^* \in A\}$. That z satisfies the conclusion of Theorem 3.5 follows from

3.6. Lemma. Let H be a subset of $B(\kappa)$ such that either H is homeomorphic to $B(\kappa)$ or H is closed and $\{h^*: h \in H\}$ is stationary in κ. Then there is a set C cub in κ such that $C \cap CF\ \omega \subset \{h^*: h \in H\}$.

Proof. We will define $\theta: \Sigma_\kappa \to \Sigma_\kappa$ by induction on dom σ to satisfy

i) $\sigma \subset \tau$ implies $\theta(\sigma) \subset \theta(\tau)$; $\operatorname{dom} \sigma \leq \operatorname{dom}\theta(\sigma)$

ii) $\sigma^* \leq \theta(\sigma)^*$

iii) for all $f \in B(\kappa)$, $\cap_{n\in\omega}[\theta(f|n)] \subset H$.

Note that by iii) we must require at least that

iv) range $\theta \subset \{\sigma \in \Sigma_\kappa: H \cap [\sigma] \neq \emptyset\}$.

If $H \cong B(\kappa)$, then H has a complete metric, d. To satisfy iii) it suffices to additionally require that the d-diameter of $H \cap [\theta(\sigma)]$ is less than $1/\operatorname{dom} \sigma$. It is easy to satisfy i) and ii) because every nonempty open subset of $B(\kappa)$, hence also of H, has cellularity κ, i.e. contains a family of κ disjoint open sets.

If H is closed and $\{h^*: h \in H\}$ is stationary, the usual metric of $B(\kappa)$ is complete on H, so to satisfy iii) it suffices to require iv). It is not true in general that every nonempty open subset of H has cellularity κ--H might have isolated points, for example--but we will find enough such open sets.

It will suffice to choose $\theta(\sigma) \in T$, defined in the next paragraph.

For $\sigma \in \Sigma_\kappa$ let $A_\sigma = \{h^*: \sigma \subset h \in H\}$. Let $T = \{\sigma: A_\sigma$ is stationary$\}$. By hypothesis, $\emptyset \in T$. We claim that if $\sigma \in T$, dom $\sigma = n$, then $\{\tau^*: \sigma \subset \tau \in T\}$ is unbounded in κ. If not, then it is bounded -- by β, say. For every δ in the stationary set $S = \text{Lim} \cap A_\sigma - (\beta+1)$ there is $h_\delta \in [\sigma] \cap H$ with $h_\delta^* = \delta$. Then there is $n_\delta \in \omega$ such that $\beta < (h_\delta|n_\delta)^* < \delta$. Define $f: S \to \kappa - (\beta+1)$ by $f(\delta) = (h_\delta|n_\delta)^*$. Since f preses down, for some γ $f^{\leftarrow}\{\gamma\}$ is stationary. Because $|\{\tau: \tau^* = \gamma\}| < \kappa$, there is τ such that $\{\delta: h_\delta|n_\delta = \tau\}$ is stationary. Then $\sigma \subset \tau \in T$ and $\tau^* > \beta$, contradiction.

Having defined θ, we define $g: \kappa \to \kappa$ by $g(\eta) = \sup\{\theta(\sigma)^*: \sigma \in \Sigma_\eta\}$. By the Sup Function Lemma, g has a cub set C of fixed points. We must verify that C satisfies the conclusion of the lemma. Let $\gamma \in C \cap CF\ \omega$; let $f: \omega \to \gamma$ be cofinal in γ. By i) $\cap_{n\in\omega}[\theta(f|n)]$ is a singleton, $\{h\}$; by iii) $h \in H$. Because γ is a fixed point of g, $h^* \leq \gamma$; because of ii) and the fact that f is cofinal in γ, we have $\gamma \leq h^*$. Thus there is $h \in H$ with $h^* = \gamma$, establishing the lemma.

Another application of Lemma 3.6 is to find two Baire spaces whose product is not Baire.

3.7. Definition. A space X is a *Baire* space iff the intersection of countably many dense open subsets of X is dense in X.

3.8. Theorem. (P.E. Cohen). There are two Baire spaces whose product is not Baire.

Proof. Let A and B be disjoint stationary subsets of ω_1.

Let $X = \{f \in B(\omega_1): f^* \in A\}$; $Y = \{g \in B(\omega_1): g^* \in B\}$. Every dense G_δ of $B(\kappa)$ is homeomorphic to $B(\kappa)$, so by Corollary 3.7, X and Y are Baire spaces.

For $n \in \omega$, let

$$U_n = \{(f,g) \in X \times Y: f(n) < g^* \text{ and } g(n) < f^*\}.$$

It is easy to verify that each U_n is a dense open subset of $X \times Y$. We claim that $\cap_{n\in\omega} U_n = \emptyset$.

Suppose that $(f,g) \in \cap_{n\in\omega} U_n$. For each $n \in \omega$ $f(n) < g^*$ because $(f,g) \in U_n$; thus $f^* = \sup_{n\in\omega} f(n) \leq g^*$. Similarly, $g^* \leq f^*$, so $f^* = g^* \in A \cap B = \emptyset$. Contradiction.

Our final example of the section is the construction of a normal collectionwise Hausdorff, nonmetrizable Moore space.

3.9. Definition. A space X is *collectionwise Hausdorff* iff whenever Y is a closed discrete subset of X there is a disjoint family of open sets $\{U_y: y \in Y\}$ such that $y \in U_y$. A space X is *collectionwise normal* iff whenever $\mathcal{Y} = \{Y_i: i \in I\}$ is a discrete family of closed sets there is a disjoint family of open sets $\{U_i: i \in I\}$ such that $Y_i \subset U_i$. Clearly, metrizable spaces are collectionwise normal and T_1 collectionwise normal spaces are collectionwise Hausdorff.

This construction requires some extra set theoretic hypotheses. First, we need a normal Moore space M of the form $M = U \cup Y$, where U is an open metrizable subspace of M, and Y is a closed discrete subspace of M which shows that M is not collectionwise Hausdorff. Second, we need an E set.

3.10. Definition. $E(\kappa)$ is the assertion that κ is a regular cardinal greater than ω_1, and that there is a set $E \subset CF\omega \cap \kappa$ satisfying

i) E is stationary in κ

ii) for all $\alpha < \kappa$, $E \cap \alpha$ is not stationary in α.

$\exists\kappa E(\kappa)$ is the assertion that $E(\kappa)$ is true for some κ.

We need $E(\kappa)$ where $\kappa > |Y|$. For concreteness, let us assume that $\kappa = \omega_2$ and $|Y| = \omega_1$. By Lemma 2.10, we can partition E into ω_1 disjoint stationary subsets, $E = \cup\{E_y: y \in Y\}$. Let

$$W = \{(m,f) \in M \times B(\omega_2): m \in U \text{ or } m \in Y \text{ and } f^* \in E_m\}.$$

W is a Moore space because products of Moore spaces are Moore spaces. It is a not difficult, purely topological task to verify that W is normal. We will verify below that W is collectionwise Hausdorff and not collectionwise normal.

Towards showing that W is collectionwise Hausdorff, let Z be a closed discrete subset of W. $Z \cap (Y \times B(\omega_2))$ and $Z \cap (U \times B(\omega_2))$ are disjoint closed sets in the normal space W, and $Z \cap (U \times B(\omega_2))$ is contained in the metrizable space $U \times B(\omega_2)$. It suffices, therefore, to consider the case $Z \subset Y \times B(\omega_2)$.

We will use the following consequence of the discreteness of Z and normality of W.

* If for each $n \in \omega$, $K_n \subset Z$ and K_n can be separated, then $\cup_{n\in\omega} K_n$ can be separated.

We show by induction on $\beta \leq \omega_2$ that $Z_\beta = \{(y,f) \in z: f^* < \beta\}$ can be separated. If $\beta = \delta + 1$ or cf $\beta = \omega$, the assertion follows from induction hypothesis and *. If cf $\beta > \omega$, we must first establish that $A_\beta = \{f^*: (y,f) \in Z_\beta\}$ is is not stationary in β. For $\beta < \omega_2$, this follows from $E(\omega_2)$ because $A_\beta \subset E \cap \beta$. Towards showing that A_{ω_2} is not statioanry in ω_2, for each $y \in Y$, set $Wy = \{y\} \times B(\omega_2)$. For each $y \in Y$,

$Z_y = Z \cap W_y$ is a closed discrete subset of the metrizable space W_y. We can separate Z_y in W_y by basic open sets; say, $(y,f) \in \{y\} \times [\sigma_f]$. Note that $\sigma_f^* < f^*$. Since $\beta < \kappa$ implies $|\Sigma_\beta| < \kappa$, the contrapositive of the Pressing Down Lemma yields that A_{ω_2} is not stationary.

Returning to the case where cf $\beta > \omega$, let $C_\beta = (\gamma_\nu: \nu < \text{cf } \beta)$ be a cub subset of β disjoint from A_β, enumerated in increasing order. For each $\nu < \text{cf } \beta$, let $S_\nu = \{(m,f) \in Z: \gamma_\nu < f^* < \gamma_{\nu+1}\}$; for $(m,f) \in S_\nu$, let $n(m,f)$ be the least element of ω such that $\gamma_\nu < f(n(m,f))$. Note that if $n(m,f) = n(m',f')$, $(m,f) \in S_\nu$, $(m',f') \in S_{\nu'}$, and $\nu \neq \nu'$, then $[f\ n(m,f)] \cap [f'\ n(m',f')] = \emptyset$. Thus the induction hypothesis gives that for each $n \in \omega$, $K_n = \{(m,f) \in Z: n(m,f) = n\}$ can be separated, and * gives that Z_β can be separated. We have shown that W is collectionwise Hausdorff.

To show that W is not metrizable, we will show that the discrete family of closed sets $\{W_y: y \in Y\}$ cannot be separated in W. For each $y \in Y$, let U_y be an open subset of $M \times B(\omega_2)$ containing W_y. By Lemma 3.6 there is C_y cub in ω_2 such that $f \in B(\omega_2)$ and $f^* \in C_y$ imply that $(y,f) \in U_y$. Let $\delta \in CF\omega \cap \cap_{y \in Y} C_y$; choose $g \in B(\omega_2)$, $g^* = \delta$. For each $y \in Y$, $(y,g) \in U_y$, an open set, so there is V_y an open subset of M such that $y \times \{g\} \subset U_y$. Because Y cannot be separated in M, there is a point $u \in U \subset M$ which is in more than one V_y. Hence $\{U_y \cap W: y \in Y\}$ is not disjoint and W is not metrizable.

4. *Applications to combinatorial set theory*

The theory of stationary sets and the Pressing Down Lemma can have indirect applications to set theoretic topology in the following way. They can be used to give neat proofs of other useful combinatorial lemmas. We begin this section with two examples.

4.1. *Definition.* A family $\mathcal{S}$ of sets is called a *Δ-system* iff there is a set R such that whenever $S, S' \in \mathcal{S}$, $S \neq S'$, then $S \cap S' = R$.

4.2. *Delta System Lemma.* (Erdös-Rado). Let $\mu < \kappa$ be two regular cardinals such that $\nu < \mu$, $\lambda < \kappa$ implies $\lambda^\nu < \kappa$. Let $\mathcal{S}$ be a family of sets of cardinality less than μ, $|\mathcal{S}| = \kappa$. Then there is a Δ-system $\mathcal{S}' \subset \mathcal{S}$, $|\mathcal{S}'| = |\mathcal{S}| = \kappa$.

Remark. Two important special cases of this lemma are when $\mathcal{S}$ is a family of ω_1 finite sets and when $\mathcal{S}$ is a family of $(2^\omega)^+$ countable sets.

Proof. Well order $\mathcal{S}$, $\mathcal{S} = (S_\alpha : \alpha < \kappa)$. $|\cup\mathcal{S}| \leq \kappa \cdot \mu = \kappa$, so we may assume that $\cup\mathcal{S} \subset \kappa$. The set $E = \{\alpha \in \kappa : \text{cf}\ \alpha = \mu\}$ is stationary in κ. For each $\alpha \in E$ define $f(\alpha) = \sup(S_\alpha \cap \alpha)$. Because $|S_\alpha| < \mu = \text{cf}\ \alpha$, $f(\alpha) < \alpha$. By the Pressing Down Lemma $f^{\leftarrow}\{\beta\}$ is stationary for some $\beta < \kappa$. For each $\nu < \mu$ there are $|\beta|^\nu < \kappa$ subsets of β of cardinality ν, so there are $<\kappa$ subsets of β of cardinality $<\mu$. Thus there is a stationary set $E' \subset E$ and a set R, $R \subset \beta$, $|R| < \mu$, such that for all $\alpha \in E'$, $S_\alpha \cap \alpha = R$.

For $\gamma < \kappa$ define $g(\gamma) = \sup \cup_{\alpha<\gamma} S_\alpha$. By the Sup Function Lemma, there is a cub set C of fixed points of g. Set $\mathcal{S}' = \{S_\alpha : \alpha \in E' \cap C\}$. We must verify that if $\alpha < \beta$ and $\alpha, \beta \in E' \cap C$, then $S_\alpha \cap S_\beta = R$. Since $S_\alpha \cap \alpha = R = S_\beta \cap \beta$, we have $R \subset S_\alpha \cap S_\beta$. If $\gamma < \beta$ and $\gamma \notin R$ then $\gamma \notin S_\beta$. If $\gamma \geq \beta$, $\gamma \notin S_\alpha$. So $S_\alpha \cap S_\beta = R$.

4.3. Definition. For A a set and κ a cardinal, set $[A]^{\kappa} = \{B \subset A: |B| = \kappa\}$ and $[A]^{<\kappa} = \{B \subset A: |B| < \kappa\}$.

4.4. Definition. If κ, ρ, and μ are cardinals, let $\kappa \to (\rho,\mu)^2_2$ mean that whenever $F: [\kappa] \to 2$, then either there is $H_0 \in [\kappa]^{\rho}$ such that $F^{\to}[H_0]^2 = \{0\}$ or there is $H_1 \in [\kappa]^{\mu}$ such that $F^{\to}[H_1]^2 = \{1\}$.

4.5. Theorem. (Erdös-Rado). Let $\mu < \kappa$ be regular cardinals such that $\nu < \mu$, $\lambda < \kappa$ implies that $\lambda^{\nu} < \kappa$. Then $\kappa \to (\kappa,\mu)^2_2$.

Remark. Two important special cases are $\kappa = \omega_1$, $\mu = \omega$ and $\kappa = (2^{\omega})^+$, $\mu = \omega_1$. The proof below can be easily modified to give $\kappa \to (\kappa,\mu)^2_{<\mu}$.

Proof. Let $F: [\kappa]^2 \to 2$ be given. We assume that there is no $H_1 \in [\kappa]^{\mu}$ such that $F^{\to}[H_1]^2 = \{1\}$, and we will find $H_0 \in [\kappa]^{\kappa}$ such that $F^{\to}[H_0]^2 = \{0\}$. Set $E = \{\alpha \in \kappa: \text{cf}\,\alpha = \mu\}$.

For each $\alpha < \kappa$, let S_α be a subset of α maximal with respect to $F^{\to}[S_\alpha \cup \{\alpha\}]^2 = \{1\}$. Note that our assumption gives $|S_\alpha| < \mu$.

For $\alpha \in E$ define $f(\alpha) = \sup S_\alpha$. Since $|S_\alpha| < \mu = \text{cf}\,\alpha$, f presses down on the stationary set E, and thus there is such that $f^{\leftarrow}\{\beta\}$ is stationary. The cardinal arithmetic hypotheses imply $|[\beta]^{<\mu}| < \kappa$, so there is a set $R \in [\beta]^{<\mu}$ and a stationary set $H_0 \subset f^{\leftarrow}\{\beta\}$ such that for all $\alpha \in H_0$, $S_\alpha = R$.

We must verify that if $\alpha < \beta$, $\alpha,\beta \in H_0$ then $F(\{\alpha,\beta\}) = 0$. Since $\alpha \in \beta - S_\beta$ and $\{\beta\} \cup S_\beta$ is a subset of β maximal with respect to $F^{\to}[\{\beta\} \cup \beta]^2 = \{1\}$, there must be $\delta \in S_\beta \cup \{\beta\}$ such that $F(\{\alpha,\delta\}) = 0$. $S_\alpha = S_\beta$ so $F^{\to}[\{\alpha\} \cup S_\beta]^2 = \{1\}$, so that δ must be β.

There are many other applications of stationary sets in combinatorial set theory, but we have space here only to mention a few. Stationary sets play a central role in Silver's beautiful theorem on the continuum hypothesis at singular cardinals. The simplest instance of this theorem is

4.6. Theorem. (Silver). Let S be a stationary subset of ω_1. If for all $\alpha \in S$, $2^{\omega_\alpha} = \omega_{\alpha+1}$, then $2^{\omega_{\omega_1}} = \omega_{\omega_1+1}$.

Stationary sets can be generalized from the power set of κ to the family of subsets of λ of cardinality less than κ ($P_\kappa(\lambda)$ or $[\lambda]^{<\kappa}$). The key definitions: $C \subset P_\kappa(\lambda)$ is *unbounded* iff for all $S \in P_\kappa(\lambda)$ there is $S' \in C$ such that $S \subset S'$. $C \subset P_\kappa(\lambda)$ is *closed* iff for all $\rho < \kappa$, all increasing (with respect to $\subset$) sequences $(S_\alpha : \alpha < \rho)$ of elements of C, $\cup_{\alpha<\rho} S_\alpha \in C$. $A \subset P_\kappa(\lambda)$ is *stationary* iff A meets every closed and unbounded subset of $P_\kappa(\lambda)$. There is a pressing down lemma.

4.7. Lemma. Let κ be an uncountable regular cardinal, and A a stationary subset of $P_\kappa(\lambda)$, where $\lambda \geq \kappa$. If f is a function from A to λ such that $f(S) \in S$, for all $S \in A$, then there is a stationary $A' \in A$ and an $\alpha \in \lambda$ such that $f(S) = \alpha$ for all $S \in A'$.

The definition of weakly Mahlo cardinal (2.10) is only the simplest contribution of stationary sets to large cardinals. The notion of normality is the abstraction from a fundamental property of stationary sets. A filter is normal iff it is closed under diagonal intersections. Equivalently, an ideal is normal if it satisfies a pressing down lemma--a function which presses down on a set not in the ideal is constant on a set not in the ideal. The notion of normal ideal is a key tool in the theory of large cardinals.

Van Douwen [vD] has used stationary sets (or, more generally, normal ideals) to construct a space X with interesting generalized compactness properties. Let κ be a regular cardinal, and let $\beta\kappa$ be the Stone-Čech compactification of the discrete space of cardinality κ. Let X be the subspace of $\beta\kappa$ consisting of ultrafilters on which contain a nonstationary subset of κ.

Shelah and Baumgartner have developed an extension of Martin's Axiom. The key definition is that a poset P is a *proper* if, for every λ, every stationary subset of $P_\kappa(\lambda)$ remains stationary after forcing with P. Countable chain condition posets are proper. The axiom asserts that if P is a proper poset, and $\mathcal{D}$ is a family of ω_1 dense subsets of P, then there is a filter G on P which meets every member of $\mathcal{D}$.

5. *The Diamond Constellation*

Perhaps the most frequent way that stationary sets have been used in recent set theoretic topology is in the formulation combinatorial principles to be assumed as extra axioms of set theory. We have already seen such an axiom in 3.10, E(κ). The most important of these combinatorial principles is Jensen's $\diamondsuit$. A small book could be written about the applications of $\diamondsuit$ to set theoretic topology, so here we only sketch the constellation of axioms associated with $\diamondsuit$.

The implications among the following axioms is illustrated in Diagram 1. Below we formulate them and list some typical applications. Generalizing these axioms by adding parameters suggests itself and is useful, but for simplicity we only state the simplest case.

$\diamond^+$: There is a sequence $(W_\alpha : \alpha < \omega_1)$ satisfying

1. Each W_α is a countable family of subsets of α
2. For every $X \subset \omega_1$, $\{\alpha : X \cap \alpha \in W_\alpha\}$ contains a cub set $C(X)$
3. If $\gamma \in C(X)$, then $C(X) \cap \gamma \in W_\gamma$.

Consequences of $\diamond^+$: There is a Kurepa tree [J_2].

There is a first countable space containing a homeomorph of ω_1 with no outer base of cardinality ω_1 [F_3].

$\diamond^*$: Same as $\diamond^+$ without 3.

Consequences of $\diamond^*$: Every normal T_2 space of character $\leq 2^\omega$ is $\leq 2^\omega$-collectionwise Hausdorff [Sh].

There is a nonnormal Aronszajn tree with no stationary antichain [DS_1].

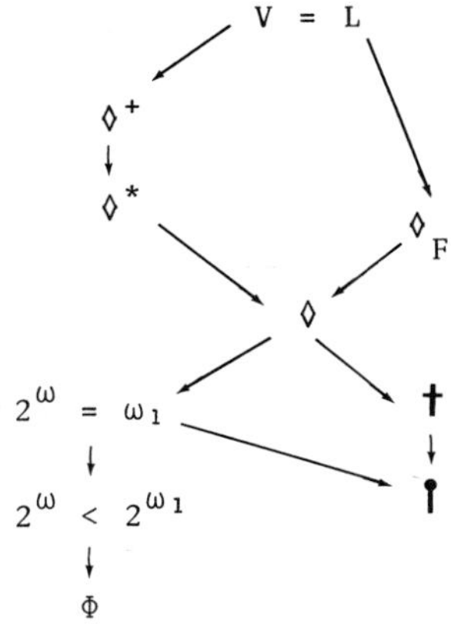

Diagram 1: The Diagram Constellation

$\diamondsuit_F$: If A is a function from the set of functions from ω_2 to ω_1 to the set of subsets of ω_1 satisfying

1. For all f, A(f) is stationary
2. If $f|\alpha = g|\alpha$ then $A(f) \cap (\alpha+1) = A(g) \cap (\alpha+1)$

then there is $(h_\alpha: \alpha < \omega_1)$ such that

3. For all f, $\{\alpha \in A_f: h_\alpha = f|\alpha\}$ is stationary.

Consequences of $\diamondsuit_F$: Every normal T_2 space of character $\leq 2^\omega$ is $\leq 2^\omega$-collectionwise Hausdorff [F_4].

$\diamondsuit$: There is a sequence $(\Gamma_\alpha: \alpha < \omega_1)$ such that for every $X \subset \omega_1$ $\{\alpha: \Gamma_\alpha = X \cap \alpha\}$ is stationary.

Consequences of $\diamondsuit$: There is a Souslin line.

There is a countably compact perfectly normal hereditarily separable space which is not Lindelöf [Os].

$\dagger$: There is a sequence $(S_\lambda: \lambda \in \omega_1 \cap \text{LIM})$ of infinite sets satisfying

1. For each λ, S_λ is a cofinal subset of λ
2. For each uncountable subset X of ω_1, there is $\lambda \in \omega_1 \cap \text{LIM}$ such that $S_\lambda \subset X \cap \lambda$.

Consequences of $\dagger$: There is a Dowker space of cardinality ω_1 [dC].

$\clubsuit$: Same as $\dagger$ without 1.

Consequences of $\clubsuit$: There is a regular hereditarily separable space which is not Lindelöf.

Consequences of $2^\omega = \omega_1$: There is a regular countably compact hereditarily separable space which is not Lindelöf [JKR].

Every separable countably paracompact Moore space is metrizable [F_2].

Consequences of $2^{\omega} < 2^{\omega_1}$: Every separable normal Moore space is metrizable [Jo].

Φ: Let H be the family of functions of the form $h: \delta \to 2$, where δ is a countable ordinal. For every $F: H \to 2$ there is a function $g: \omega_1 \to 2$ satisfying

for all $f: \omega_1 \to 2$, $\{\alpha \in \omega_1: F(f|\alpha) = g(\alpha)\}$ is stationary.

Consequences of Φ: Every special Aronszajn tree (considered as a Moore space) is not normal [DS_2].

Two implications not indicated in Diagram 1 should be noted. First, $2^{\omega} = \omega_1$ plus † implies ◊ [D]. Second, $2^{\omega} = \omega_1$ plus $2^{\omega_1} = \omega_2$ implies the following variant of ◊*.

There is a sequence $(W_{\alpha}: \alpha < \omega_2)$ satisfying

1. Each W_{α} is a family of subsets of α, $|W_{\alpha}| \leq \omega_1$
2. For every $X \subset \omega_2$ there is a cub set $C(X)$ such that $\alpha \in C(X) \cap CF\omega$ implies that $X \cap \alpha \in W_{\alpha}$.

This axiom plus $E(\omega_2)$ implies that there is an ω_2-Souslin tree [Gr].

6. *Historical and Bibliographic Notes*

The term stationary is due to Bloch [Bl].

Gillman and Hendriksen [GH] proved that a linearly ordered topological space is paracompact iff every gap is a Q-gap. Engelking and Lutzer [EL] proved Theorem 1.6 by showing that a linearly ordered topological space has a non-Q-gap iff it contains a closed homeomorph of a stationary set.

The special case of Lemma 2.2 where $S = \omega_1 - \{0\}$ is in Aleksandrov and Urysohn [AU]. The Pressing Down Lemma (weak form) is Neumer's Theorem [Ne]; the Pressing Down Lemma (strong form) is Fodor's Theorem [Fo].

Ulam [Ul] proved that there is no real valued measurable cardinal less than the first weakly inaccessible cardinal. His proof shows that every stationary subset of a successor cardinal κ can be split into κ disjoint stationary subsets. His paper contains the trick of exchanging indices (e.g. defining $h_n(\delta) = \eta_\delta(n)$). Solovay's Theorem (2.12) is in [So], where he shows that the first real valued measurable cardinal must be greater than the first weakly Mahlo cardinal.

Nonseparable Borel theory was introduced (or perhaps, first pursued) by A. Stone and his students. Stationary sets where introduced to this study by Pol $[P_1]$, $[P_2]$. For a bibliography for this topic, see those two papers and $[F_1]$.

Theorem 3.9 was proved by Oxtoby in [Ox] assuming $2^\omega = \omega_1$. P.E. Cohen [Co] gave the first proof without extra assumptions. The spaces used here to prove Theorem 3.9 are from [FK] where more examples of this technique are given. For example, a Baire space X such that X^2 is first category in itself, and for each cardinal κ, a family of spaces $\{Y_\alpha : \alpha < \kappa\}$ such that the product of all of them is not Baire, but the product of all but any one of them is Baire.

The example of a normal collectionwise Hausdorff nonmetrizable Moore space is from $[F_2]$. Axiom $\exists\kappa E(\kappa)$ is consistent with almost everything else. It is quite hard to make it false. It is true, for example in every forcing extension of L, the class of constructible sets. Axiom $\neg E(\omega_2)$ implies that ω_2 is Mahlo in L [HS]; $\neg\exists\kappa E(\kappa)$ implies that 0# exists (a much stronger assetion)--see the footnote at the end of [BKM]. The existence of normal, not $\leq\omega_1$ collectionwise Hausdorff Moore spaces follows from Martin's Axiom plus $2^\omega > \omega_1$, and is consistent with the Generalized Continuum Hypothesis [DS_2]. On the other hand, assuming there is a real valued measurable cardinal, there are no such spaces [Ny].

Theorems 4.2 and 4.5 are due to Erdös and Rado. The proof given here, using the Pressing Down Lemma, is due, so far as I know, to Mate. See the appendix of [Ju] these and more theorems with different proofs. See Jech [J_1] for an introduction to $P_\kappa(\lambda)$. For large cardinals and normal ideals, see [Dr], [SRK], [Ba] and [BTW]. See [BP] for an easy exposition of Theorem 4.6. Its bibliography contains references to Silver's original theorem and the many generalizations.

Section 5 already contains references to the topological results. For more details about the axioms themselves and how to make them true or false, the reader is referred to [De].

References

[AU] P. Aleksandrove and P. Urysohn, *Memoire sur les espaces compacts*, Ver. Kon. Acad. Amsterdam 1 (1929).

[BKM] J. Barwise, M. Kaufmann and M. Makkai, *Stationary logic*, Ann. Math. Logic 13 (1978), 171-224.

[Ba] J. Baumgartner, *Ineffiability properties of cardinal I*, Proc. Intl. Colloq. Infinite and Finite sets, (1973), 109-130.

[BP] J. Baumgartner and K. Prikry, *Singular cardinals and the generalized continuum hypothesis*, Monthly 84 (1977), 108-113.

[BTW] J. Baumgartner, A. Taylor and S. Wagon, *Structural properties of ideals*, to appear Diss. Math.

[Bl] G. Bloch, *Sur les ensembles stationaries de nombres ordinaux et les suites distingueés de functions regressives*, C.R. Acad. Sci. Paris 236 (1953), 265-267.

[Co] P.E. Cohen, *Products of Baire spaces*, Proc. AMS 55 (1976), 119-124.

[dC] P. deCaux, *A collectionwise normal, θ-refinable Dowker spaces which is neither irreducible nor realcompact*, Top. Proc. 1 (1976), 67-78.

[D] K. Devlin, *Variations on ◊*, JSL 44 (1979), 51-58.

[DS_1] K. Devlin and S. Shelah, *Souslin properties and tree topologies*, Proc. London Math. Soc., to appear.

[DS_2] ________, *A note on the normal Moore space conjecture*, Canad. J. Math., to appear.

[DS_3] ________, *A weak version of ◊ which follows from* $2^{\aleph_0} < 2^{\aleph_1}$, Isr. Jour. Math. 29 (1978), 239-247.

[Dr] F. Drake, *Set theory, an introduction to large cardinals*, North Holland Pub., Co., Amsterdam, 1974.

[EL] R. Engelking and D. Lutzer, *Paracompactness in ordered spaces*, Fund. Math. 94 (1977), 49-58.

[F_1] W. Fleissner, *An axiom for nonseparable Borel theory*, Trans. AMS, 251 (1979), 309-328.

[F_2] ________, *Separation properties in Moore spaces*, Fund. Math. 98 (1978), 279-286.

[F_3] ________, *The character of* ω_1 *in first countable spaces*, Proc. AMS 62 (1977), 149-155.

[F_4] ________, *Normal Moore spaces in the constructible universe*, Proc. AMS 46 (1974), 294-298.

[FK] W. Fleissner and K. Kunen, *Barely Baire spaces*, Fund. Math. 101 (1978), 229-240.

[Fo] G. Fodor, *Eine Benierkung zur theorie der regressiven Fuktionen*, Acta Scientarium Mathematicam 17 (1956), 139-142.

[GH] L. Gillman and M. Hendriksen, *Concerning rings of continuous functions*, Trans. AMS 77 (1954), 340-362.

[Gr] J. Gregory, *Higher Souslin trees*, JSL 41 (1976), 663-671.

[HS] L. Harrington and S. Shelah, *Exact equiconsistency results*, Not. AMS, 78T-E74, 25 (1978), A-602.

[J_1] T. Jech, *Some combinatorial problems concerning uncountable cardinals*, Ann. Math. Logic 5 (1973), 165-198.

[J_2] ________, *Trees*, JSL 36 (1971), 1-14.

[Jo] F. Jones, *Concerning normal and completely normal spaces*, Bull. AMS 43 (1937), 671-677.

[Ju] I. Juhász, *Cardinal functions in topology*, Mathematical centre, Amsterdam (1971).

[JKR] I. Juhász, K. Kunen and M. Rudin, *Two more hereditarily separable non-Lindelöf spaces*, Canad. J. Math. 28 (1976), 998-1005.

[Ne] W. Neumer, *Verallgemeinerung eines Satzes von Alexandroff und Urysohn*, Mathematische Zeitschrift 54 (1951), 254-261.

[Ny] P. Nyikos, *A provisional solution to the normal Moore space problem*, Proc. AMS, to appear.

[OS] A. Ostaszewski, *On countably compact perfectly normal spaces*, Jour. London Math. Soc. 14 (1976), 193-204.

[Ox] J. Oxtoby, *Catesian products of Baire spaces*, Fund. Math. 49 (1961), 157-161.

[P_1] R. Pol, *Note on decomposition of metrizable spaces, I*, Fund. Math. 95 (1977), 95-103.

[P_2] ________, *Note on decomposition of metrizable space, II*, Fund. Math. 100 (1978), 129-143.

[Sh] S. Shelah, *unpublished.*

[SRK] R. Solovay, W. Reinhardt and A. Kanamori, *Strong axioms of infinity and elementary embeddings*, Ann. Math. Logic 13 (1978), 73-116.

[So] R. Solovay, *Real-valued measurable cardinals*, AMS Symp. Pure Math. 13, I (1971), 397-428.

[Ul] S. Ulam, *Zur Masstheorie in der algemeinen Mengenlehre*, Fund. Math. 16 (1930), 140-150.

[vD] E.K. van Douwen, *Compactness-like properties and non-normality of the space of non-stationary ultrafilters*, to appear.

SURVEYS IN
GENERAL TOPOLOGY

THREE COVERING PROPERTIES

Heikki J.K. Junnila
Ohio University and
University of Pittsburgh

This paper gives a survey of results obtained in research on the properties of metacompactness, submetacompactness and subparacompactness. Each of these three properties is weaker than the property of paracompactness. Of the many generalizations of paracompactness introduced in the last thirty years, we consider here only three because these three seem to us to be the most important of those generalizations of paracompactness which can be studied by modifying techniques invented to prove results on paracompactness; some generalizations for which these techniques do not seem to apply, for example, the para-Lindelöf property, are potentially interesting, but the theory connected with these properties is still very fragmentary (e.g. it is not known whether every regular para-Lindelöf space is paracompact; see [T_1] and [FR]).

Instead of just giving a list of the most important results dealing with the covering properties studied in this paper, we also demonstrate some of the techniques of proof that have been used to analyze these properties. The most important of these techniques, illustrated in the proofs of Proposition 1.4, Theorem 1.6 and Proposition 2.6 below, are modifications of techniques earlier invented by A.H. Stone to

ISBN: 0-12-584960-5

establish the equivalence of paracompactness and full normality ([St]), by P.S. Alexandrov and P. Urysohn to characterize compactness of a space in terms of the existence of finite subcovers for all monotone covers of the space ([AU]) and by E.A. Michael and K. Nagami to prove that every collectionwise normal metacompact space is paracompact ([Mi_2] and [N_1]).

Besides presenting a sketch of the development of the "internal" theory dealing with metacompactness, submetacompactness and subparacompactness, we also try to indicate some of the connections that these properties have to other kinds of topological properties. Many of the generalizations of paracompactness considered in the literature are associated with certain generalizations of metrizable spaces; for example, subparacompactness first appeared as a property of developable spaces. In the following, we indicate some generalized metric spaces that have one or more of the three covering properties under consideration. We refer the reader to [BuL] for a survey on generalized metric spaces.

To keep this work in reasonable size, we have to leave parts of the theory dealing with the three covering properties out of consideration. We do not consider the invariance of these properties under topological operations. For some results and counterexamples on invariance, we refer the reader to [G_2] and [Ch_2] (open mappings), [Wo_2], [Ju_5] and [Bu_1] (closed mappings), [Ho], [Ju_2], [Ju_5] and [Bu_1] (sums), [AE] and [Pr] (products). Also, we do not consider such special topics as the properties of Borel measures on spaces satisfying covering properties (for this topic, we refer the reader to [P]).

For expositions and surveys on the theory of paracompactness and certain other covering properties, see [E], [Na], [Si], [T_1], [Bu_3] and [J].

0. Preliminaries

In this section, we explain the terminology and notation to be used below (for the meaning of concepts used without definition here, see [E] or [Na]). We also state some elementary results dealing with point-finite and locally finite families of sets.

Throughout the following, X denotes a topological space. The word "space" always refers to a topological space, and the word "iff" is used as an abbreviation of the phrase "if and only if".

The set $\{1,2,...\}$ of the natural numbers is denoted by $\mathbb{N}$. An infinite sequence whose nth term is x_n (for $n \in \mathbb{N}$) is denoted by $\langle x_n \rangle$. Ordinals are sets of smaller ordinals, and cardinals are initial ordinals. The symbols ω, ω_1 and ω_2 denote the first infinite cardinal and the first and second uncountable cardinals, respectively. The cardinality of a set A is denoted by $|A|$.

Let L be a family of subsets of X. For each $A \subset X$, the family $\{L \in L | L \cap A \neq \emptyset\}$ is denoted by $(L)_A$; if $A = \{x\}$, then we write $(L)_x$ in place of $(L)_A$. As usual, $St(A,L) = \cup(L)_A$. The family consisting of all finite unions of sets from L is denoted by L^F. We say that L is a *directed family* if L^F is a refinement of L. Let N be a family of subsets of X. The family N is a *partial refinement* of the family L if for each $N \in N$, there exists $L \in L$ such that $N \subset L$. Let $x \in X$. We say that N is a *pointwise (local)*

W-*refinement* of L *at* x if there exists a finite subfamily K of L (and a neighborhood U of x) such that the family $(L)_x$ (the family $(L)_U$) is a partial refinement of K; if we can choose K to consist of a single member of L, then we say that N is a *pointwise (local) star-refinement* of L *at* x. If $\cup N = \cup L$ and N is a pointwise (local) W-refinement of L at each point of X, then we say that N is a *pointwise (local) W-refinement* of L.

The concepts of a pointwise W-refinement and a local W-refinement were introduced by J.M. Worrell, Jr. who characterized metacompactness and paracompactness in terms of the existence of such refinements ([Wo_2] and [Wo_4]). The importance of these concepts is due to the fact that pointwise W-refinements and local W-refinements provide a simultaneous generalization of the concepts of a point-finite refinement and a pointwise star-refinement and the concepts of a locally finite refinement and a local star-refinement, respectively. In this paper, we take Worrell's characterization of submetacompactness in terms of the existence of certain W-refinements ([Wo_3]) as the starting point of our exposition, and we show how that characterization can be used to derive other results on submetacompactness; we then indicate how results obtained for submetacompact spaces can be used to derive results dealing with metacompact, subparacompact and paracompact spaces.

We now show that open pointwise W-refinements can be generated from certain other kinds of refinements. Recall that a cover L of X is *semi-open* ([Ju_1]) provided that for each $x \in X$, the set $St(x,L)$ is a neighborhood of x. Besides open covers, all closure-preserving closed covers are semi-open.

Lemma 0.1 [Ju_2]. If an open cover of a space has a point-finite semi-open refinement, then the cover has an open pointwise W-refinement.

Proof. If $\mathcal{L}$ is a point-finite semi-open refinement of an open cover $\mathcal{U}$ of X, and for each $L \in \mathcal{L}$, $L \subset U_L \in \mathcal{U}$, then the sets $[\text{Int St}(x,\mathcal{L})] \cap [\cap\{U_L | L \in (\mathcal{L})_x\}]$, $x \in X$, form an open pointwise W-refinement for $\mathcal{U}$. □

The same proof can be used to establish a "local" version of the above result. For locally finite semi-open covers, we also have the following result [Ju_2]: if $\mathcal{L}$ is a locally finite semi-open cover of a space, then $\mathcal{L}^F$ has a locally finite closed refinement.

A family $\mathcal{L}$ of subsets of X is *interior-preserving* [Ju_1] if for each $\mathcal{K} \subset \mathcal{L}$, we have $\text{Int} \cap \mathcal{K} = \cap\{\text{Int } L | L \in \mathcal{K}\}$. Note that $\mathcal{L}$ is interior-preserving iff the family $\{X \sim L | L \in \mathcal{L}\}$ is closure-preserving and that $\mathcal{L}$ is interior-preserving and open iff for each $x \in X$, the set $\cap(\mathcal{L})_x$ is a neighborhood of x.

Lemma 0.2 [Ju_3]. An interior-preserving open cover $\mathcal{U}$ of X has an interior-preserving open pointwise W-refinement iff the cover $\mathcal{U}^F$ has a closure-preserving closed refinement.

Proof. If $\mathcal{F}$ is a closure-preserving closed refinement of $\mathcal{U}^F$, then the sets $[\cap(\mathcal{U})_x] \cap [X \sim \cup(\mathcal{F} \sim (\mathcal{F})_x)]$, $x \in X$, form an interior-preserving open pointwise W-refinement for $\mathcal{U}$.

If $\mathcal{V}$ is an interior-preserving open pointwise W-refinement of $\mathcal{U}$, then the sets $\{x \in X | \text{St}(x,\mathcal{V}) \subset G\}$, $G \in \mathcal{U}^F$, form a closure-preserving closed refinement for $\mathcal{U}^F$. □

A point-finite family is an interior-preserving pointwise W-refinement of itself; consequently, the following result obtains.

Corollary 0.3. If $\mathcal{U}$ is a point-finite open cover of a space, then $\mathcal{U}^F$ has a closure-preserving closed refinement.

We end these preliminaries with a lemma which shows that every point-finite family of open subsets of X is countable provided that either X has countable spread or X is a Baire-space and X satisfies the countable chain condition; the assertion concerning Baire-spaces that satisfy ccc was proved by F. Tall in [T_2] (for related results, see [Ar_4], [Ju_4] and [FL_2]).

Lemma 0.4. Let $\mathcal{U}$ be a point-finite family of open subsets of X. Then there exists a σ-discrete subspace A of X and a disjoint family $\mathcal{G}$ of open subsets of X such that for each $U \in \mathcal{U}$, we have $U \cap A \neq \emptyset$ and if U is of the second category, then we have $G \subset U$ for some $G \in \mathcal{G}$.

Proof. For each $n \in \omega$, let $H_n = \{x \in X \mid |(\mathcal{U})_x| = n\}$.

Let A be maximal among subsets B of X having the property that for all distinct $b \in B$ and $c \in B$, the families $(\mathcal{U})_b$ and $(\mathcal{U})_c$ are distinct. Then $\mathcal{U} = (\mathcal{U})_A$, and A is σ-discrete since for every $n \in \omega$, the subspace $A \cap H_n$ is discrete.

For every $n \in \omega$, let $\mathcal{G}_n = \{(\cap \mathcal{V}) \cap \operatorname{Int} H_n \mid \mathcal{V} \subset \mathcal{U}$ and $|\mathcal{V}| = n\}$. Then the family $\mathcal{G} = \bigcup_{n\in\omega} \mathcal{G}_n$ is disjoint and open. If $U \in \mathcal{U}$ is of the second category, then there exists $n \in \omega$ such that $U \cap \operatorname{Int} H_n \neq \emptyset$ and consequently, there exists $G \in \mathcal{G}_n$ such that $G \subset U$. □

1. Submetacompactness

Before defining submetacompactness, let us recall the definitions of three other covering properties.

Definition 1.1. A space is *paracompact* (*metacompact, subparacompact*) if every open cover of the space has a locally finite open (point-finite open, σ-discrete closed) refinement.

For Hausdorff-spaces, both metacompactness and subparacompactness are weaker properties than paracompactness. Looking for a simultaneous generalization of those two generalizations of paracompactness, we can make the following observation (see Lemma 2.1): an open cover of a space X has a σ-discrete closed refinement iff the cover has a sequence $\langle \mathcal{V}_n \rangle$ of open refinements such that for each $x \in X$, there exists $n \in \mathbb{N}$ such that only one member of $\mathcal{V}_n$ contains x. The following definition gives a generalization of such sequences and a corresponding covering property.

Definition 1.2. A sequence $\langle \mathcal{L}_n \rangle$ of covers of X is a *θ-sequence* if for each $x \in X$, there exists $n \in \mathbb{N}$ such that the family $\mathcal{L}_n$ is finite at x. The space X is *submetacompact* if every open cover of X has a θ-sequence of open refinements.

Submetacompact spaces were introduced by J.M. Worrell, Jr. and H.H. Wicke in 1965 [WoW] under the name "θ-refinable"; the name "submetacompact" for these spaces was suggested in [Ju_5].

So far submetacompactness has remained more or less just a generalization of other properties: no particularly interesting "characteristic" subclass of submetacompact spaces has yet emerged (for a candidate for such a subclass, see

Problem 1.20). However, the redundancy caused by the introduction of this property is compensated by the existence of certain "transition theorems" (Theorems 2.7, 3.3 and 4.1) that allow us to derive results dealing with the other three covering properties from those proved for submetacompactness.

a) Characterizations

One characterization of submetacompactness is contained in the following simple, but important observation.

Lemma 1.3 [WoW]. A cover L of X has a θ-sequence of open refinements iff there exists an open refinement $\cup_{n\in\mathbb{N}} V_n$ of L and a closed cover $\{F_n | n \in \mathbb{N}\}$ of X such that for each $n \in \mathbb{N}$, $F_n \subset \cup V_n$ and V_n is point-finite on F_n.

A highly non-trivial characterization for submetacompactness was obtained by Worrell in 1967; to prove Worrell's result we need the following result on the existence of θ-sequences of refinements of a cover.

Proposition 1.4. A cover L of X has a θ-sequence of open refinements if there exists a sequence $\langle U_n \rangle$ of open refinements of L such that for each $x \in X$, there exists a sequence $\langle t(n) \rangle$ of natural numbers such that for each $n \in \mathbb{N}$, $U_{t(n+1)}$ is a pointwise W-refinement of $U_{t(n)}$ at x.

Proof. Represent L in the form $L = \{W_\alpha | \alpha < \gamma\}$ where γ is an ordinal number. For each $U \in \cup_{k\in\mathbb{N}} U_k$, let $\alpha(U)$ be the least ordinal $\alpha < \gamma$ such that $U \subset L_\alpha$. For $n \in \mathbb{N}$ and $V \in \cup_{k\in\mathbb{N}} U_k$, we say that the family U_n is "precise" at the set V provided that $\{\alpha(U) | V \subset U \in U_n\} \subset \{\alpha(V)\}$. For all $n \in \mathbb{N}$ and $k \in \mathbb{N}$, let $W_{n,k} = \{V \in U_k | U_n \text{ is precise at } V\}$.

For all $n \in \mathbb{N}$ and $k \in \mathbb{N}$, denote by $L_{n,k}$ the set consisting of all those points of X at which $\mathcal{U}_k$ is a pointwise W-refinement of $\mathcal{U}_n$. For every $h > 1$ and for every $s \in \mathbb{N}^h$, let $L_s = \bigcap_{i=1}^{n-1} L_{s(i),s(i+1)}$, and let $H_s = \{x \in L_s | St(x,\mathcal{U}_{s(h)}) \subset \cup(\bigcup_{i=1}^{h-1} \mathcal{W}_{s(i),s(i+1)})\}$. We show that the sets H_s, $s \in \bigcup_{h>1} \mathbb{N}^h$, cover X. Let $x \in X$. Then there exists a sequence $\langle t(n)\rangle$ of natural numbers and a sequence $\langle \mathcal{Q}_n\rangle$ of finite families of sets such that for each $n \in \mathbb{N}$, $\mathcal{Q}_n \subset (\mathcal{U}_{t(n)})_x$ and $(\mathcal{U}_{t(n+1)})_x$ is a partial refinement of $\mathcal{Q}_n$. For every $n > 1$, let $\mathcal{R}_n = \{Q \in \mathcal{Q}_n | Q \not\subset \cup(\bigcup_{i=1}^{n-1} \mathcal{W}_{t(i),t(i+1)})\}$. We show that there exists $n > 1$ such that $\mathcal{R}_n = \emptyset$. Assume on the contrary that $\mathcal{R}_n \neq \emptyset$ for every $n > 1$. For each $n > 1$, let $\alpha(n)$ be the maximum of the ordinals numbers $\alpha(R)$, $R \in \mathcal{R}_n$. Note that for each $n > 1$, $\mathcal{R}_{n+1}$ is a partial refinement of $\mathcal{R}_n$, and hence $\alpha_{n+1} \leq \alpha_n$. It follows that there exists $k > 3$ such that $\alpha_k = \alpha_{k-1} = \alpha_{k-2}$. Let $R \in \mathcal{R}_k$ be such that $\alpha(R) = \alpha_k$. We show that the family $\mathcal{U}_{t(k-1)}$ is precise at R. Let $U \in \mathcal{U}_{t(k-1)}$ be such that $R \subset U$. Then $U \in (\mathcal{U}_{t(k-1)})_x$ and consequently there exists $Q \in \mathcal{Q}_{k-2}$ such that $U \subset Q$. We have $R \subset Q$ and it follows that $Q \in \mathcal{R}_{k-2}$; hence $\alpha(Q) \leq \alpha_{k-2}$. Since $R \subset U \subset Q$, we have $\alpha_k = \alpha(R) \leq \alpha(U) \leq \alpha(Q) \leq \alpha_{k-2}$, and it follows that $\alpha(U) = \alpha(R)$. We have shown that $\mathcal{U}_{t(k-1)}$ is precise at R. Consequently, $R \in \mathcal{W}_{t(k-1),t(k)}$; this, however, is a contradiction since $R \in \mathcal{R}_k$. It follows that there exists $n > 1$ such that $\mathcal{R}_n = \emptyset$. If we set $r = \langle t(1),\dots,t(n+1)\rangle$, then $x \in H_r$. We have shown that the sets H_s, $s \in \bigcup_{n>1} \mathbb{N}^h$, cover X.

For all $n \in \mathbb{N}$ and $k \in \mathbb{N}$, let $V_{\alpha,n,k} = \cup\{W \in \mathcal{W}_{n,k} | \alpha(W) = \alpha\}$ for each $\alpha < \gamma$, and let $\mathcal{V}_{n,k} = \{V_{\alpha,n,k} | \alpha < \gamma\}$. We show that for all $n \in \mathbb{N}$ and $k \in \mathbb{N}$, the family $\mathcal{V}_{n,k}$ is point-finite on the set $L_{n,k}$. Let $x \in L_{n,k}$. Then there exists a finite subfamily $\mathcal{Q}$ of $\mathcal{U}_n$ such that $(\mathcal{U}_k)_x$ partially refines $\mathcal{Q}$. Let $A = \{\alpha(Q) | Q \in \mathcal{Q}\}$. To show that $\mathcal{V}_{n,k}$ is finite at x, it suffices to show that $\{\alpha < \gamma | x \in V_{\alpha,n,k}\} \subset A$. Let $\alpha < \gamma$ be such that $x \in V_{\alpha,n,k}$. Then there exists $W \in \mathcal{W}_{n,k}$ such that $x \in W$ and $\alpha(W) = \alpha$. We have $W \in (\mathcal{U}_k)_x$ and it follows that there exists $Q \in \mathcal{Q}$ such that $W \subset Q$. The family $\mathcal{U}_n$ is precise at W and hence $\alpha(W) = \alpha(Q)$; consequently, $\alpha \in A$. We have shown that the family $\mathcal{V}_{n,k}$ is finite at x.

For every $h > 1$, and for each $s \in \mathbb{N}^h$, let $\mathcal{V}_s = \bigcup_{i=1}^{h-1} \mathcal{V}_{s(i),s(i+1)}$, let $\mathcal{U}_s = \{U \in \mathcal{U}_{s(h)} | U \not\subset \cup\mathcal{V}_s\}$ and let $\mathcal{O}_s = \mathcal{U}_s \cup \mathcal{V}_s$; note that $\mathcal{O}_s$ is an open refinement of $\mathcal{L}$. To complete the proof, it suffices to show that for each $s \in \bigcup_{h>1} \mathbb{N}^h$, the family $\mathcal{O}_s$ is point-finite on the set H_s. Let $s \in \mathbb{N}^h$, with $h > 1$, and let $x \in H_s$. Then $St(x,\mathcal{U}_{s(h)}) \subset \cup(\bigcup_{i=1}^{h-1} \mathcal{W}_{s(i),s(i+1)})$, that is, $St(x,\mathcal{U}_{s(h)}) \subset \cup\mathcal{V}_s$. Consequently, $x \notin \cup\mathcal{U}_s$. We have $(\mathcal{O}_s)_x = (\mathcal{V}_s)_x = \bigcup_{i=1}^{h-1} (\mathcal{V}_{s(i),s(i+1)})_x$ and it follows from the preceding part of the proof that the family $(\mathcal{O}_s)_x$ is finite. □

The central idea of the above proof, which we formalized in the concept of "preciseness", was originally used by Worrell in [Wo_1] to prove an important characterization of metacompactness (see Theorem 3.4 below).

Worrell's characterization of submetacompactness is an easy consequence of the result above. To state the characterization, we employ the following terminology: a sequence of covers of X is said to be a *pointwise W-refining (pointwise star-refining) sequence for a cover* $\mathcal{L}$ of X provided that for each $x \in X$, some member of the sequence is a pointwise

W-refinement (a pointwise star-refinement) of L at x.

Theorem 1.5 [Wo_3]. A space is submetacompact iff every open cover of the space has a pointwise W-refining sequence by open covers of the space.

In the above characterization, the condition that the required refinements from θ-sequences was relaxed; we now use that characterization to derive another one in which the requirement that *all* open covers have θ-sequences of open requirements is relaxed.

Recall that a family L of sets is *monotone* if the partial order of set-inclusion is a linear order on L; if this order is a well-order, then we say that L is *well-monotone*. Note that if L is well-monotone and $L' \subset L$, then $\cap L' \in L$. Consequently, every well-monotone family of subsets of a topological space is interior-preserving.

Theorem 1.6 [Ju_5]. A space is submetacompact iff every well-monotone open cover of the space has a θ-sequence of open refinements.

Proof. Necessity is obvious. To prove sufficiency, assume that every well-monotone open cover of X has a θ-sequence of open refinements. For each cardinal number κ, denote by P_κ the proposition that every open cover of X of cardinality κ has a pointwise W-refining sequence by open covers of X. We use transfinite induction on κ to show that P_κ holds for every cardinal κ; it then follows from Theorem 1.5 that X is submetacompact. For κ finite, P_κ is trivially true. Let κ be an infinite cardinal such that P_λ holds for every cardinal $\lambda < \kappa$. To show that P_κ holds, let U be an open cover of X such that $|\mathcal{U}| = \kappa$. Represent $\mathcal{U}$ in the form $\mathcal{U} = \{U_\alpha | \alpha < \kappa\}$. The family $\{\cup_{\beta\leq\alpha} U_\beta | \alpha < \kappa\}$ is a well-monotone open cover of X and hence this family has a θ-sequence, say $\langle \mathcal{V}_n \rangle$, of open refinements. For each $V \in \cup_{n\in\mathbb{N}} \mathcal{V}_n$, let $\alpha(V) < \kappa$ be such that $V \subset \cup\{U_\beta | \beta \leq \alpha(V)\}$. For all $\alpha < \kappa$ and $n \in \mathbb{N}$, let $P_{\alpha,n} = \cup\{V \in \mathcal{V}_n | \alpha(V) > \alpha\}$ and let $\mathcal{U}_{\alpha,n} = \{U_\beta | \beta \leq \alpha\} \cup \{P_{\alpha,n}\}$. Note that $\mathcal{U}_{\alpha,n}$ is an open cover of X and $|\mathcal{U}_{\alpha,n}| < \kappa$; hence, by the induction hypothesis, $\mathcal{U}_{\alpha,n}$ has a pointwise W-refining sequence, say $\langle \mathcal{W}_{\alpha,n,k} \rangle_{k=1}^{\infty}$, by open covers of X.

For all $n \in \mathbb{N}$ and $h \in \mathbb{N}$, denote by $F_{n,h}$ the closed set $\{x \in X | |(\mathcal{V}_n)_x| \leq h\}$. For every $n \in \mathbb{N}$, let $H_n = \cup_{h\in\mathbb{N}} F_{n,h}$, and for each $x \in H_n$, let $\alpha(x,n)$ be the maximum of the ordinal numbers $\alpha(V)$, $V \in (\mathcal{V}_n)_x$. For all $n \in \mathbb{N}$ and $k \in \mathbb{N}$, and for every $x \in H_n$, let $W_{n,k}(x)$ be a member of the family $\mathcal{W}_{\alpha(x,n),n,k}$ containing x; further, let $G_{n,k}(x) = (\cap_{i\leq k} W_{n,i}(x)) \cap (\cap(\mathcal{V}_n)_x)$. Then, for all $n \in \mathbb{N}$ and $k \in \mathbb{N}$, the family $\mathcal{G}_{n,k} = \{G_{n,k}(x) | x \in H_n\} \cup \{X \sim F_{n,k}\}$ is an open cover of X. Let $x \in X$. We show that there exist $n \in \mathbb{N}$ and $k \in \mathbb{N}$ such that $\mathcal{G}_{n,k}$ is a pointwise W-refinement of $\mathcal{U}$ at x. Let $n \in \mathbb{N}$ be such that $x \in H_n$, and let $A = \{\alpha(V) | V \in (\mathcal{V}_n)_x\}$. For each $\alpha \in A$, there exists $k(\alpha) \in \mathbb{N}$, and a finite subfamily $\mathcal{R}_\alpha$ of $\mathcal{U}_{\alpha,n}$ such that the family $(\mathcal{W}_{\alpha,n,k(\alpha)})_x$ is a partial refinement of $\mathcal{R}_\alpha$. Let $\mathcal{R} = \cup_{\alpha\in A} (\mathcal{R}_\alpha \cap \mathcal{U})$ and let k be the maximum of the numbers $|(\mathcal{V}_n)_x|$ and $k(\alpha)$, $\alpha \in A$. We show that $(\mathcal{G}_{n,k})_x$ is a partial refinement of $\mathcal{R}$. Let $G \in (\mathcal{G}_{n,k})_x$. We have $x \in F_{n,k}$ and it follows that there exists $y \in H_n$ such that $G = G_{n,k}(y)$. Let $\alpha = \alpha(y,n)$, and let $V \in (\mathcal{V}_n)_y$ be such that $\alpha(V) = \alpha$. We have $x \in G_{n,k}(y) \subset V$ and consequently $\alpha \in A$. Let $j = k(\alpha)$. Since $j \leq k$, we have $G_{n,k}(y) \subset W_{n,j}(y)$ and hence $x \in W_{n,j}(y)$. It follows that there exists $R \in \mathcal{R}_\alpha$ such that $W_{n,j}(y) \subset R$. We have $y \notin P_{\alpha,n}$ and it follows, since $y \in R \in \mathcal{U}_{\alpha,n}$, that $R \in \mathcal{U}$. By the foregoing, $R \in \mathcal{R}$ and $G \subset R$. Since $\mathcal{R}$ is a finite subfamily of $\mathcal{U}$, we have shown that $\mathcal{G}_{n,k}$ is a pointwise W-refinement of $\mathcal{U}$ at x. This completes the proof that P_K holds. □

In [Mi_3], E. Michael showed that a Hausdorff-space is paracompact, iff every open cover of the space has a closure-preserving closed refinement; we'll now show that submetacompactness admits a similar characterization.

Theorem 1.7 [Ju_5]. The following conditions are mutually equivalent for a topological space:

(i) The space is submetacompact.

(ii) Every interior-preserving directed open cover of the space has a σ-closure-preserving closed refinement.

(iii) Every directed open cover of the space has a σ-closure-preserving closed refinement.

Proof. (i) ⇒ (iii): Use Lemma 1.3 and Corollary 0.3.
(ii) ⇒ (i): Assume that X satisfies condition (ii). To show that X is submetacompact, it suffices, by Theorem 1.6, to show that every interior-preserving open cover of X has a θ-sequence of open refinements. Note that for every interior-preserving open cover $\mathcal{V}$ of X, the family $\mathcal{V}^F$ is an interior-preserving directed open cover of X and hence this family has a σ-closure-preserving closed refinement; a modification of the proof of Lemma 0.2 then shows that $\mathcal{V}$ has a pointwise W-refining sequence by interior-preserving open covers of X. It follows that for every interior-preserving open cover $\mathcal{U}$ of X, there exists a countable collection $\mathcal{C}$ of interior-preserving

open refinements of $\mathcal{U}$ such that for all $x \in X$ and $\mathcal{V} \in \mathcal{C}$, some member of $\mathcal{C}$ is a pointwise W-refinement of $\mathcal{V}$ at x; by Proposition 1.4, $\mathcal{U}$ has a θ-sequence of open refinements. □

In [Mi_4], Michael showed that a Hausdorff-space is paracompact iff every open cover of the space has a cushioned refinement (recall that a family $\mathcal{N}$ of subsets of X is *cushioned* in a family $\mathcal{L}$ of subsets of X provided that there exists a map $\Phi: \mathcal{N} \to \mathcal{L}$ such that $\overline{\cup\mathcal{M}} \subset \cup\Phi(\mathcal{M})$ for every $\mathcal{M} \subset \mathcal{N}$). It is not known whether the analogue of Michael's result holds for submetacompact spaces.

Problem 1.8 [Ka_2]. Does a space have to be submetacompact if every directed open cover of the space has a σ-cushioned refinement?

Using the following result, we could state Problem 1.8 in another form.

Lemma 1.9 [Ju_5]. A cover of a space has a σ-cushioned refinement iff the cover has a pointwise star-refining sequence by semi-open covers of the space.

In light of this result, the following characterization of submetacompactness can be considered as a partial solution to Problem 1.8; this characterization is a direct corollary to Theorem 3.2 of [Ju_5].

Theorem 1.10. A space is submetacompact iff every open cover of the space has a pointwise W-refining sequence by semi-open covers of the space.

b) Properties of submetacompact spaces

Recall that a space is *countably metacompact* if every countable open cover of the space has a point-finite open refinement. It is easily seen that a countable open cover of a space has a point-finite open refinement provided that the cover has a θ-sequence of open refinements; consequently, the following result, due to R. Gittings, holds.

Proposition 1.11 [G_1]. Every submetacompact space is countably metacompact.

In 1950, A. Arens and J. Dugundji introduced the concept of a metacompact space, and they showed that every countably compact metacompact space is compact [AD]; to prove this result, they first showed that every point-finite cover of a set contains an irreducible subcover (recall that a cover of a set is *irreducible* if the cover has no proper subcover). Using this last-mentioned result and Lemma 1.3, one can easily show that if a cover of a space has a θ-sequence of open refinements, then the cover has an open irreducible refinement; consequently, we have the following result, obtained by Worrell and Wicke in 1965.

Proposition 1.12 [WoW]. In a submetacompact space, every open cover has an irreducible open refinement.

Corollary 1.13 [WoW]. Let X be a submetacompact space.

i) If X is countably compact, then X is compact.

ii) If X is $\aleph_1$-compact, then X is a Lindelöf-space.

Recently, various weak covering properties have been introduced, and it has been shown that the above proposition, or at least part of the corollary, holds for mnay of these weak properties; see e.g.[Bo_2], [Smi_1], [Aq], [WW_1],[Au], [Ch_1], [D] and [WW_3], and see [vDW] for an example that is relevant for this line of research.

Our next result deals with certain partitions of submetacompact spaces. We need some preliminaries.

Definition 1.14 [W], [WW_2]. A partition $\mathcal{P}$ of a space is *scattered* provided that whenever $\mathcal{P}'$ is a non-empty subfamily of $\mathcal{P}$, there exists $P \in \mathcal{P}'$ and an open set U such that $P \subset U$ and $[\cup(\mathcal{P}' \sim \{P\})] \cap U = \emptyset$.

Note that X is a scattered space iff the identity partition $\{\{x\} | x \in X\}$ of X is scattered.

Lemma 1.15 [WW_2]. A partition $\mathcal{P}$ of a space is scattered iff there exists a well-ordering $\leq$ on $\mathcal{P}$ such that for each $P \in \mathcal{P}$, the set $\cup\{P' \in \mathcal{P}\ P' \leq P\}$ is open.

Proof. Sufficiency is obvious. Necessity: Let $\mathcal{P}$ be a scattered partition of X. For each non-empty $\mathcal{Q} \subset \mathcal{P}$, let $P(\mathcal{Q}) \in \mathcal{Q}$ be such that for some open set U, we have $P(\mathcal{Q}) \subset U$ and $[\cup(\mathcal{Q} \sim \{P(\mathcal{Q})\})] \cap U = \emptyset$. Let A be a set not in $\mathcal{P}$, and let $P(\emptyset) = A$. Let κ be a cardinal number bigger than $|\mathcal{P}|$. Define a map $R: \kappa \to \mathcal{P} \cup \{A\}$ be the condition that for each $\alpha < \kappa$, $R(\alpha) = P(\mathcal{P} \sim \{R(\beta) | \beta < \alpha\})$. Let γ be the least ordinal in κ such that $R(\gamma) = A$. Then $\mathcal{P} = \{R(\alpha) | \alpha < \gamma\}$ and this representation induces the desired well-ordering on $\mathcal{P}$. □

It follows from the above result that every open cover of a space is refined by some scattered partition of the space.

The following result has been obtained by J. Chaber and this author; in light of Lemma 1.15, the result can be seen to be a generalization of a part of Theorem 2 of [ChZ] (see Theorem 2.18 below).

Proposition 1.16. In a submetacompact space, every scattered partition of the space into G_δ-sets has a σ-discrete closed refinement.

Proof. Let X be a submetacompact space, and let $\mathcal{P}$ be a scattered partition of X such that every member of $\mathcal{P}$ is a G_δ-subset of X. By Lemma 1.15, there exists an ordinal γ and a representation of $\mathcal{P}$ in the form $\mathcal{P} = \{P_\alpha | \alpha < \gamma\}$ such that for each $\alpha < \gamma$, the set $\bigcup_{\beta \leq \alpha} P_\beta$ is open. Let $P_\gamma = \emptyset$. For each $\alpha \leq \gamma$, let $U_\alpha = \bigcup_{\beta \leq \alpha} P_\beta$ and let $\langle G_{\alpha,n} \rangle$ be a sequence of open sets such that $P_\alpha = \bigcap_{n \in \mathbb{N}} G_{\alpha,n}$ and for each $n \in \mathbb{N}$, $G_{\alpha,n} \subset U_\alpha$. For each $x \in X$, let $\alpha(x) < \gamma$ be such that $x \in P_{\alpha(x)}$.

Let $\mathcal{V}$ be an open cover of X. For each $x \in X$, let $\beta(x) \leq \gamma$ be such that $St(x,\mathcal{V}) \subset U_\beta$. We show that there exists a countable collection $\mathcal{C}$ of open covers of X such that for each $x \in X$, if $\beta(x) > \alpha(x)$, then there exists $\delta < \beta(x)$ and $\mathcal{W} \in \mathcal{C}$ such that $St(x,\mathcal{W}) \subset U_\delta$. For each $V \in \mathcal{V}$, let $\alpha(V)$ be the least ordinal $\alpha \leq \gamma$ such that $V \subset U_\alpha$. For all $n \in \mathbb{N}$ and $\alpha \leq \gamma$, let $V_{\alpha,n} = G_{\alpha,n} \cap (\bigcup\{V \in \mathcal{V} | \alpha(V) \geq \alpha\})$. For every $n \in \mathbb{N}$, let $\mathcal{V}_n = \{V_{\alpha,n} | \alpha \leq \gamma\}$ and note that this open family covers X since for each $x \in X$, we have $x \in V_{\alpha(x),n}$; consequently, $\mathcal{V}_n$ has a θ-sequence, say $\langle \mathcal{W}_{n,k} \rangle_{k=1}^{\infty}$, of open refinements. Let $\mathcal{C} = \{\mathcal{W}_{n,k} | n \in \mathbb{N} \text{ and } k \in \mathbb{N}\}$. To show that the collection $\mathcal{C}$ has the desired property, let $x \in X$ be such that $\beta(x) > \alpha(x)$. Then $x \notin P_{\beta(x)}$ and hence there exists $n \in \mathbb{N}$ such that $x \notin G_{\beta(x),n}$. Note that $x \notin \bigcup\{V_{\alpha,n} | \alpha \geq \beta(x)\}$. Let $k \in \mathbb{N}$ be such that the family $(\mathcal{W}_{n,k})_x$ is finite. Then, for some $\delta < \beta(x)$, we have $St(x,\mathcal{W}_{n,k}) \subset \bigcup_{\alpha \leq \delta} V_{\alpha,n} \subset U_\delta$.

It follows from the foregoing that there exists a countable collection $\mathcal{D}$ of open covers of X such that for each $x \in X$, there exists $\mathcal{O} \in \mathcal{D}$ such that $St(x,\mathcal{O}) \subset U_{\alpha(x)}$. Let $\mathcal{D} = \{\mathcal{O}_n | n \in \mathbb{N}\}$, and for all $n \in \mathbb{N}$ and $\alpha < \gamma$, let

$F_{\alpha,n} = \{x \in P_\alpha | St(x,\mathcal{O}_n) \subset U_\alpha\}$. The family $\bigcup_{n\in\mathbb{N}} \{F_{\alpha,n} | \alpha < \gamma\}$ is a σ-discrete closed refinement of $\mathcal{P}$. □

Note that if a partition of a space has a σ-discrete closed refinement, then the partition consists of G_δ-sets.

Proposition 1.16 has the following corollaries. The result of the first corollary was obtained by P. Nyikos in 1977; with "subparacompact" substituted for "submetacompact", the result was earlier obtained by R. Telgarsky [Te_1].

Corollary 1.17 [Ny_1]. If X is a submetacompact scattered space such that every singleton subset of X is a G_δ-set, then X is σ-discrete.

Corollary 1.18. In a submetacompact space, every cover consisting of ω_1 open F_σ-sets has a σ-discrete closed refinement.

Proof. If $\{U_\alpha | \alpha < \omega_1\}$ is a cover of X by open F_σ-sets, then the partition $\{U_\alpha \sim \bigcup_{\beta<\alpha} U_\beta | \alpha < \omega_1\}$ of X is scattered and for each $\alpha < \omega_1$, the difference of the open set U_α and the F_σ-set $\bigcup_{\beta<\alpha} U_\beta$ is a G_δ-set. □

The result of Corollary 1.18 also follows from the result recently obtained by D. Burke [Bu_5] that in a submetacompact space, every locally countable family of closed subsets of the space has a σ-closure-preserving closed refinement.

c) Subclasses

To indicate one class of submetacompact spaces that is not contained in any class determined by a stronger covering property, we employ the following terminology. We say that a family $\mathcal{L}$ of subsets of a space is a *refiner for directed open covers* of the space provided that every directed open cover of the space has a refinement by members of $\mathcal{L}$. It is easily seen that whenever F is a family of closed subsets of X such that F is closed under finite intersections, the family F is a refiner for directed open covers of X iff for each $x \in X$, the set $\cap(F)_x$ is compact and the family $(F)_x$ is a network around the set $\cap(F)_x$.

In 1969, K. Nagami defined strong Σ-spaces (see Section 2.c), and in 1971, Michael introduced the wider class of strong $\Sigma\#$-spaces [Mi_6]. In the terminology introduced above, a space is a *strong* $\Sigma\#$*-space* iff the space has a σ-closure-preserving closed refiner for directed open covers. This characterization of strong $\Sigma\#$-spaces and Theorem 1.7 establish the following result.

Proposition 1.19 [Ju_5]. Every strong $\Sigma\#$-space is submetacompact.

The importance of strong $\Sigma\#$-spaces for the theory of submetacompact spaces is due to the fact that any countable product of strong $\Sigma\#$-spaces is again a strong $\Sigma\#$-space and hence submetacompact. Unfortunately, it is not known whether other covering properties have nice multiplicative properties in the class of strong $\Sigma\#$-spaces.

One class of spaces in which all four of the covering properties studied in this paper are countably multiplicative is the class of p-spaces, introduced by Arhangel'skii in 1963 [Ar_2]. In 1966, Arhangel'skii defined strict p-spaces [Ar_3]. Strict p-spaces form a particularly interesting subclass of p-spaces; in 1969, D. Burke and R. Stoltenberg [BuS] obtained the result that a Tychonoff-space X is a *strict p-space* iff there exists a sequence $\langle U_n \rangle$ of open covers of X such that for each $x \in X$, the set $K(x) = \bigcap_{n \in \mathbb{N}} St(x, U_n)$ is compact and the sequence $\langle St(x,\mathcal{U}_n)\rangle$ is a neighborhood-base around the set $K(x)$.

In 1970, Burke showed that every submetacompact p-space is a strict p-space [Bu_2]; whether the converse of this result holds is an open problem.

Problem 1.20 [ChJ]. Is every strict p-space submetacompact?

Note that in a strict p-space, every directed open cover has a pointwise star-refining sequence by open covers; hence, in light of the result of Lemma 1.9, the above problem is related to Problem 1.8.

Problem 1.20 can be stated in another form by means of the following result.

Proposition 1.21 [Ju_5]. A strict p-space is submetacompact iff the space is a strong $\Sigma\#$-space.

A partial solution to the problem is given in the following result.

Proposition 1.22 [ChJ]. Every locally compact strict p-space is submetacompact.

2. *Subparacompactness*

Most of the basic ideas in the theory of subparacompact spaces were introduced by R.H. Bing in 1951 [B]. Many would agree that Bing's 1951-paper on metrization of topological spaces is the most important single paper in the literature on that topic; here we'll see that the concepts and results appearing in the paper are of central importance also in the theory of covering properties. In an implicit form, the concept of a subparacompact space first appeared in that paper: a key result used by Bing in the proof of the factorization of metrizability into developability and collectionwise normality was that in a developable space, every open cover has a σ-discrete closed refinement. An explicit definition of subparacompact spaces was given by L.F. McAuley in 1958 [Mc]; McAuley called these spaces "F_σ-screenable" (screenability is a covering property introduced by Bing in [B]; a space is screenable provided that every open cover of the space has a σ-disjoint open refinement). McAuley generalized Bing's result on developable spaces by showing that every semi-metrizable space is F_σ-screenable. In 1966, Arhangel'skii defined a covering property under the name "σ-paracompactness" [Ar_3]. By 1969, it had become evident that the conditions defining F_σ-screenability and σ-paracompactness are actually equivalent (see [Wo_3], [Co_1], [BuS] and [Bu_1]). In [Bu_1], D.K. Burke established the equivalence of several different conditions, including those defining F_σ-screenability and σ-paracompactness, and he gave the name "subparacompact" to spaces that satisfy these conditions.

a) Relation to submetacompactness

We start with a characterization of subparacompactness that illustrates the relationship between this property and submetacompactness. To state the characterization, we introduce the following notation: when $\mathcal{L}$ is a family of sets, we denote the family $\{L \sim \cup(\mathcal{L} \sim \{L\}) | L \in \mathcal{L}\}$ by $M(\mathcal{L})$. Note that $x \in \cup M(\mathcal{L})$ iff $|(\mathcal{L})_x| = 1$. The characterization is a consequence of the following observation.

Lemma 2.1. Let $\mathcal{U}$ be an open cover of X, and let $\mathcal{F}$ be a family of subsets of X. Then $\mathcal{F}$ is a closed and discrete partial refinement of $\mathcal{U}$ iff there exists an open refinement

$\mathcal{V}$ of $\mathcal{U}$ such that $F \subset M(\mathcal{V})$.

Proposition 2.2 [Bu_2]. A space is subparacompact iff for every open cover $\mathcal{U}$ of the space, there exists a sequence $\langle \mathcal{V}_n \rangle$ of open refinements of $\mathcal{U}$ such that the family $\bigcup_{n \in \mathbb{N}} M(\mathcal{V}_n)$ covers the space.

Note that if $\mathcal{U}$ and $\langle \mathcal{V}_n \rangle$ are as above, then $\langle \mathcal{V}_n \rangle$ is simultaneously a pointwise star-refining sequence for $\mathcal{U}$ and a θ-sequence.

In particular, as observed by Worrell and Wicke in [WoW], every subparacompact space is submetacompact. The converse is not true; however, by Corollary 1.18, the following result holds.

Proposition 2.3. Every submetacompact, completely regular space of weight ω_1 is subparacompact.

The weight of the space cannot be increased in the result above: in 1966, Worrell gave an example of a metacompact, non-subparacompact space of weight ω_2 [Wo_2], and in 1969, Burke independently gave an example of a similar space that is also locally compact.

Example 2.4 [Bu_1]. Let $Y = [(\omega_2+1) \times (\omega_2+1)] \sim \{\langle \omega_2, \omega_2 \rangle\}$. For every $\alpha \in \omega_2$, let $L_\alpha = \{\alpha\} \times (\omega_2+1)$ and $K_\alpha = (\omega_2+1) \times \{\alpha\}$. Topologize Y so that the points in $\omega_2 \times \omega_2$ are isolated, the points $\langle \alpha, \omega_2 \rangle$ have a neighborhood basis by sets of the form $L_\alpha \sim A$, where A is finite, and the points $\langle \omega_2, \alpha \rangle$ have a similar neighborhood base, with K_α replacing L_α.

It is easy to see that Y is a metacompact, locally compact Hausdorff-space of weight ω_2. Let $\mathcal{U} = \{\omega_2 \times (\omega_2+1), (\omega_2+1) \times \omega_2\}$. Then $\mathcal{U}$ is an open cover of Y, and it can be shown that whenever $\langle \mathcal{V}_n \rangle$ is a sequence of open refinements of $\mathcal{U}$, there exists $y \in Y$ such that for every $n \in \mathbb{N}$, $y \notin \cup M(\mathcal{V}_n)$. By Proposition 2.2, Y is not subparacompact. □

To be able to use results dealing with submetacompact spaces in the theory of subparacompact spaces, we need to find some conditions under which a submetacompact space is subparacompact. In 1955, E. Michael and K. Nagami showed, independently, that every collectionwise normal metacompact space is paracompact, and in 1958, McAuley showed that every collectionwise normal subparacompact space is paracompact ([Mi_2], [N_1], [Mc]); a simultaneous generalization of these results was obtained by Worrell and Wicke in 1965 in the theorem that every collectionwise normal submetacompact space is paracompact [WoW]. In light of the last-mentioned result, it seems possible that some suitable generalization of collectionwise normality could characterize subparacompact spaces in the class of submetacompact spaces; this is indeed the case, as has been shown by Y. Katuta and J. Chaber. In 1975, Katuta defined the concept of a subexpandable space and he proved that a space is subparacompact iff the space is submetacompact and subexpandable. Katuta also considered a generalization of subexpandability that he called "discrete subexpandability" [Ka_2]. In 1979, Chaber generalized Katuta's result by showing that every discretely subexpandable submetacompact space is subparacompact, and he renamed discretely subexpandable spaces "collectionwise subnormal" [Ch_3]. We'll now show that a generalization of collectionwise subnormality characterizes subparacompactness in the class of submetacompact spaces.

To state the following definition, let us recall that an *expansion* of a family $\mathcal{L}$ of subsets of X is a family $\{E(L) | L \in \mathcal{L}\}$ of subsets of X such that $L \subset E(L)$ for each $L \in \mathcal{L}$. An expansion $E(L)$ $L \in \mathcal{L}$ of $\mathcal{L}$ is a *disjoint expansion* of $\mathcal{L}$ provided that $E(L) \cap E(L') = \emptyset$ whenever L and L' are distinct members of $\mathcal{L}$. Open expansion, a G_δ-expansion etc. are defined in the obvious way.

Definition 2.5. A space is *collectionwise δ-normal* provided that every discrete family of closed subsets of the space has a disjoint G_δ-expansion.

Proposition 2.6. In a collectionwise δ-normal space, every point-finite open cover has a σ-discrete closed refinement.

Proof. Assume that X is collectionwise δ-normal. We start the proof by showing that for every point-finite open cover $\mathcal{V}$ of X, there exists a σ-discrete closed partial refinement $\mathcal{F}$ of $\mathcal{V}$ and a G_δ-subset A of X such that $\cup M(\mathcal{V}) \subset A \subset \cup\mathcal{F}$. Let $\mathcal{V}$ be a point-finite open cover of X. For every $V \in \mathcal{V}$, let $M(V) = V \sim \cup(\mathcal{V} \sim \{V\})$. Then $M(\mathcal{V}) = \{M(V) | V \in \mathcal{V}\}$. Let $\{B(V) | V \in \mathcal{V}\}$ be a disjoint G_δ-expansion of the discrete family $M(\mathcal{V})$. We may assume that for each $V \in \mathcal{V}$, we have $B(V) \subset V$, and then we can represent B(V) in the form $B(V) = \bigcap_{n\in\mathbb{N}} G_n(V)$ so that for each $n \in \mathbb{N}$, the set $G_n(V)$ is open and $G_{n+1}(V) \subset G_n(V) \subset V$. For every $n \in \mathbb{N}$, let $G_n = \cup\{G_n(V) | V \in \mathcal{V}\}$ and $S_n = X \sim G_n$. Let $K = \cup M(\mathcal{V})$. For every $n \in \mathbb{N}$, the family $\{K, S_n\}$ is discrete and hence there are open sets $Q_{n,k}$ and $W_{n,k}$, for $k \in \mathbb{N}$, such that $K \subset \bigcap_{k\in\mathbb{N}} Q_{n,k}$, $S_n \subset \bigcap_{k\in\mathbb{N}} W_{n,k}$ and $(\bigcap_{k\in\mathbb{N}} Q_{n,k}) \cap (\bigcap_{k\in\mathbb{N}} W_{n,k}) = \emptyset$. For all $n \in \mathbb{N}$ and $k \in \mathbb{N}$, the family $\mathcal{G}_{n,k} = \{W_{n,k} \cap V | V \in \mathcal{V}\} \cup \{G_n(V) | V \in \mathcal{V}\}$ is an open refinement of $\mathcal{V}$; consequently, the family $M(\mathcal{G}_{n,k})$ is a closed and discrete partial refinement of $\mathcal{V}$. Let $\mathcal{F} = \cup\{\mathcal{M}(\mathcal{G}_{n,k}) | n \in \mathbb{N}$ and $k \in \mathbb{N}\}$ and $A = \cap\{Q_{n,k} | n \in \mathbb{N}$ and $k \in \mathbb{N}\}$. Then A is a G_δ-set and $\cup\mathcal{M}(\mathcal{V}) \subset A$. To show that $A \subset \cup\mathcal{F}$, let $x \in A$. Let $\mathcal{N} = \{V \in (\mathcal{V})_x | x \notin B(V)\}$. There exists $m \in \mathbb{N}$ such that $x \notin \cup\{G_m(V) | V \in \mathcal{N}\}$. Note that $x \in G_m(V)$ for at most one $V \in \mathcal{V}$. Since $x \in A$, we have $x \in \bigcap_{k\in\mathbb{N}} Q_{m,k}$. It follows that there exists $j \in \mathbb{N}$ such that $x \notin W_{m,j}$. We have $x \in \cup\mathcal{M}(\mathcal{G}_{m,j})$, and hence $x \subset \cup\mathcal{F}$, as required.

To complete the proof, let $\mathcal{U}$ be a point-finite open cover of X. For every $n \in \mathbb{N}$, let $K_n = \{x \in X \,|\, |(\mathcal{U})_x| < n\}$. We use induction on n to show that for every $n \in \mathbb{N}$, there exists a G_δ-set A_n and a σ-discrete closed partial refinement $\mathcal{F}_n$ of $\mathcal{U}$ such that $K_n \subset A_n \subset \cup\mathcal{F}_n$. Choose $A_1 = \mathcal{F}_1 = \emptyset$. Assume that A_n and $\mathcal{F}_n$ have been defined. To define A_{n+1} and $\mathcal{F}_{n+1}$, represent A_n in the form $A_n = \bigcap_{k\in\mathbb{N}} Q_k$ where each Q_k is open. Let $\mathcal{W} = \{\cap\mathcal{V} | \mathcal{V} \subset \mathcal{U}$ and $|\mathcal{V}| = n\}$, and for each $k \in \mathbb{N}$, let $\mathcal{W}_k = \mathcal{W} \cup \{Q_k \cap U | U \in \mathcal{U}\}$. Note that for each $k \in \mathbb{N}$, the family $\mathcal{W}_k$ is an open refinement of $\mathcal{U}$ and $K_{n+1} \sim Q_k \subset \cup\mathcal{M}(\mathcal{W}_k)$. By the first part of this proof there exists, for each $k \in \mathbb{N}$, a G_δ-set B_k and a σ-discrete closed partial refinement $\mathcal{K}_k$ of $\mathcal{W}_k$ such that $\cup\mathcal{M}(\mathcal{W}_k) \subset B_k \subset \cup\mathcal{K}_k$. Let $A_{n+1} = \bigcap_{k\in\mathbb{N}} (Q_k \cup B_k)$ and $\mathcal{F}_{n+1} = \mathcal{F}_n \cup (\bigcup_{k\in\mathbb{N}} \mathcal{K}_k)$. Then A_{n+1} is a G_δ-set and $\mathcal{F}_{n+1}$ is a σ-discrete closed partial refinement of $\mathcal{U}$. For each $k \in \mathbb{N}$, we have $K_{n+1} \sim Q_k \subset \cup\mathcal{M}(\mathcal{W}_k) \subset B_k$ and hence $K_{n+1} \subset Q_k \cup B_k$; it follows that $K_{n+1} \subset A_{n+1}$. To show that $A_{n+1} \subset \cup\mathcal{F}_{n+1}$, let $x \in A_{n+1}$. If $x \in A_n$, then $x \in \cup\mathcal{F}_n \subset \cup\mathcal{F}_{n+1}$. Assume that $x \notin A_n$. Then there is $j \in \mathbb{N}$ such that $x \notin Q_j$. Since $A_{n+1} \subset Q_j \cup B_j$, we have $x \in B_j$ and hence $x \in \cup\mathcal{K}_j \subset \cup\mathcal{F}_{n+1}$. We have shown that $A_{n+1} \subset \cup\mathcal{F}_{n+1}$. This completes the proof for the induction. Since $\bigcup_{n\in\mathbb{N}} K_n = X$, the σ-discrete closed partial refinement $\bigcup_{n \in \mathbb{N}} \mathcal{F}_n$ of $\mathcal{U}$ is a refinement of $\mathcal{U}$. □

In 1955, Michael proved that in a collectionwise normal space, every point-finite open cover has a locally finite open refinement [Mi_2]. Note that Michael's result can be quite easily derived using Proposition 2.6.

We can now decompose subparacompactness into two other properties.

Theorem 2.7. A space is subparacompact iff the space is submetacompact and collectionwise δ-normal.

Proof. Necessity. Assume that X is subparacompact. We already know that X is then submetacompact. Let $\mathcal{F}$ be a nonempty discrete family of closed subsets of X. Then the family $\mathcal{G} = \{X \sim \cup(\mathcal{F} \sim \{F\}) \mid F \in \mathcal{F}\}$ is an (interior-preserving) open cover of X. Let $\mathcal{K}$ be a σ-discrete closed refinement of $\mathcal{G}$, and for each $F \in \mathcal{F}$, let $A(F) = X \sim \cup\{K \in \mathcal{K} \mid K \cap F = \emptyset\}$. The family $\{A(F) \mid F \in \mathcal{F}\}$ is a disjoint G_δ-expansion of $\mathcal{F}$.
Sufficiency. Use Proposition 2.6 together with the observation that collectionwise δ-normality is hereditary with respect to closed subsets. □

Collectionwise δ-normality and collectionwise subnormality [Ch_3] are related to a generalization of normality in the same way as collectionwise normality is related to normality. In [Kra], T.R. Kramer called a space subnormal if every finite open cover of the space has a countable closed refinement. In [Ch_3], Chaber independently, and with a definition differing from that used by Kramer, introduced subnormal spaces. We use a characterization of subnormality given by Chaber to define this property.

Definition 2.8 [Kra], [Ch_3]. A space is *subnormal* provided that whenever S and F are disjoint closed subsets of the space, there are disjoint G_δ-sets A and B in the space such that $S \subset A$ and $F \subset B$.

The property defined above could also be called δ-normality. Even though "subnormality" and "δ-normality" would just be different names for one concept, the difference in viewpoint leads to different "collectionwise" versions of this concept (see [Ch_3]).

A natural question to ask is whether the result of Theorem 2.7 would remain valid with "subnormal" substituted for "collectionwise δ-normal" (note that the space of Example 2.4 is not subnormal). The answer is in the negative, except for spaces with a small Lindelöf-number. In [Bu_4], Burke shows that Bing's famous normal, non-collectionwise normal space (Example G of [B]), if chosen "big enough," has a closed metacompact subspace Y that is not subparacompact. The Lindelöf-number of Y is bigger than $\underline{c}$ $(=2^\omega)$; the next results show that the number cannot be $\underline{c}$ or less.

Lemma 2.9. In a subnormal space, every discrete and closed family consisting of no more than $\underline{c}$ sets has a disjoint G_δ-expansion.

Proof. Let X be a subnormal space, and let $\mathcal{F}$ be a discrete family of closed subsets of X such that $|\mathcal{F}| \leq \underline{c}$. For every $\mathcal{K} \subset \mathcal{F}$, the sets $\cup\mathcal{K}$ and $\cup(\mathcal{F} \sim \mathcal{K})$ are closed and mutually disjoint; it follows that there is a family $\{A(\mathcal{K}) | \mathcal{K} \subset \mathcal{F}\}$ of G_δ-subsets of X such that for each $\mathcal{K} \subset \mathcal{F}$, we have $\cup\mathcal{K} \subset A(\mathcal{K})$ and $A(\mathcal{K}) \cap A(\mathcal{F} \sim \mathcal{K}) = \emptyset$. Since $|\mathcal{F}| \leq \underline{c}$, there is a collection $\{\mathcal{F}_n | n \in \mathbb{N}\}$ of subfamilies of $\mathcal{F}$ such that for each $F \in \mathcal{F}$, we have $\cap\{\mathcal{F}_n | n \in \mathbb{N}$ and $F \in \mathcal{F}_n\} = \{F\}$. Let $\mathcal{A} = \{\mathcal{A}(\mathcal{F}_n) | n \in \mathbb{N}\} \cup \{\mathcal{A}(\mathcal{F} \sim \mathcal{F}_n) | n \in \mathbb{N}\}$. It is easily seen that if we set $A(F) = \cap\{A \in \mathcal{A} | F \subset A\}$ for each $F \in \mathcal{F}$, then the family $\{A(F) | F \in \mathcal{F}\}$ is a disjoint G_δ-expansion of $\mathcal{F}$. □

The argument employed in the above proof, that every set of cardinality $\leq c$ has a countable "point-separating" family of subsets, has been used to prove results on covering properties of normal spaces in [RZ] and [Gr].

The result of Lemma 2.9 and the proof of Proposition 2.6 establish the following result.

Proposition 2.10. In a subnormal space, every point-finite open cover of cardinality $\leq c$ has a σ-discrete closed refinement.

Corollary 2.11. In a subnormal submetacompact space, every open cover of cardinality $\leq c$ has a σ-discrete closed refinement.

Burke's example shows that a subnormal metacompact space may fail to be subparacompact. However, the following problem is open.

Problem 2.12. Is every first countable subnormal submetacompact space subparacompact?

The problem seems to remain open even if "subnormal" is omitted; it was stated in the above form since there does not seem to be any reason why all first countable submetacompact spaces would have to be subparacompact.

To obtain his famous "provisional" solution to the Normal Moore Space Problem, Nyikos showed in [Ny_2] that if the Product Measure Extension Axiom (PMEA) holds, then every first countable normal space is collectionwise normal. In [Ju_7], a modification of Nyikos's proof is used to show that under PMEA, every first countable subnormal space is collectionwise δ-normal; it follows, by Theorem 2.7, that under PMEA, Problem 2.12 has a positive solution. This result makes it seem unlikely that there would exist any "real" example of a first countable subnormal submetacompact space that is not subparacompact; so far, not even any "consistent" example of such a space has been found.

Before turning to consider other kinds of characterizations of subparacompactness, we mention a necessary and sufficient condition, different from that given in Theorem 2.7, for a submetacompact space to be subparacompact. The following result has recently been obtained by Burke; the result is a consequence of a result mentioned after Corollary 1.18 above and a characterization of subparacompactness earlier obtained by Burke (see Theorem 2.15 below).

Theorem 2.13 [Bu_5]. A space is subparacompact iff the space is submetacompact and every open cover of the space has a σ-locally countable closed refinement.

Note that for the ordinal space ω_1, the family $\{\{\alpha\} | \alpha < \omega_1\}$ is locally countable and closed; hence we cannot omit "submetacompact" from the above theorem.

It is an easy consequence of Lemma 0.4 that if X is a submetacompact space and X has countable spread locally, then every open cover of X has a locally countable refinement; hence, by Theorem 2.13, a regular submetacompact space is subparacompact provided that the space has countable spread locally.

b) Characterizations

The following characterizations of subparacompactness follow easily from Theorem 2.7 and the corresponding characterizations of submetacompactness (Theorems 1.5, 1.7 and 1.10, respectively; for the third result below, use also Lemma 1.9).

Theorem 2.14. A space is subparacompact iff every open cover of the space has a pointwise star-refining sequence by open covers of the space.

The necessity of the above condition for subparacompactness was proved, independently, by Čoban [Co_1] and Burke and Stoltenberg [BuS], and the sufficiency by Burke ([Bu_1]; essentially the same result was announced by Worrell in [Wo_3]).

Note that Theorem 2.14 gives as a corollary Bing's result on subparacompactness of developable spaces.

Theorem 2.15. A space is subparacompact iff every interior-preserving open cover of the space has a σ-closure-preserving closed refinement.

In the above form, the result of Theorem 2.15 was obtained by this author in [Ju_5]; with "interior-preserving" omitted, the result was earlier obtained by Burke [Bu_1]. As observed by Burke in [Bu_1], it follows from the above result that a space is subparacompact iff every open cover of the space has a σ-locally finite closed refinement.

Theorem 2.16 [Ju_5]. A space is subparacompact iff every open cover of the space has a σ-cushioned refinement.

Tamano's Theorem [Ta] and the parallels between the theory of paracompactness and the theory of subparacompactness motivate the following problem.

Problem 2.17. If X is a Tychonoff-space and the product space $X \times \beta X$ is subnormal, is X then necessarily subparacompact?

c) Subclasses

We now indicate some special subclasses of the class of subparacompact spaces. First, we consider perfectly subparacompact spaces, i.e. spaces that are both perfect and subparacompact. The following result shows that there is no need to talk about perfectly submetacompact spaces.

Theorem 2.18. The following conditions are mutually equivalent for a topological space:

(i) The space is perfectly subparacompact.

(ii) The space is submetacompact and locally perfect.

(iii) Every scattered partition of the space has a σ-discrete closed refinement.

Proof. (ii) $\Rightarrow$ (iii): Assume that X is submetacompact and locally perfect. Let $\mathcal{P}$ be a scattered partition of X. By Lemma 1.15, there exists a well-order $\leq$ on $\mathcal{P}$ such that for each $P \in \mathcal{P}$, the set $\cup\{P' \in \mathcal{P} | P' \leq P\}$ is open. It is easily seen that there exists an ordinal number γ and an open cover $\mathcal{V} = \{V_\alpha | \alpha < \gamma\}$ of X such that each member of $\mathcal{V}$ is perfect as a subspace of X and for each $P \in \mathcal{P}$, there exists $\alpha < \gamma$ such that $\bigcup_{\beta<\alpha} V_\beta = \cup\{P' \in \mathcal{P} | P' \leq P\}$. Denote by $\mathcal{D}$ the scattered partition $\{V_\alpha \sim \bigcup_{\beta<\alpha} V_\beta | \alpha < \gamma\}$ of X and note that $\mathcal{D}$ is a refinement of $\mathcal{P}$. For each $\alpha < \gamma$, the closed subset $V_\alpha \sim \bigcup_{\beta<\alpha} V_\beta$ of V_α is a G_δ-set in V_α and hence in X. By Theorem 1.16, $\mathcal{D}$ has a σ-discrete closed refinement.

(iii) $\Rightarrow$ (i): Assume that X satisfies condition (iii). Then X is subparacompact. If U is an open subset of X, then the family $\{U, X \sim U\}$ is a scattered partition of X; if $\mathcal{F}$ is a

σ-discrete closed refinement of this partition, then $U = \cup\{F \in \mathcal{F} | F \subset U\}$ and hence U is an F_σ-set. □

The equivalence of conditions (i) and (iii) above was established in [ChZ]; the equivalence of (i) and (ii), with "locally" omitted from (ii), follows from the lemma in [WoW].

Note that not all perfect spaces, and not even all perfectly normal spaces are subparacompact; in [Po], there is an example of a non-subparacompact space that is perfectly normal and locally metrizable.

We already know from McAuley's result that every semimetrizable space is subparacompact; an even larger class of (perfectly) subparacompact spaces is indicated in the following result; the result was obtained by G. Creede and J.A. Kofner who, independently, introduced the concept of a semi-stratifiable space.

Theorem 2.19 [Cr], [Ko]. Every semi-stratifiable space is subparacompact.

Among regular spaces, an interesting class of subparacompact spaces is formed by the strong Σ-spaces of Nagami [N_2]. Strong Σ-spaces can be characterized as those spaces that have a σ-locally finite closed refiner for directed open covers (see Section 1.c); from this characterization and from Burke's results on subparacompact spaces, we can infer the following result.

Proposition 2.20 [Mi_6]. Every regular strong Σ-space is subparacompact.

Note that it follows from Proposition 2.20 that every regular strong Σ-space has a σ-discrete closed refiner for directed open covers.

The principal reason for the importance of strong Σ-spaces is that these spaces have nice multiplicative properties: the product of countably many regular (paracompact, metacompact) strong Σ-spaces is a (paracompact, metacompact) strong Σ-space, and the product of a regular strong Σ-space with a perfectly (sub)paracompact regular space is perfectly (sub)paracompact (see [N_2] and [L]). The class of strong Σ-spaces is rather large: it contains all compact spaces, all σ-spaces and all subparacompact p-spaces.

3. *Metacompactness*

The concept of a metacompact space was introduced by Arens and Dugundji in 1950 [AD] and, independently, by Bing in 1951 [B] (Bing called metacompact spaces "pointwise paracompact"). Arens and Dugundji showed that every countably compact metacompact space is compact. Bing gave an example of a metacompact developable space that is not paracompact. Among the early results on metacompact spaces, the most important one is the theorem, obtained independently by Michael and Nagami in 1955 [Mi_2], [N_1], that every collectionwise normal metacompact space is paracompact. The first significant results on metacompactness as such, and not on the relations of this property to other properties, were obtained by Worrell in 1966 [Wo_1] and [Wo_2]. Worrell showed that metacompactness admits a characterization similar to that given by A.H. Stone for paracompactness. In 1971, W.B. Sconyers characterized metacompactness by a condition similar to a condition used by Alexandrov and Urysohn to characterize compactness [S].

Among metacompact spaces, those that are developable are of special interest. Metacompact developable T_1-spaces constitute a natural generalization of metrizable spaces; these spaces can be characterized as the continuous images of metrizable spaces under open, compact mappings (Hanai, 1961; Arhangel'skii, 1962) or as the spaces with a uniform base (Alexandrov, 1960). In 1964, R.W. Heath gave several results and examples on the relationship between metacompactness and screenability in developable spaces [He]. Metacompactness has had its role in the attempts to solve the Normal Moore Space Problem. Heath showed in [He] that if every metacompact normal Moore space is metrizable, then so is every separable normal Moore space. By subsequent results, it followed from Heath's result that the existence of a non-metrizable metacompact normal Moore space is consistent with the usual axioms of set-theory [T_1].

a) Characterizations

The following result can be used to construct point-finite and locally finite open refinements for certain open covers.

Proposition 3.1. Let $\langle \mathcal{U}_n \rangle$ be a sequence of open covers of a space such that for each $n \in \mathbb{N}$, $\mathcal{U}_{n+1}$ is a pointwise (local) W-refinement of $\mathcal{U}_n$. Then $\mathcal{U}_1$ has a point-finite (locally finite) open refinement.

The "pointwise" part of the above result extends a result implicit in Worrell's proof of his characterization of metacompactness ([Wo_1]; see also [Wo_4]).

The following result was obtained by Sconyers in 1971 (for a proof of the result, see [Ju_3]).

Theorem 3.2 [S]. A space is metacompact iff every well-monotone open cover of the space has a point-finite open refinement.

Using the above results, we could derive several characterizations for metacompactness in the same way as we derived characterizations for submetacompactness from Proposition 1.4 and Theorem 1.6; however, the characterizations can also be derived from those obtained for submetacompactness if we first find a necessary and sufficient condition for a submetacompact space to be metacompact. Such a condition has been found by J.C. Smith and L. Krajewski in 1971 [SmiK]. In 1973, J. Boone characterized metacompactness in the class of submetacompact spaces by a condition weaker than that used by Smith and Krajewski (Boone's result can be obtained from the result of Smith and Krajewski by using Proposition 1.11 and a result from [SmiK]).

Theorem 3.3 [Bo_1]. A space is metacompact iff the space is submetacompact and every discrete family of closed subsets of the space has a point-finite open expansion.

(A *point-finite expansion* of a family $\mathcal{L}$ of sets is an expansion $\{E(L) | L \in \mathcal{L}\}$ of $\mathcal{L}$ such that for every point x, the family $\{L \in \mathcal{L} | x \in E(L)\}$ is finite; locally finite expansion is defined similarly).

The next two theorems follow easily from Theorem 3.3 and the corresponding results from Section 1.a. The first result was obtained by Worrell in 1966; the second result was obtained by this author.

Theorem 3.4 [Wo_1]. A space is metacompact iff every open cover of the space has an open pointwise W-refinement.

By the above theorem and Lemma 0.1, we have the following result [Ju_2]: a space is metacompact if every open cover of the space has a point-finite semi-open refinement. A more general result follows from Theorems 3.3 and 1.10: a space is metacompact if every open cover of the space has a semi-open pointwise W-refinement.

It is not known whether a space has to be metacompact if every directed open cover of the space has a cushioned refinement (this problem was raised in [Ka_2]); however, the following result obtains (to prove that (i) ⇒ (iii), use Corollary 0.3).

Theorem 3.5 [Ju_3]. The following conditions are mutually equivalent for a topological space:

(i) The space is metacompact.

(ii) Every interior-preserving directed open cover of the space has a closure-preserving closed refinement.

(iii) Every directed open cover of the space has a closure-preserving closed refinement.

According to Tamano's Theorem [Ta], a Tychonoff-space X is paracompact iff the product space $X \times \beta X$ is normal. In [Ju_3], the result of Theorem 3.5 is used to obtain the following analogue of Tamano's Theorem: A Tychonoff-space X is metacompact iff the product space $X \times \beta X$ is orthocompact (a space is *orthocompact* if every open cover of the space has an interior-preserving open refinement; see [FL_1]). In [Sc_1], B. Scott characterized metacompactness of a space X in terms of orthocompactness of the product of X with certain compact spaces that are not necessarily compactifications of X; many of the results obtained by Scott in [Sc_1] are analogues of earlier results concerning normality of products with a paracompact factor.

In 1962, Arhangel'skii showed that the continuous image of a paracompact space under an open, compact mapping is metacompact [Ar_1]; in [Ju_2], a result mentioned after Theorem 3.4 was used to show that Arhangel'skii's result remains valid if "pseudo-open" is substituted for "open" in the statement of the result. There does not yet exist any characterization of those spaces that can be represented as continuous images of paracompact spaces under (pseudo-) open compact mappings; in particular, the following problem remains unsolved.

Problem 3.6. Is every metacompact space the continuous image of a paracompact space under a (pseudo-) open compact mapping?

For open mappings, the problem was first stated by Arhangel'skii. Some results relating to the problem can be found in [Ar_1], [H], [Co_2] and [Ju_6].

One of the best-known examples of a non-paracompact metacompact space is the product space formed by the irrationals and the space known as the "Michael line" (see [Mi_5], or Example V.2 of [Na]). Even for this simple space it seems to remain unknown whether the space can be represented as the continuous image of a paracompact space under an open compact mapping (however, it is easily seen that the space is the continuous image of a paracompact space under a pseudo-open finite-to-one mapping).

Before turning to consider some special metacompact spaces, we indicate certain properties implied by metacompactness.

According to the result of Corollary 1.13(i), every countably compact submetacompact space is compact. A stronger result has recently been obtained for metacompact (Tychonoff) spaces by Scott [Sc_2]: every pseudocompact metacompact space is compact.

It is easily seen that if a partition of a metacompact space has a σ-discrete closed refinement, then the partition has a point-finite open expansion; consequently, by Proposition 1.16, the following result obtains.

Proposition 3.7. In a metacompact space, every scattered partition of the space into G_δ-sets has a point-finite open expansion.

b) Some metacompact spaces

A strengthening of the conclusion of Proposition 3.7 characterizes those spaces in which every subspace is metacompact.

Proposition 3.8. A space is hereditarily metacompact iff every scattered partition of the space has a point-finite open expansion.

In the particular case of a scattered space, the above result was obtained by Nyikos [Ny_1].

In the introduction to this chapter, we mentioned some characterizations of metacompact developable spaces. For some further characterizations of these spaces, see [Ar_5], [A] and [Co_1].

For metacompact semi-stratifiable spaces, the following result obtains.

Proposition 3.9 [Ju_1]. Let $\mathcal{U}$ be an interior-preserving open cover of a metacompact semi-stratifiable space. Then there exists a point-finite open cover $\mathcal{V}$ of the space such that for every $U \in \mathcal{U}$, $U = \cup\{V \in \mathcal{V} | V \subset U\}$.

Note that every space for which the conclusion of Proposition 3.9 holds is hereditarily metacompact; however, as wittnessed by the Sorgenfrey line, the conclusion does not hold for all hereditarily metacompact spaces.

Next we indicate a base property that implies metacompactness. We say that a family $\mathcal{L}$ of sets is of *point-finite rank* if for each $x \in \cup\mathcal{L}$ there exists $n \in \mathbb{N}$ such that for every $\mathcal{L}' \subset (\mathcal{L})_x$, if $|\mathcal{L}'| > n$, then $\mathcal{L}'$ contains two members that are related by inclusion; the family $\mathcal{L}$ is of *sub-infinite rank* provided that for every infinite subfamily $\mathcal{L}'$ of $\mathcal{L}$, if $\cap\mathcal{L}' \neq \emptyset$, then $\mathcal{L}'$ contains two members that are related by inclusion [GrN].

The following result is due to G. Gruenhage and P. Nyikos.

Proposition 3.10 [GrN]. If a T_1-space has a base of point-finite rank, then the space is metacompact.

Note that every uniform base [Al] is of sub-infinite rank, this observation makes the following problem interesting.

Problem 3.11 [GrN]. If a T_1-space has a base of sub-infinite rank, is the space then necessarily metacompact?

We consider one more subclass of metacompact spaces. The following result, obtained by Y. Katuta and, independently, by H. Potoczny and this author, can be easily derived from the result of Theorem 3.5.

Proposition 3.12 [Ka_1] and [PotJu]. If a space has a closure-preserving cover by compact, closed subsets, then the space is metacompact.

For the sake of brevity, let us say that a family of subsets of a space is a *cpc-family* (*cpf-family*) if the family is closure-preserving and it consists of compact (finite) closed sets. Spaces that have a cpc-cover have been found to have many interesting properties; many of the results dealing with these spaces follow from certain structural properties of cpc-families that were discovered by Potoczny in [Pot]. The following remarkable result was obtained by Telgársky [Te_1]: if a regular paracompact space Y has a σ-cpc-cover, then for every regular paracompact space X, the product space Y × X is paracompact. It is not known whether Telgársky's result remains valid if "paracompact" is replaced "metacompact." In [Y_1], Y. Yajima shows that if a regular paracompact space has a σ-cpc-cover, then every base of the sapce contains a locally finite cover. Using Yajima's technique of proof and Proposition 3.12, one can show that the following result holds: if a space has a cpc-cover, then every base of the space contains a point-finite cover.

Among spaces which have a cpc-cover, we have all locally compact metacompact spaces (by Corollary 0.3) and, more generally, all metacompact spaces that admit cpc-covers locally (to establish this result, use the complementary interior-preserving open families and note that any point-finite sum of interior-preserving open families is interior-preserving). Telgársky has shown that every hereditarily paracompact scattered space has a cpf-cover [Te_2]; from a result of Nyikos [Ny_1] mentioned after Proposition 3.8, it follows (see [Ju_6]) that every hereditarily metacompact scattered space has a cpf-cover. Note that for any space X, the Pixley-Roy hyperspace $\mathcal{F}[X]$ of X (see [vD]) has a cpf-cover. For characterizations of spaces that admit a cpf-cover, see [Y_2] and [Ju_6].

4. *Relations to paracompactness*

Besides compactness, paracompactness is the most important of all topological covering properties. Consequently, research on various generalization of paracompactness is to a large extent motivated by the fact that results obtained for these more general properties are often helpful in establishing paracompactness of certain spaces (however, as we hope the results presented in this paper demonstrate, generalizations of paracompactness are also of independent interest). In this section, we indicate some conditions under which a submetacompact, subparacompact, or metacompact space is paracompact.

We start with a characterization of paracompactness in the class of submetacompact spaces.

Theorem 4.1 [Kr]. A space is paracompact iff the space is submetacompact and every locally finite family of closed subsets of the space has a locally finite open expansion.

The above result is due to Krajewski. Spaces with the property that every locally finite family of closed subsets of the space has a locally finite open expansion were first considered by Katětov [K], who showed that a normal space has this property iff the space is collectionwise normal and countable paracompact. Since every normal submetacompact space is countably paracompact (this follows from the result of Proposition 1.11), the results of Krajewski and Katětov can be used to derive the result, stated by Worrell and Wicke in [WoW], that every collectionwise normal submetacompact space is paracompact. The result of Worrell and Wicke can also be derived from Theorem 2.7 and the result of McAuley [Mc] that every collectionwise normal subparacompact space is paracompact.

For some extensions of the result of Theorem 4.1, see [SmiK] and [Smi_2].

With the help of Theorem 4.1, one can derive several characterizations for paracompactness from those given above for submetacompactness. For example, many of the characterizations of paracompactness obtained by Michael in [Mi_1], [Mi_3] and [Mi_4], as well as the extensions of some of Michael's results obtained by this author in [Ju_3], can be easily derived this way. From Theorems 4.1 and 1.5, we obtain the result, announced by Worrell in [Wo_4], that a space is paracompact iff every open cover of the space has an open local W-refinement. Alternatively, to derive these characterizations, one can use Theorem 3.1 and the result of J. Mack that a space is paracompact iff every well-monotone open cover of the space has a locally finite open refinement [M]. Of course, many characterizations of paracompactness, e.g. those given in terms of the existence of continuous pseudometrics or partitions of unity, do not fall within the framework of the theory presented here, and these characterizations cannot be obtained as direct corollaries to results dealing with more general properties.

In the remaining results of this section, we indicate some sufficient conditions for paracompactness. The first result follows from Lemma 0.4 and Yu M. Smirnov's result that a regular space is (strongly) paracompact provided that every open cover of the space has a star-countable open refinement [Sm].

Proposition 4.2. A regular metacompact space is paracompact provided that either the space has countable spread locally or the space is a Baire-space and it satisfies ccc locally.

Note that a metacompact space satisfying ccc is not necessarily paracompact (see [PR]), even if the space is normal (see [PrT] and [Ju_4]).

The part of the above result dealing with Baire-spaces is due to Tall [T_2]. Both parts of Proposition 4.2 imply the result of Arhangel'skii that every perfectly normal, locally compact metacompact space is paracompact [Ar_4]. Arhangel'skii's result motivates the following question, raised by Tall.

Problem 4.3. Is every normal, locally compact metacompact space paracompact?

The following result, obtained recently by Gruenhage, is considerably deeper than the result of Proposition 4.2. The result shows that if there exists a space giving a negative solution to Problem 4.3, then such a space cannot be locally connected.

Theorem 4.4 [Gr]. A locally connected normal space is paracompact provided that the space is either submetacompact and locally compact or subparacompact and rim-compact.

(Recall that a space is *rim-compact* if the space has a base by sets whose boundaries are compact).

Theorem 4.4 extends certain results on paracompactness of locally connected, perfectly normal subparacompact spaces obtained by G.M. Reed and P.L. Zenor in [RZ] and by Chaber and Zenor in [ChZ]. The result of the theorem leaves the following question unanswered.

Problem 4.5 [Gr]. Is every locally connected, normal, submetacompact rim-compact space paracompact?

In Theorem 4.4, the assumptions of subparacompactness and submetacompactness cannot be omitted; this follows from an example, constructed under CH by M.E. Rudin and Zenor in [RZ], of a non-paracompact, perfectly normal locally Euclidean space. However, for perfectly normal spaces, we can replace the assumptions in question by a set-theoretic assumption: Rudin showed in [Ru] that under MA + ¬CH, every perfectly normal locally Euclidean space is paracompact, and subsequently, D. Lane modified Rudin's proof to show that under MA + ¬CH, every perfectly normal, locally connected, locally compact space is paracompact. Both Rudin's proof of her result and Gruenhage's proof of the result of Theorem 4.4 are based on techniques devised by Reed and Zenor in [RZ].

ACKNOWLEDGEMENTS. This paper is based on talks that the author gave in seminars at Ohio University and the University of Pittsburgh during the academic year 1978-79. The comments made by the students and faculty attending those seminars helped the author much in the task of writing this paper.

While writing this paper, the author was visiting the University of Pittsburgh as an Andrew Mellon Postdoctoral Fellow.

References

[A] C.C. Alexander, *Semi-developable spaces and quotient images of metric spaces,* Pacific J. Math. 37 (1971), 277-293.

[Al] P.S. Alexandrov, *On metrization of topological spaces,* Bull. Acad. Polon. Sci. Sér. Math. 8 (1960), 135-140. (In Russian).

[AU] P.S. Alexandrov and P. Urysohn, *Mémoire sur les espaces topologiques compacts,* Verh. Nederl. Akad. Wetensch. (Amsterdam) 14 (1929), 1-96.

[AE] K. Alster and R. Engelking, *Subparacompactness and product spaces,* Bull. Acad. Polon. Sci. Sér. Math. 20 (1972), 763-767.

[Aq] G. Aquaro, *Point-countable open coverings in countably compact spaces,* General Topology and its Relations to Modern Analysis and Algebra II; Proc. Second Prague Topology Symposium, 1966 (Academie, Prague; Academic Press, New York, 1967), 39-41.

[AD] R. Arens and J. Dugundji, *Remark on the concept of compactness,* Portugal. Math. 9 (1950), 141-143.

[Ar_1] A.V. Arhangel'skii, *On mappings of metric spaces,* Soviet Math. Dokl. 3 (1962), 953-956.

[Ar_2] ________, *On a class of spaces containing all metric and all locally bicompact spaces,* Soviet Math. Dokl. 4 (1963), 1051-1053.

[Ar_3] ________, *Mappings and spaces,* Russian Math. Surveys, 21 (1966), 115-162.

[Ar_4] ________, *The property of paracompactness in the class of perfectly normal, locally bicompact spaces,* Soviet Math. Dokl. 12 (1971), 1253-1257.

[Ar_5] ________, *The intersection of topologies, and pseudo-open compact mappings,* Soviet Math. Dokl. 17 (1976), 160-163.

[Au] C.E. Aull, *A generalization of a theorem of Aquaro*, Bull. Austral. Math. Soc. 9 (1973), 105-108.

[B] R.H. Bing, *Metrization of topological spaces*, Canad. J. Math. 3 (1951), 175-186.

[Bo_1] J.R. Boone, *A characterization of metacompactness in the class of θ-refinable spaces*, Gen. Topology and Appl. 3 (1973), 253-264.

[Bo_2] ________, *On irreducible spaces II*, Pacific J. Math. 62 (1976), 351-357.

[Bu_1] D.K. Burke, *On subparacompact spaces*, Proc. Amer. Math. Soc. 23 (1964), 655-663.

[Bu_2] ________, *On p-spaces and ωΔ-spaces*, Pacific J. Math. 35 (1970), 285-296.

[Bu_3] ________, *Subparacompact spaces*, Proc. Washington State Univ. Conference on General Topology, 1970 (Pi Mu Epsilon, Washington Alpha Center, 1970), 39-49.

[Bu_4] ________, *A note on R.H. Bing's Example G*, Topology Conference, Virginia Polytechnic Institute and State University, 1973 (Lecture Notes in Math. 375, Springer-Verlag, Berlin, 1974), 47-52.

[Bu_5] ________, *Refinements of locally countable collections*, preprint [Announcement in Notices Amer. Math. Soc. 26 (1979), A-394].

[BuL] D.K. Burke and D.J. Lutzer, *Recent advances in the theory of generalized metric spaces*, Topology, Proc. Memphis State Univ. Conference (Marcel Dekker, Inc., New York, 1976), 1-70.

[BuS] D.K. Burke and R.A. Stoltenberg, *A note on p-spaces and Moore spaces*, Pacific J. Math. 30 (1969), 601-608.

[Ch_1] J. Chaber, *Conditions which imply compactness in countably compact spaces*, Bull. Acad. Polon. Sci. Sér. Math. 24 (1976), 993-998.

[Ch_2] ________, *Metacompactness and the class MOBI*, Fund. Math. 91 (1976), 211-217.

[Ch_3] ________, *On subparacompactness and related properties*, Gen. Topology and Appl. 10 (1979), 13-17.

[ChJ] J. Chaber and H. Junnila, *On θ-refinability of strict p-spaces*, to appear in Gen. Topology and Appl.

[ChZ] J. Chaber and P. Zenor, *On perfect subparacompactness and a metrization theorem for Moore spaces*, Topology Proc. 2 (1977), 401-407.

[Co_1] M.M. Čoban, *On σ-paracompact spaces*, Moscov Univ. Math. Bull. 24 (1969), 11-14.

[Co_2] ________, *On the theory of p-spaces*, Soviet Math. Dokl. 11 (1970), 1257-1260.

[Cr] G.D. Creede, *Concerning semi-stratifiable spaces*, Pacific J. Math. 32 (1970), 47-54.

[D] S.W. Davis, *A cushioning-type weak covering property*, to appear in Pacific J. Math.

[vD] E.K. van Douwen, *The Pixley-Roy topology in spaces of subsets*, Set Theoretic Topology (Academic Press, Inc., New York, 1977), 111-134.

[vDW] E.K. van Douwen and H.H. Wicke, *A real, weird topology on the reals*, Houston J. Math. 3 (1977), 141-152.

[E] R. Engelking, *General Topology* (PWN, Warszawa, 1977).

[FR] W.G. Fleissner and G.M. Reed, *ParaLindelöf spaces and spaces with a σ-locally countable base*, Topology Proc. 2 (1977), 89-110.

[FL_1] P. Fletcher and W.F. Lindgren, *Transitive quasi-uniformities*, J. Math. Anal. and Appl. 39 (1972), 397-405.

[FL_2] ________, *A note on spaces of second category*, Arch. Math. 24 (1973), 186-187.

[G_1] R.F. Gittings, *Some results on weak covering conditions*, Canad. J. Math. 26 (1974), 1152-1156.

[G_2] ________, *Open mapping theory*, Set Theoretic Topology (Academic Press, Inc., New York, 1977), 141-191.

[Gr] G. Gruenhage, *Paracompactness in normal, locally connected, locally compact spaces*, preprint [Announcement in Notices Amer. Math. Soc. 26 (1979), A-285].

[GrN] G. Gruenhage and P.J. Nyikos, *Spaces with bases of countable rank*, Gen. Topology and Appl. 8 (1978), 233-257.

[H] S. Hanai, *On open mappings II*, Proc. Japan Acad. 37 (1961), 233-238.

[He] R.W. Heath, *Screenability, pointwise paracompactness and metrization of Moore spaces*, Canad. J. Math. 16 (1964), 763-770.

[Ho] R.E. Hodel, *Sum theorems for topological spaces*, Pacific J. Math. 30 (1969), 59-65.

[J] P. Jain, *Subparacompact spaces*, Math. Student 40 (1972), 231-249.

[Ju_1] H.J.K. Junnila, *Neighbornets*, Pacific J. Math. 76 (1978), 83-108.

[Ju_2] ________, *Paracompactness, metacompactness, and semi-open covers*, Proc. Amer. Math. Soc. 73 (1979), 244-248.

[Ju_3] ________, *Metacompactness, paracompactness, and interior-preserving open covers*, Trans. Amer. Math. Soc. 249 (1979), 373-385.

[Ju_4] ________, *On countability of point-finite families of sets*, to appear in Canad. J. Math.

[Ju_5] ________, *On submetacompactness*, Topology Proc. 3 (1978), 375-405.

[Ju_6] ________, *Stratifiable pre-images of topological spaces*, to appear in Proc. Budapest Topology Colloq., 1978.

[Ju_7] ________, *Some topological consequences of the Product Measure Extension Axiom*, in preparation.

[K] M. Katětov, *Extension of locally finite coverings*, Colloq. Math. 6 (1958), 145-151. (In Russian).

[Ka_1] J. Katuta, *On spaces which admit closure-preserving covers by compact sets*, Proc. Japan Acad. 50 (1974), 826-828.

[Ka_2] ________, *Expandability and its generalizations*, Fund. Math. 87 (1975), 231-250.

[Ko] J.A. Kofner, *On pseudostratifiable spaces*, Fund. Math. 70 (1971), 25-47. (In Russian).

[Kr] L.L. Krajewski, *Expanding locally finite collections*, Canad. J. Math. 23 (1971), 58-68.

[Kra] T.R. Kramer, *A note on countably subparacompact spaces*, Pacific J. Math. 46 (1973), 209-213.

[L] D.J. Lutzer, *Another property of the Sorgenfrey line*, Compos. Math. 24 (1972), 359-363.

[M] J. Mack, *Directed covers and paracompact spaces*, Canad. J. Math. 19 (1967), 649-654.

[Mc] L.F. McAuley, *A note on complete collectionwise normality and paracompactness*, Proc. Amer. Math. Soc. 9 (1958), 796-799.

[Mi_1] E.A. Michael, *A note on paracompact spaces*, Amer. Math. Soc. 4 (1953), 831-838.

[Mi_2] E.A. Michael, *Point-finite and locally finite coverings*, Canad. J. Math. 7 (1955), 275-279.

[Mi_3] ________, *Another note on paracompact spaces*, Proc. Amer. Math. Soc. 8 (1958), 822-828.

[Mi_4] ________, *Yet another note on paracompact spaces*, Proc. Amer. Math. Soc. 10 (1959), 309-314.

[Mi_5] ________, *The product of a normal space and a metric space need not be normal*, Bull. Amer. Math. Soc. 69 (1963), 375-376.

[Mi_6] ________, *On Nagami's Σ-spaces and some related matters*, Proc. Washington State Univ. Conference on General Topology, 1970 (Pi Mu Epsilon, Washington Alpha Center, 1970), 13-19.

[N_1] K. Nagami, *Paracompactness and strong screenability*, Nagoya Math. J. 8 (1955), 83-88.

[N_2] ________, *Σ-spaces*, Fund. Math. 65 (1969), 169-192.

[Na] J. Nagata, *Modern General Topology*, North-Holland Publ. Co., Amsterdam, 1974.

[Ny_1] P.J. Nyikos, *Covering properties on σ-scattered spaces*, Topology Proc. 2 (1977), 509-542.

[Ny_2] ________, *A provisional solution of the Normal Moore Space Problem*, to appear in Proc. Amer. Math. Soc.

[P] W.K. Pfeffer, *Integrals and measures*, Marcel Dekker, Inc., New York, 1977.

[PR] C. Pixley and P. Roy, *Uncompletable Moore spaces*, Proc. Auburn Topology Conference, 1969 (Auburn, Alabama), 75-85.

[Po] R. Pol, *A perfectly normal locally metrizable non-paracompact space*, Fund. Math. 97 (1977), 37-42.

[Pot] H. Potoczny, *Closure-preserving families of compact sets*, Gen. Topology and Appl. 3 (1973), 243-248.

[PotJu] H. Potoczny and H.J.K. Junnila, *Closure-preserving families and metacompactness*, Proc. Amer. Math. Soc. 53 (1975), 523-529.

[Pr] T.C. Przymusiński, *Normality and paracompactness in finite and countable cartesian products*, to appear in Fund. Math.

[PrT] T.C. Przymusiński and F.D. Tall, *The undecidability of the existence of a non-separable normal Moore space satisfying the countable chain condition*, Fund. Math. 85 (1974), 291-297.

[RZ] G.M. Reed and P.L. Zenor, *Metrization of Moore spaces and generalized manifolds*, Fund. Math. 91 (1976), 203-210.

[Ru] M.E. Rudin, *The undecidability of the existence of a perfectly normal non-metrizable manifold*, preprint.

[RuZ] M.E. Rudin and P.L. Zenor, *A perfectly normal non-metrizable manifold*, Houston J. Math. 2 (1976) 129-134.

[S] W.B. Sconyers, *Metacompact spaces and well-ordered open coverings*, Notices Amer. Math. Soc. 18 (1970), 230.

[Sc_1] B.M. Scott, *Toward a product theory for orthocompactness*, Studies in Topology (Academic Press, New York, 1975), 517-537.

[Sc_2] ________, *Pseudocompact, metacompact spaces are compact*, preprint.

[Si] M.K. Singal, *Some recent work on paracompact spaces*, Math. Student 38 (1970), 139-164.

[Sm] Yu M. Smirnov, *On strongly paracompact spaces*, Izv. Akad. Nauk SSSR, Math. Ser. 20 (1956), 253-274. (In Russian).

[Smi_1] J.C. Smith, *A remark on irreducible spaces*, Proc. Amer. Math. Soc. 57 (1976), 133-139.

[Smi_2] ________, *On Ⓗ-expandable spaces*, Glasnik Mat. 11 (1976), 335-346.

[SmiK] J.C. Smith and L.L. Krajewski, *Expandability and collectionwise normality*, Trans. Amer. Math. Soc. 160 (1971), 437-451.

[St] A.H. Stone, *Paracompactness and product spaces*, Bull. Amer. Math. Soc. 54 (1948), 977-982.

[T_1] F.D. Tall, *Set-theoretic consistency results and topological theorems concerning the Normal Moore Space Conjecture and related problems*, Thesis, University of Wisconsin, 1969 [Published with some changes in Dissertationes Math. 148 (1977).]

[T_2] ________, *The countable chain condition versus separability--applications of Martin's Axiom*, Gen. Topology and Appl. 4 (1974), 315-339.

[Ta] H. Tamano, *On paracompactness*, Pacific J. Math. 10 (1969), 1043-1047.

[Te_1] R. Telgársky, *Spaces defined by topological games*, Fund. Math. 88 (1975), 193-223.

[Te_2] ________, *Concerning two covering properties*, Collog. Math. 46 (1976), 57-61.

[W] H.H. Wicke, *A functional characterization of primitive base*, preprint. [Announcement in Notices Amer. Math. Soc. 25 (1978), A-138.]

[WW_1] H.H. Wicke and J.M. Worrell, Jr., *Point-countability and compactness*, Proc. Amer. Math. Soc. 55 (1976), 427-431.

[WW_2] ________, *Spaces which are scattered with respect to collections of sets*, Topology Proc. 2 (1977), 281-307.

[WW_3] ________, *A covering property which implies isocompactness, I and II*, Notices Amer. Math. Soc. 26 (1979), A-124 and A-324.

[Wo_1] J.M. Worrell, Jr., *A characterization of metacompact spaces*, Portugal. Math. 25 (1966), 171-174.

[Wo_2] ________, *The closed continuous images of metacompact topological spaces*, Portugal. Math. 25 (1966), 175-179.

[Wo_3] ________, *Some properties of full normalcy and their relations to Čech completeness*, Notices Amer. Math. Soc. 14 (1967), 555.

[Wo_4] ________, *Paracompactness as a relaxation of full normalcy*, Notices Amer. Math. Soc. 15 (1968), 661.

[WoW] J.M. Worrell, Jr. and H.H. Wicke, *Characterizations of developable topological spaces*, Canad. J. Math. 17 (1965), 820-830.

[Y_1] Y. Yajima, *Solution of R. Telgársky's problem*, Proc. Japan. Acad. 52 (1976), 348-350.

[Y_2] ________, *On spaces which have a closure-preserving cover by finite sets*, Pacific J. Math. 69 (1977), 571-578.

Added in proof: The result of Proposition 1.22 has been independently obtained by N.K. Shamgunov.

SURVEYS IN
GENERAL TOPOLOGY

ORDERED TOPOLOGICAL SPACES

David J. Lutzer
Texas Tech University

1. *Introduction*

Begin with a linearly ordered set $(X,<)$ and use the family of all open half-spaces $(\leftarrow,b) = \{x \in X: x < b\}$ and $(a,\rightarrow) = \{x \in X | x > a\}$ as a subbase. The resulting topology $\mathcal{T} = \mathcal{T}(<)$ is the <u>open interval topology</u> of the order $<$ and the triple $(X,<,\mathcal{T})$ is a <u>linearly ordered topological space</u> (abbreviated LOTS). Next take a subspace Y of X and consider the relative topology $\mathcal{T}_Y$ on Y and the order $<_Y$ obtained by restricting $<$ to Y. Even in the simplest cases, $\mathcal{T}_Y$ can fail to be the open interval topology of $<_Y$; the most one can say without futher information about Y is that $\mathcal{T}(<_Y) \subset \mathcal{T}_Y$. The triple $(Y,<_Y,\mathcal{T}_Y)$ is called a <u>generalized ordered space</u> (abbreviated GO space). There is another (equivalent - see [Lu_3]) way to obtain GO spaces: start with a linearly ordered set $(Y,<)$ and equip Y with any topology which contains $\mathcal{T}(<)$ and has a base of open sets each of which is order convex. (A set S in Y is called <u>order convex</u> if $x \in S$ for every point x lying between two points of S.) Historically, it is the linearly ordered spaces which have received the most attention, but recent experience indicates that it is the larger class of GO spaces which deserves to be studied.

ISBN: 0-12-584960-5

In this paper, I will present a few topics from the theory of GO spaces which have interested me over the years. I will emphasize areas in which there has been recent progress and will try to show the reader how researchers in ordered spaces view the subject. I have tried to make this paper reflect the general ideas we study, even though this led to the omission of some fine, but highly technical, work that has been done recently.

Undefined terms--and there are many--can usually be found in [En], [Lu_3] or [BkL] and I have tried to be liberal with bibliographical references to enable readers to find details elsewhere. The only conventions needed are that all spaces are at least T_1 and that cardinals are initial ordinals. Finally, the phrase "if and only if" is abbreviated "iff" and the usual space of real numbers is denoted by $\mathbb{R}$.

2. *Orderability*

As defined above, a LOTS or a GO space is a topological space already equipped with a compatible ordering. Over the years, some effort has been devoted to giving a characterization of those topological spaces for which some compatible ordering can be constructed. Results of that type are called orderability theorems. Characterizations of the arc, Cantor set and space of irrationals might be viewed as orderability theorems (e.g. any compact, separable connected, locally connected space with at most two non-cut points is homeomorphic to [0,1] and is therefore orderable). However, the first results dealing directly with orderability seem to be the theorems of Eilenberg [Ei] who proved

2.1. *Theorem.* Let (X,T) be connected and locally connected. Then there is a linear ordering $<$ of X such that $(X,T,<)$ is a LOTS iff $(X\times X) - \{(x,x): x \in X\}$ is not connected.

Other orderability theorems for connected locally connected spaces can be found in [Kk]; as a sample, we reproduce a result due to Kowalsky [Ko].

2.2. *Theorem.* A connected, locally connected space is orderable iff whenever A_1, A_2, and A_3 are connected proper subsets of X, there exist distinct $i,j \in \{1,2,3\}$ having $A_i \cup A_j \neq X$.

Removing the connectedness hypotheses (both known to be necessary) from (2.1) and (2.2) was the major stumbling block in characterizing orderable spaces. One step in that direction was to characterize spaces having a compatible dense ordering $<$ (i.e. if $a < b$, some $c \in X$ has $a < c < b$), a condition known to be much weaker than connectedness in ordered spaces. Banaschewski [Bn] proved:

2.2. *Theorem.* A topological space X is orderable by a dense ordering iff there is a uniformity $\mathcal{U}$ (viewed as a family of entourages), compatible with the topology of X, such that

(a) if $U \in \mathcal{U}$ and if $U^{n+1} = U^n \circ U$ for each $n \geq 1$, then $\cup\{U^n: n \geq 1\} = X\times X$;

(b) if $x,y \in X$ and if there are two sequences $x = x_0, x_1, \ldots, x_n = y$ and $y = y_0, y_1, \ldots, y_m = x$ in X such that for some sets $V_1, \ldots, V_n \in \mathcal{U}$, we have $V_i(x_i) \cap V_{i+1}(x_{i+1}) \neq \emptyset$ and $V_i(y_i) \cap V_{i+1}(y_{i+1}) \neq \emptyset$ for $1 \leq i \leq n$, then for some j $V_j(x_j) \cap V_j(y_j) \neq \emptyset$.

Banaschewski's result, while not dealing directly with connected orderable spaces, is still far from a general orderability theorem because if $(X,\mathcal{T},<)$ is a LOTS with a dense ordering, then the usual Dedekind compactification of X is connected.

If one considers spaces which are antipodal to connected spaces, namely 0-dimensional spaces, there are nice orderability theorems. The next result was obtained by Lynn [Ly_1] for separable spaces and the general case is due to Herrlich [Hr]. (A proof also appears in [En, pp. 457-8].)

2.3. Theorem. Any strongly zero-dimensional metric space is linearly orderable.

The general orderability problem--with no connectedness, disconnectedness or metrizability hypothesis, was solved by van Dalen and Wattel [vDW] and by Deak [De]. If we call a collection $\mathcal{N}$ of sets a <u>nest</u> whenever $\mathcal{N}$ is linearly ordered by inclusion, then we have

2.4. Theorem ([vDW]). A topological space $(X,\mathcal{T})$ admits a linear ordering < such that $(X,\mathcal{T},<)$ is a generalized ordered space iff $(X,\mathcal{T})$ is T_1 and has a subbase which is the union of two nests.

The characterization of linearly orderable spaces is slightly more complicated. Let us say that a collection $\mathcal{S}$ is <u>interlocking</u> if, whenever $S_0 \in \mathcal{S}$ satisfies

$$S_0 = \cap\{T \in \mathcal{S} \mid T \supset S_0,\ T \neq S_0\}$$

then

$$S_0 = \cup\{T \in \mathcal{S} \mid T \subset S_0 \text{ and } T \neq S_0\}.$$

2.5. *Theorem* [vDW]. A topological space $(X,\mathcal{T})$ admits a linear ordering $<$ such that $(X,\mathcal{T},<)$ is a LOTS iff X is a T_1 space and has a subbase $\mathcal{S} = \mathcal{N}_1 \cup \mathcal{N}_2$ such that each $\mathcal{N}_i$ is an interlocking nest.

As a corollary, one obtains a result of de Groot and Schnare [dGS].

2.6. *Theorem*. A compact space $(X,\mathcal{T})$ admits a linear ordering $<$ such that $(X,\mathcal{T},<)$ is a LOTS iff there is a collection $\mathcal{S} \subset \mathcal{T}$ which T_1-separates points of X and which is the union of two nests.

To derive (2.6) from (2.4) one considers $\mathcal{T}'$, the topology having $\mathcal{S}$ as a subbase. Then $(X,\mathcal{T}')$ is a GO space and is Hausdorff. But then the identity map i: $(X,\mathcal{T}) \to (X,\mathcal{T}')$ is a continuous bijection from a compact space to a Hausdorff space, so $\mathcal{T} = \mathcal{T}'$ and $(X,\mathcal{T},<)$ is a compact GO space, whence a LOTS.

While the two-nest subbase theorem does solve the orderability problem, it is not always clear how to apply it in specific situations. For example, we know that the square of the space of rationals with the half-open interval topology, being a topological copy of the usual space of rationals, is orderable. But it is not so clear how to find a subbase for this space having the properties given in (2.5).

Another solution of the general orderability problem, in terms of the existence of special "directional structures" compatible with the topology of a space, has been given by Deak [De]. Orderability of metric spaces has been studied by Purisch [Pu]. The results of Deak and Purisch are too technical to reproduce here.

There is a second orderability problem which asks "given a GO space $(X,\mathcal{T},<)$, when is there a linear ordering $<<$ of X having $\mathcal{T}$ as its open-interval topology?" Historically this problem has been studied under the guise "Given a LOTS $(Y,\mathcal{I},<)$ and a subset X of Y, when is the relative topology $\mathcal{I}_X$ on X orderable with respect to the restricted order $<_X$ and when is it orderable with respect to any ordering?" M.E. Rudin [R_1] gave a very complicated solution to this problem and special cases of the problem were studied by Purisch [Pu]. See also [Ly_1]. However these results are often hard to use and some interesting orderability questions remain unsolved. For example, Galvin (private communication) asked whether each subspace of a scattered LOTS X must be orderable (possibly under an order different from the ordering inherited from X.)

3. *Normality*

It is an easy matter to prove that every generalized ordered space X is normal. The traditional direct proof depends on the fact that each open set U can be decomposed into a pairwise disjoint family of maximal convex subsets, called <u>convex components</u> of U, and on an easy lemma.

3.1. Lemma. Suppose X is a GO space and that, for each γ in an index set Γ, we have nonvoid sets $A_\gamma \subset U_\gamma$. If the sets U are pairwise disjoint convex, open sets, and if $\cup\{A_\gamma: \gamma \in \Gamma\}$ is closed in X, then the collection $\{U_\gamma: \gamma \in \Gamma\}$ is locally finite in X.

3.2. Corollary. Any GO space is hereditarily normal.

Proof. Since the GO-class is hereditary, it is enough to show that each GO-space is normal. To that end, suppose $A \subset U$ where A is closed in X and U is open. Let $\{U_\gamma\colon \gamma \in \Gamma\}$ be the family of convex components of U. Let $A_\gamma = A \cap U_\gamma$ for each $\gamma \in \Gamma$. Because each A_γ is closed in X, it is easy to see that there is an open convex set V_γ having $A_\gamma \subset V_\gamma \subset \mathrm{cl}_X(V_\gamma) \subset U_\gamma$. According to (3.1), the collection $\{U_\gamma\colon A_\gamma \neq \emptyset\}$ is locally finite so that the open set $V = \cup\{V_\gamma\colon A_\gamma \neq \emptyset\}$ satisfies $A \subset V \subset \mathrm{cl}_X(V) \subset U$. □

The use of the axiom of choice in the proof of (3.2) is clear; whether (3.2) is provable without that axiom is not known.

A similar, but notationally more complicated, proof can be used to show that any GO space is hereditarily collectionwise normal (cf. [St]) but today we know other proofs which are either prettier or which give even stronger conclusions. One such proof (unpublished) was given by Eric van Douwen who has given me permission to include it here. The main steps are as follows;

(1) Any GO space can be embedded in a compact LOTS [Lu_3] so it will be enough to prove that every compact LOTS X is hereditarily collectionwise normal.

(2) Let X' be the lexicographic product $X \times \{0,1\}$ equipped with the open interval topology of the lexicographic order. The natural map $p\colon X' \to X$ given by $p(x,i) = x$ is a closed continuous surjection, so that if we prove that X' is hereditarily collectionwise normal, then the same will be true for X.

(3) It is easy to see that X' is a compact, totally disconnected LOTS. We claim that any closed subspace C of X' is a retract of X'. Write the set X' - C as a union of a pairwise disjoint collection of convex open sets. Then, X' being compact, each of these sets is an interval (or half line) which must have at least one endpoint in C. Because X' is totally disconnected, each of these intervals can be retracted onto its endpoint(s) which lie in C and that gives a retraction from X' onto C.

(4) In order to show that X' is hereditarily collectionwise normal, it will be enough to show that whenever $\mathcal{D}$ is a collection of sets such that, in the subspace $Y = \cup\mathcal{D}$, $\mathcal{D}$ is a relatively closed, relatively discrete collection, then there are pairwise disjoint open sets $\{U(D): D \in \mathcal{D}\}$ in X' having $D \subset U(D)$ for each $D \in \mathcal{D}$. We let $Z = cl_{X'}(Y)$ and observe that if we define $V(D) = Z - cl_{X'}(Y-D)$ then $\{V(D): D \in \mathcal{D}\}$ is a pairwise disjoint collection of relatively open subsets of Z with $D \subset V(D)$. Let $r: X' \to Z$ be a retraction; then $\{r^{-1}[V(D)]: D \in \mathcal{D}\}$ is the required pairwise disjoint collection of open subsets of X'.

Over the years, researchers have discovered that there are certain topological properties, known to be stronger than collectionwise normality, which every GO space possesses. I have in mind the properties called monotone normality and the Dugundji Extension Property.

A space X is monotonically normal if there is a function H which assigns to each pair (p,C), where C is a closed set and $p \in X - C$, an open set H(p,C) satisfying:

(a) $p \in H(p,C) \subset X - C$;

(b) if $C \subset D$ are closed and $p \in X - D$, then $H(p,D) \subset H(p,C)$; and

(c) if $p \neq q$ then $H(p,\{q\}) \cap H(q,\{p\}) = \emptyset$.

(Other characterizations of monotone normality show more clearly its relation to normality, e.g. X is monotonically normal iff there is a function G which assigns to every pair (A,U), where A is closed, U is open and $A \subset U$, an open set $G(A,U)$ in such a way that

(a) $A \subset G(A,U) \subset \overline{G(A,U)} \subset U$; and

(b) if $(A,U) \subset (A',U')$ -- meaning that $A \subset A'$, $U \subset U'$ --then $G(A,U) \subset G(A',U')$.)

It is known [HLZ] that every monotonically normal space is hereditarily collectionwise normal and in [HLZ] we prove

3.3. Theorem. Any GO space is monotonically normal.

Proof. Let X be a GO space. Let W be any well-ordering of the underlying set of X. If C is closed and $p \in X - C$, let $I(p,C)$ be the convex component of $X - C$ to which p belongs. If the set $I_0(p,C) = \{y \in I(p,C): p < y\}$ is nonvoid let $x(p,C)$ be its W-first point. If $I_1(p,C) = \{y \in I(p,C): p < y\}$ is nonempty let $y(p,C)$ be its W-first point. Let

$$H(p,C) = \begin{cases} (x(p,C),\ y(p,C)) & \text{if } I_0(p,C) \neq \emptyset \neq I_1(p,C) \\ [p,\ y(p,C)) & \text{if } I_0(p,C) = \emptyset \neq I_1(p,C) \\ (x(p,C),p] & \text{if } I_0(p,C) \neq \emptyset = I_1(p,C) \\ \{p\} & \text{if } I_0(p,C) = \emptyset = I_1(p,C) \end{cases}$$

Then H satisfies the three requirements of the above definition. □

A space X has the <u>Dugundji Extension Property</u> [HL_1] if for every closed set $A \subset X$ there is a function $e: C^*(A) \to C^*(X)$ such that:

(a) e is linear;

(b) if $f \in C^*(A)$ then $e(f)$ extends f;

(c) if $f \in C^*(A)$ then the range of $e(f)$ is contained in the closed convex hull of the range of f.

(Here $C^*(Y)$ denotes the Banach space if all bounded, continuous, real-valued functions on a space Y, equipped with the sup-norm topology.)

In [HL_1] we proved that any space with the Dugundji Extension Property must be collectionwise normal. Since the class of GO-spaces is hereditary, if we could prove that every GO space has the Dugundji Extension Property, we would have another proof that every GO space is hereditarily collectionwise normal. (It should be pointed out that in [vD_1], van Douwen has proved directly that any space having the Dugundji Extension Property is hereditarily collectionwise normal.) The result needed to complete the proof appears in [HL_2]:

3.4. Theorem. Any GO space has the Dugundji Extension Property.

The proof of (3.4) is too technical to reproduce here, but its central idea is easy to describe. Take a closed subset A of a GO space X and consider the family $\{U_\gamma: \gamma \in \Gamma\}$ of convex components of X - A. Certain U_γ's will have endpoints belonging to A. We let $\Gamma_1 = \{\gamma \in \Gamma; U_\gamma$ has exactly one end point in A$\}$, $\Gamma_2 = \{\gamma \in \Gamma; U_\gamma$ has two distinct endpoints in A$\}$ and $\Gamma_0 = \Gamma - (\Gamma_1 \cup \Gamma_2)$. Given any bounded continuous $f: A \to \mathbb{R}$ we may extend f over $A \cup (\cup\{U_\gamma: \gamma \in \Gamma_1\})$ by defining the extended function's value on U_γ to be the value of f at the unique endpoint of U_γ lying in A. Then, using a family of monotonic continuous functions from the sets U_γ to [0,1], we can extend f over $A \cup (\cup\{U_\gamma: \gamma \in \Gamma_1\}) \cup (\cup\{U_\gamma: \gamma \in \Gamma_2\})$ by taking appropriate linear combinations of the values of f at the two endpoints of U_γ which lie in A. The sets U_γ for $\gamma \in \Gamma_0$ are the only ones which cause trouble and extension over these sets is accomplished using Banach limits. Consider a set U_γ for $\gamma \in \Gamma_0$. Let $A'_\gamma = \{a \in A: a < x \text{ for each } x \in U_\gamma\}$. If F denotes the family of all bounded revalued functions on A'_γ, it is possible to find a real-valued, linear function Ψ on F such that, using obvious interpretations for the limits involved,

$$\lim \inf\{f(a): a \in A'_\gamma\} \leq \Psi(f) \leq \lim \sup\{f(a): a \in A'_\gamma\}$$

for each $f \in F$. (Such a Ψ is called a Banach limit.) This function γ allows us to assign a number to each $f \in F$ which can play the role of the value of f at the lower endpoint of U_γ lying in A (which of course doesn't really exist because $\gamma \in \Gamma_0$). An analogous choice of a suitable value for each $f \in C^*(A)$ at the (nonexistent) right end point of U_γ is made and then the function is extended over U_γ by forming suitable linear combinations of these values, as was done for the case where $\gamma \in \Gamma_2$.

Now let me turn to covering properties which are possessed by every GO space. There is a characterization of normality which makes normality look like a covering property: a space X is normal iff given any point-finite open cover $\mathcal{U} = \{U_\gamma: \gamma \in \Gamma\}$ of X there is an open cover $\mathcal{V} = \{V_\gamma: \gamma \in \Gamma\}$ of X having $V_\gamma \subset \overline{V}_\gamma \subset U_\gamma$. (See [En]). The collection $\mathcal{V}$ is called a shrinking of $\mathcal{U}$. If a shrinking exists for an open cover $\mathcal{U}$

we say that $\mathcal{U}$ can be shrunk. It is known that point-finiteness of $\mathcal{U}$ cannot be omitted in the above characterization of normality-consider any Dowker space--so that the next result, (due to Fleischman [Fl]) gives another stronger-than-normality property of ordered spaces.

3.5. Theorem. Any open cover of a GO space can be shrunk.

Countable paracompactness is a covering property which has beeen studied in conjunction with normality since Dowker's paper [Do] because the space X × [0,1] is normal if X is both normal and countably paracompact [R_6] so that the next theorem (when combined with (3.2)) yields another way in which GO spaces are more than normal. The result was obtained in the mid-1950's by Ball [Ba] and by Gillman and Henriksen [GH]; we will give a simple proof below.

3.6. Theorem. Any GO space is countably paracompact.

Cohen [Co] introduced the notion of $\aleph_0$-full normality: a space X is $\aleph_0$-fully normal if every open cover $\mathcal{U}$ of X has an open refinement $\mathcal{V}$ such that whenever $\mathcal{W}$ is a countable subcollection of $\mathcal{V}$ with $\cap\mathcal{W} \neq \emptyset$, then $\cup\mathcal{W}$ is contained in a single member of $\mathcal{U}$. Cohen proved that any $\aleph_0$-fully normal space is collectionwise normal, and in 1957 Mansfield [Ma] proved

3.7. Theorem. Every GO space is hereditarily $\aleph_0$-fully normal (and hence hereditarily collectionwise normal).

Today we can reorganize and slightly shorten Mansfield's proof using "the method of coherent collections". We describe this approach at the end of the current section since it gives a good insight into our methods for manipulating covers of ordered spaces. However, before we discuss coherent collections, one final covering property of every GO space deserves to be mentioned. A space X is orthocompact if every open cover $\mathcal{U}$ of X admits an open refinement $\mathcal{V}$ such that for every $p \in X$ the set $\cap\{V \in \mathcal{V}: p \in V\}$ is open. A nice proof of the next theorem appears in [Sc]; the theorem was announced in [Fl].

3.8. Theorem. Every GO space is hereditarily orthocompact.

Let me now describe a general technique (the promised "method of coherent collections") which can be used to give a unified treatment of Mansfield's theorem (3.7), Fleischman's theorems (3.5) and (3.8), the theorem of Ball, Gillman and Henriksen (3.6) and which can be the basis for most of the results on paracompactness in §4. First let me explain the terminology "coherent collection". A collection $\mathcal{C}$ of subsets of X is coherent if whenever $\mathcal{C}'$ is a proper, non-empty subcollection of $\mathcal{C}$, then some members $C_1 \in \mathcal{C}'$ and $C_2 \in \mathcal{C} - \mathcal{C}'$ have $C_1 \cap C_2 \neq \emptyset$. (Such collections were used by Ball in [Ba].) Coherent collections can be obtained via Zorn's lemma (see [BeL_1]) but Engelking [En] pointed out a better approach leading to the same result. Today all that remains of the coherent collections approach is its name, as the reader will now see. We begin with the following lemmas, whose statements are as long as their proofs.

3.9. Lemma. Let $\mathcal{U}$ be a cover of a GO space X by convex, open sets. For $p \in X$ define $S_0(p) = \{p\}$,

$S_{n+1}(p) = \cup\{U \in \mathcal{U} | U \cap S_n(p) \neq \emptyset\}$, and $S(p) = \cup\{S_n(p): n \in \omega\}$. Then:

(a) for $p,q \in X$ either $S(p) = S(q)$ or $S(p) \cap S(q) = \emptyset$;

(b) each $S(p)$ is a convex open subset of X;

(c) for each $p \in X$ there are sequences (possibly finite) $x(m)$ and $y(n)$ in $S(p)$ having

(i) $x(m+1) < x(m) < \ldots < x(0) = p = y(0) < \ldots < y(n) < y(n+1)$;

(ii) $x(m+1) \notin S_1(x(m))$ and $S_1(x(m+1)) \cap S_1(x(m)) \neq \emptyset$;

(iii) $y(n+1) \notin S_1(y(n))$ and $S_1(y(n+1)) \cap S_1(y(n)) \neq \emptyset$;

(iv) $S(p) = \cup\{S_1(x(m)) \cup S_1(y(n)): m,n \geq 0\}$.

3.10. Lemma. With notation as in (3.9), if the sequence $y(0) < y(1) \ldots$ is infinite, then there is a subcollection of $\mathcal{U}$ which covers $[p,\rightarrow) \cap S(p)$ and is locally finite in X, and if the sequence $y(0) < y(1) < \ldots$ is finite with last term $y(N)$, then

(a) there is a finite subcollection of $\mathcal{U}$ which covers $[p,y(N)]$ and

(b) $S_1(y(N)) \cap [y(N),\rightarrow) = S(p) \cap [y(N),\rightarrow)$.

(Analogous statements about the sequence $x(0) > x(1) > \ldots$ also hold.)

It follows from (3.9) and (3.10) that in attempting to refine open coverings of GO spaces, one need only consider convex open coverings $\mathcal{U}$ and then only the convex open pieces $S(p)$ into which $\mathcal{U}$ divides X and, more precisely, the special subcollections $\mathcal{U}(p) = \{U \in \mathcal{U} | U \subset S(p)\}$ which cover $S(p)$. This reduction of the problem, and the use of (3.10), is what has come to be known as "the method of coherent collections".

To help readers visualize what (3.9) and (3.10) give, here are the two relevant pictures of $[p,\rightarrow) \cap S(p)$ showing certain members of the covers $\mathcal{U}(p)$ and the points $y(n)$.

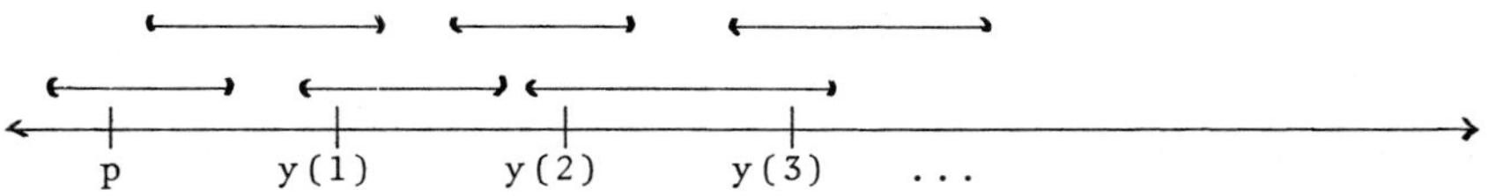

* *

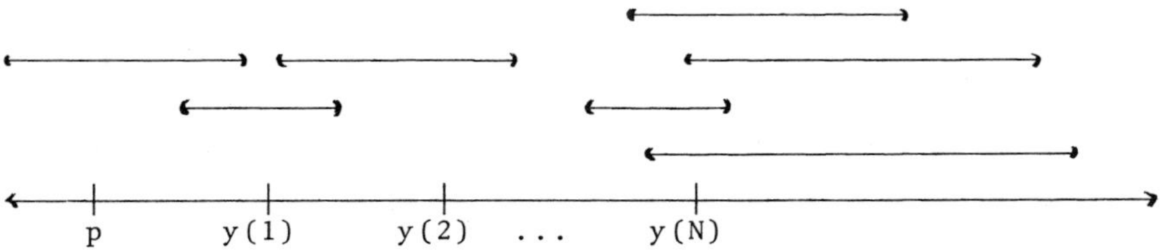

The first diagram represents the situation in which $\langle y(n) \rangle$ is an infinite sequence and shows how certain members of $\mathcal{U}(p)$ may be chosen to give a locally finite cover of $[p,\rightarrow) \cap S(p)$. The second picture represents the troublesome situation where the sequence $\langle y(n) \rangle$ has a last term $y(N)$ such that $S_1(y(N))$ reaches all the way to the right of $S(p)$. The second picture is not entirely accurate since it is possible that $y(N)$ is the right end point of $S(p)$, or that finitely many members of $\mathcal{U}(p)$ contain $y(N)$ and cover $[y(N),\rightarrow) \cap S(p)$. However, these special cases never cause problems. The difficult case arises when infinitely many (perhaps uncountably many) members of $\mathcal{U}(p)$ are needed in order to start at $y(N)$ and cover all of $[y(N),\rightarrow) \cap S(p)$. As an example of the method, I will describe the key step in the proof of Theorem (3.7) on $\aleph_0$-full normality. Start with a convex open cover and apply

(3.9) and (3.10). Fix $p \in X$ and consider $S(p) \cap [p,\rightarrow)$. If the sequence $\langle y(n)\rangle$ is infinite, then choose $I_0^+ \in \mathcal{U}(p)$ with $p \in I_0^+$ and for $n > 1$ we choose two open convex sets I_n^- and I_n^+ in such a way that

(1) $p \in I_0^+$, $y_n \in I_n^- \cap I_n^+$;

(2) $I_{n-1}^+ \cap I_n^- \neq \emptyset$ for $n \geq 1$;

(3) $y(n) \notin I_{n+1}^-$ for each $n \geq 0$.

Then the required refinement of $\mathcal{U}(p)$ consists of the sets

I_0^+, $I_1^- \cap (\leftarrow,y)$, $I_1^- \cap I_1^+$, $I_1^+ \cap (y(1),\rightarrow)$

and, in general,

$I_n^- \cap (\leftarrow,y(n))$, $I_n^- \cap I_n^+$, $I_n^+ \cap (y(n),\rightarrow)$.

In case $\langle y(n)\rangle$ is finite with last term $y(N)$, the above technique is used on $[p,y(N)]$ and we concentrate on $T = S(p) \cap (y(N),\rightarrow)$. The easy case is where there is no countable cofinal subset of T and in that case we obtain the required cover of T by using

$$\{U \cap (p,\rightarrow): y(N) \in U \in \mathcal{U}(p)\}.$$

If there is an infinite sequence $y(N) = z(0) < z(1) < \ldots$ of points of T which is cofinal in T. Choose a set $U(n) \in \mathcal{U}(p)$ with $\{p,z(n)\} \subset U(n)$. The sets required to cover T are then $U(2) \cap (z(0),z(2))$, $U(3) \cap (z(1),z(3))$, and in general $U(n) \cap (z(n-2),z(n))$. Since the case where $y(N) = \max(T)$ is trivial, that completes the proof of (3.7).

As another example of this coherent collections method, I will prove (3.6). Let $\mathcal{V}$ be any countable open cover of the GO space X. One may <u>not</u> assume that the members of $\mathcal{V}$ are convex and it is <u>not</u> necessarily true that $\mathcal{V}$ can be refined by a countable, open, convex cover. Instead we consider the collection $\mathcal{W}$ of all convex components of members of $\mathcal{V}$. Then $\mathcal{W}$ is a convex, open, point-countable cover of X which refines

$\mathcal{V}$ and we may apply the next lemma to the cover $\mathcal{W}$.

3.11. Lemma. Any point-countable open cover of a GO space X has a locally finite open refinement.

Proof. Let $\mathcal{W}$ be a point-countable open cover of X and let $\mathcal{U}$ be the family of all convex components of members of $\mathcal{W}$. Then $\mathcal{U}$ is a point-countable convex open cover of X. With notation as in (3.9) and (3.10), it is enough to find, for each p, a locally finite open cover of S(p) which refines $\mathcal{U}$ and it is obviously enough to cosider the sets $[p,\rightarrow) \cap S(p)$. The only difficulty arises when the sequence $p = y(0) < y(1) <$ is finite with last term $y(N) \neq \sup S(p)$. Then $S_1(y(N))$ contains $(y(N),\rightarrow) \cap S(p)$ and, because $\mathcal{U}$ is point-countable, there must be a countable sequence $\langle z(n)\rangle$ in S(p) with $y(N) = z(0) < z(1) \ldots$ which is cofinal in S(p) and which has the property that for each n, $[y(N),z(n)]$ is contained in some member of $\mathcal{U}$. But then the collection $\{(z(n-1),z(n+1)): n \geq 1\}$ is a locally finite refinement of $\mathcal{U}$ which covers $(y(N),\rightarrow) \cap S(p)$. □

4. *Paracompactness in GO spaces*

Every GO space is collectionwise normal (3.2) so that, in the light of general theory, many well-known covering properties, e.g. subparacompactness, θ-refinability and metacompactness, are equivalent to paracompactness in a GO space [WW, Theorem (iii); En, Thm. 5.3.2]. However, there are other properties, not generally equivalent to paracompactness in collectionwise normal spaces, which reduce to paracompactness in a GO space. The key to results of this latter type lies in specific characterizations of paracompactness in GO spaces.

The first characterization of paracompactness in GO spaces appears in [GH] and is based on a study of the ideal points which are added in forming the usual ordered compactification. By a pseudogap in a GO space $(X,\mathcal{T},<)$ we mean an ordered pair (A,B) of subsets of X satisfying

(i) $A \cup B = X$;

(ii) $a < b$ whenever $a \in A$ and $b \in B$;

(iii) either A has no right end point or B has no left end point;

(iv) $A, B \in \mathcal{T}$.

Gillman and Henriksen isolated the notion of a Q-gap. A particularly clear restatement of the Gillman-Henriksen theory was given by Faber [Fa] (see also [Lu_4, footnote 1]) and makes it unnecessary to reproduce the original definitions given in [GH].

4.1. Theorem. A GO space X is paracompact iff for each pseudogap (A,B) of X there are closed discrete subsets A',B' of X such that A' is a cofinal subset of A and B' is a coinitial subset of B.

The motivation for a theorem like (4.1) is easy to see if one recalls the method of coherent collections discussed in Section 3. Recall that, in the notation of (3.9), if we begin with a convex open cover $\mathcal{U}$, the only possible problem in obtaining locally finite refinements occurs when some point $y(N) \in S(p)$ has $[y(N),\rightarrow) \cap S(p) \subset S_1(y(N))$. If we let $A = (\leftarrow,y(N)] \cup (S(p) \cap [y(N),\rightarrow))$ and $B = X - A$, then (A,B) is a pseudogap in X. If we could find a closed discrete $A' \subset A$ which is cofinal in A, then we could choose a well-ordered increasing net $\{a(\xi): \xi < \kappa\}$, where κ is some cardinal, which is a cofinal subset of A'. For each limit ordinal $\lambda < \kappa$, let (A_λ, B_λ) be the pseudogap $A_\lambda = \{x \in X$: for some $\xi < \lambda$, $x < a(\xi)\}$ and $B_\lambda = X - A_\lambda$. Then define sets $V(\xi)$ for $\xi < \kappa$ by

$$V(\xi) = \begin{cases} (a(\xi-1), a(\xi+1)) & \text{if } \xi = \eta + 1 \\ B_\lambda \cap (\leftarrow, a(\lambda+2)) & \text{if } \xi = \lambda \text{ is a limit ordinal.} \end{cases}$$

Because A' is closed, discrete and cofinal in A, the sets $V(\xi)$ for $\xi < \kappa$ form a locally finite open cover of $S(p) \cap (y(N), \rightarrow)$ which refines the original cover $\mathcal{U}$.

A slight sharpening of (4.1) is available: it is enough to have the sets A' and B' be σ-discrete instead of discrete.

It is useful to know what happens if a GO space is not paracompact. An assertion in [GH, p. 353] has led some students to the mistaken conclusion that if X is a non-paracompact GO space, then for some regular uncountable cardinal κ, the space $[0,\kappa)$ can be topologically embedded in X. That is definitely not true, even if X is a LOTS, as examples in [Lu_4] show. However, there are special subspaces of ordinals which are embeddable in every non-paracompact GO space. Let us say that a subset S of $[0,\kappa)$ is stationary provided $S \cap C \neq \emptyset$ whenever C is a closed cofinal subset of κ. (Here κ is assumed to be a regular uncountable cardinal.) In [EnL] the following result is established:

4.2. *Theorem.* A GO space X is not paracompact iff there is a closed subspace of X which is homeomorphic to a stationary set in a regular uncountable cardinal.

The homeomorphism mentioned in (4.2) can be taken to be monotonic. Furthermore, since the GO class is hereditary,

(4.2) shows that a GO space X is hereditarily paracompact iff no subspace of X is homeomorphic to a stationary set in a regular uncountable cardinal.

Applications of (4.2) usually follow a simple pattern. Suppose we wish to show that any GO space having topological property P is paracompact. If P is known to be inherited by closed subspaces, then it is enough to show that no stationary set in a regular uncountable cardinal can have property P. That is usually accomplished by use of the next result, called the "Pressing Down Lemma" (PDL).

4.3. Theorem. Suppose S is a stationary subset of $[0,\kappa)$ where κ is a regular uncountable cardinal. If $f: S \to [0,\kappa)$ has the property that $f(\alpha) < \alpha$ for each $\alpha \in S - \{0\}$, then there is a $\beta \in [0,\kappa)$ such that $\{\alpha \in S: f(\alpha) = \beta\}$ is stationary in κ.

To give an example of this approach to paracompactness, let us consider the following property, called property D by Michael and studied by van Douwen (private communication). By a neighborhood assignment we mean a function V which assigns to each point $x \in X$ an open set $V(x)$ with $x \in V(x)$. A space X has property D iff whenever V is a neighborhood assignment for X, there is a closed discrete set $D \subset X$ such that $\cup\{V(x): x \in D\} = X$. It is easily seen that property D is closed-hereditary so that if some GO space has property D and yet is not paracompact, then some stationary set S in a regular uncountable cardinal κ has property D. Consider the neighborhood assignment $V(\alpha) = S \cap [0,\alpha+1)$ in the space S and let D be the relatively closed discrete subset of S such that $S = \cup\{V(\alpha): \alpha \in D\}$. We claim that because S is stationary, the relatively closed, discrete subspace D of S cannot be cofinal in S. But that would finish the proof because if D is not cofinal in S, neither is $\cup\{V(\alpha): \alpha \in D\}$. To prove that D cannot be cofinal in S, let $E = cl_\kappa(D) - D$ and observe that since D is closed in S, $E \cap S = \emptyset$. Since S is stationary, E cannot be cofinal in κ since E is closed in κ.

A slightly more complicated application of this technique allows us to show that a GO space X is paracompact provided each open cover of X has an open refinement $\mathcal{V} = \cup\{\mathcal{V}(n) | n \geq 1\}$ such that if $p \in X$ then for some $n \geq 1$, $1 \leq \text{ord}\ (p, \mathcal{V}(n)) \leq \omega$. (This is the property called <u>weak</u> <u>$\delta\theta$-refinability</u> by C.E. Aull.) We suppose that a stationary set S in a regular uncountable cardinal has this property. Let $\mathcal{U} = \{[0,\alpha+1) \cap S: \alpha \in S\}$ and let $\mathcal{V} = \cup\{\mathcal{V}(n): n \geq 1\}$ be a weak $\delta\theta$-refinement of $\mathcal{U}$. For each n, let $T(n) = \{\beta \in S: 1 \leq \text{ord}\ (\beta, \mathcal{V}(n)) \leq \omega\}$. Then $S = \cup\{T(n): n \geq 1\}$ so that, κ being regular and uncountable, one of the sets T(n), say $T(n_0)$, must be stationary. But then $\mathcal{W} = \{V \cap T(n_0): V \in \mathcal{V}(n_0)\}$ is a point-countable cover of the stationary set $T(n_0)$ by bounded, relatively open sets. For each nonisolated point λ of $\Gamma(n_0)$, choose $W_\lambda \in \mathcal{W}$ and then an ordinal $f(\lambda) < \lambda$ such that $T(n_0) \cap (f(\lambda),\lambda] \subset W_\lambda$. Since the set of nonisolated points of the set $T(n_0)$ is again a stationary subset of κ, the PDL can be applied to f to find an ordinal β such that $\{\lambda: f(\lambda) = \beta\}$ is cofinal in $T(n_0)$. Let γ be the first point of $T(n_0)$ which is greater than β. Since each W_λ is a bounded set, the family $\mathcal{W}$ cannot be point-countable at γ.

The next theorem lists properties which are known to imply paracompactness in any GO space. Each part of the theorem can be proved easily using (4.2) and (4.3). Further information about stationary sets can be found in [vDL].

4.4. Theorem. Let X be a GO space. Each of the following implies that X is paracompact.

(a) X is metacompact [Be_2], [Fe];

(b) X is θ-refinable [WW];

(c) X is weakly θ-refinable [BeL_2];

(d) X is metaLindelöf [Be_2];

(e) X is weakly $\delta\theta$-refinable (see above);

(f) X is subparacompact [Bk];

(g) X is Dieudonné complete [Is], [EnL];

(h) X has a G_δ-diagonal [Lu_3], [EnL];

(i) X has a σ-minimal base [BeL_3];

(j) X has property D (see above);

(k) X is perfect [Lu_3], [EnL];

(l) Every subspace of X is a β-space [BeL_4];

(m) X has a point-countable base [Be_3], [Fe];

(n) X has countable cellularity [BeL_1];

(o) Every subspace of X is a p-space [BeL_4];

(p) There is a continuous f: $X \to M$, where M is a metric space, such that $f^{-1}[\{m\}]$ is a paracompact subset of X for each $m \in M$ [Kp].

There is obviously a great deal of redundancy in the list appearing in (4.4). The proof of sufficiency for (i) is different from all the others since the property of having a σ-minimal base is not closed-hereditary. Nevertheless, (4.2) can still be used as the basis for the proof ([BeL_3]).

While the stationary set technique is useful to establish paracompactness of a GO space having a property weaker than paracompactness, it does not seem to be useful in investigating properties of GO spaces already known to be paracompact. Other techniques and (4.4) yield:

4.5. Theorem. For a GO space X, the following are equivalent:

(a) X satisfies one of the first six properties listed in (4.4);

(b) X is paracompact;

(c) X has property D [vD_2];

(d) X is strongly paracompact [Be_2];

(e) Suppose, for each $x \in X$, that $\phi(x)$ is an open neighborhood of x. Then there is a subset $D \subset X$ and an open cover $\{\psi(x): x \in D\}$ of X such that

(i) $x \in \psi(x) \subset \phi(x)$ whenever $x \in D$;

(ii) $\psi(x)$ is convex whenever $x \in D$;

(iii) if $x \in D$ then $\{y \in D | \psi(y) \cap \psi(x) \neq \emptyset\}$ has cardinality ≤ 2 [vD_2].

5. *Metrization and generalized metric classes*

The basic metrization for LOTS was proved in [Lu_1]; it asserts

5.1. Theorem. A LOTS is metrizable iff it has a G_δ-diagonal.

The short proof of that result, given in [Lu_2], proceeds by proving that a LOTS with a G_δ-diagonal is a Moore space; that, combined with (3.7), completes the proof. Unfortunately, (5.1) does not hold for arbitrary GO spaces as the example of the Sorgenfrey line shows. (Indeed, the fact that the Sorgenfrey line has a G_δ-diagonal and yet is not metrizable seems to have been the first proof that the Sorgenfrey line is not orderable.) However, the existence of a G_δ-diagonal in a GO-space does yield some special structure for the space, e.g. hereditary paracompactness, as can be seen from (4.4) plus the fact that no stationary set in a regular uncountable cardinal has a G_δ-diagonal. Alster, in [Al], gave a detailed study of the special properties and product theory of GO-spaces having G_δ-diagonals (which he called GO_δ-spaces). For example, since any paracompact space with a G_δ-diagonal is submetrizable [Bo_1], every GO_δ-space has a weaker metrizable topology. But Przymusiński proved a sharper result (see [Al]), viz.

5.2. *Theorem*. If $(X,T,<)$ is a GO-space with a G_δ-diagonal, then there is a metrizable topology $S \subset T$ such that $(X,S,<)$ is also a GO space.

Another example of special structures of GO_δ-spaces is given by Alster's result that, modulo isolated points, each GO_δ-space is perfect; to be more precise, Alster showed that if X is a GO_δ-space then the subspace X^d, the set of non-isolated points of X, is perfect [Al]. Alster used the special structure of GO_δ-spaces to obtain results in product theory, showing, for example, that the Continuum Hypothesis is equivalent to the assertion that $X \times Y$ is hereditarily sub-paracompact whenever X and Y are Lindelöf GO_δ-spaces.

The G_δ-diagonal metrization theorem for LOTS turns out to be a special case of a G_δ-diagonal metrization theorem for arbitrary GO spaces which is due to Faber [Fa].

5.3. *Theorem*. Let $(X,T,<)$ be a GO space and let I be the usual open-interval topology of $<$. Then (X,T) is metrizable iff $(X,\mathcal{T})$ has a G_δ-diagonal and the set $\{x \in X: [x,\rightarrow) \in \mathcal{T} - \mathcal{I}$ or $(\leftarrow,x] \in \mathcal{T} - \mathcal{I}\}$ is σ-discrete in $(X,\mathcal{T})$.

Using Faber's theorem, van Wouwe [vW] was able to give a direct proof of another metrization theorem for GO spaces whose original proof in [Lu_3] was much more complicated.

5.4. Theorem. A GO space is metrizable if and only if it is semi-stratifiable.

(A space X is semistratifiable [Cr] if corresponding to each open set G it is possible to choose closed sets C(G,n) such that $G = \cup\{C(G,n): n \geq 1\}$ and such that if the open sets G,H have $G \subset H$, then $C(G,n) \subset C(H,n)$ for every $n \geq 1$.)

A related metrization theorem for LOTS was given by Nedev in [Ne] and can be easily deduced from (5.4) using well-known tools from generalized metric space theory. Definitions of terms in the next theorem and proof can be found in [BkL] and [Ar_1].

5.5. Theorem. Any symmetrizable GO space is metrizable.

Proof. Any symmetrizable spaces satisfies the gf-axiom of countability, and in GO-spaces the gf-axiom is easily seen to be equivalent to first-countability. But a first-countable symmetrizable space is known to be semi-metrizable and hence semistratifiable. Now apply (5.4). □

A second branch of generalized metric space theory centers around Arhangel'skii's p-spaces and Morita's M-spaces. Each class arose in the solution of problems which are not directly related to the metrization problem, and each class has since been widely used in metrization theory. In general, the two classes are distinct--there are p-spaces which are not M-spaces, and vice-versa--but Velichko [V] and van Wouwe [vW] have shown that the properties are related in the family of LOTS. Van Wouwe's theorem is more general. (This is also an unpublished result of van Douwen.)

5.6. Theorem. Let X be a GO space. If X is a p-space, then X is an M-space.

Van Wouwe's proof is to construct, for any GO space X, two quotient spaces which he calls gX and cX and for which there is a natural closed continuous mapping from gX onto cX. Van Wouwe proves:

(a) X is a p-space iff gX is metrizable;

(b) X is an M-space iff cX is metrizable;

(c) both gX and cX are GO spaces when endowed with the quotient topology and a certain natural ordering induced by X.

Then, if X is a p-space, the space cX is a closed continuous image of the metric space gX so that, in the light of general theory, cX is at least semistratifiable. Therefore, (5.4) shows that cX is metrizable so that, from (b) above, X is an M-space.

Associated with M-spaces in the general theory are the wΔ-spaces introduced by Borges [Bo_2], and the Σ-spaces introduced by Nagami [Na]. A weak relative of p-spaces was introduced by Creede [Cr] under the name "quasi-complete spaces". In general all these classes are distinct; however things are simpler in the GO class since one can prove

5.7. Theorem: Let X be a GO space. The following are equivalent:

(a) X is a M-space;

(b) X is a wΔ-space;

(c) X is quasi-complete.

The equivalence of (a) and (b) in (5.7) is due to van Wouwe [vW]; the equivalence of (b) and (c) was established in [BeL_4].

In [Ar_2], Arhangel'skii studied the class of hereditarily p-spaces, i.e. the class of spaces X such that every subspace Y of X is a p-space. This class of spaces has a really surprising interplay with the GO class and yields a metrization theorem which does not seem to be related to the ones listed in (5.1) to (5.5). In [BeL_4] we prove

5.8. Theorem: Let X be a GO space. Then the following are equivalent:

(a) X is metrizable;

(b) X is hereditarily a p-space;

(c) X is hereditarily an M-space;

(d) X is hereditarily a wΔ-space;

(e) X is hereditarily quasi-complete.

The basis of our proof is a general metrization theorem due to Bennett [Be_4] which asserts that a space is metrizable iff it is a quasi-developable, paracompact p-space. The first step in our proof is to observe that no stationary set in a regular uncountable cardinal can be hereditarily quasi-complete so that any GO space satisfying (e) of (5.8) must be hereditarily paracompact and must therefore satisfy (b). The rest of the proof studies the way that X embeds into X^+, its usual Dedekind compactification, and constructs a quasi-development for X from this embedding.

5.9. Problem. Van Wouwe [vW] has studied GO spaces which are hereditarily Σ-spaces. Because the defining structure for a Σ-space involves locally finite closed covers (instead of open covers, as is the case with the properties in (5.8)), problems in this area seem to be considerably harder than those mentioned in (5.8). In his dissertation [vW], van Wouwe conjectured that if a GO space X is hereditarily a Σ-space, then X is metrizable. He was able to reduce that conjecture to another one, viz., if X is a Lindelöf GO space which is hereditarily a Σ-space, then X is hereditarily Lindelöf (equivalently, is perfect). To attack this second conjecture, one does not need to look at subspaces of X at all since it is known that a GO space is hereditarily Lindelöf iff it has countable cellularity [Lu_3].

Theorem 5.8 may be viewed as an attempt to mimick Theorem 4.2 for metric spaces, i.e. to isolate a certain class of easily studied spaces such that every nonmetrizable GO space must contain a subspace belonging to this class. Theorem 5.8 does not give a satisfactory solution to that problem.

A third component of generalized metric space theory involves spaces having bases of special kinds. Perhaps the first result in this direction was due to Fedorčuk [Fe] who proved the next theorem for LOTS.

5.10. Theorem. Let X be a GO space with a σ-locally countable base. Then X is metrizable.

Proof. Since no stationary set in a regular uncountable cardinal can have such a base, any GO space with a σ-locally countable base is paracompact. But now the result follows from general theory since any paracompact Hausdorff space with a σ-locally countable base is metrizable [Fe]. □

A word of warning about bases in metrizable GO spaces is in order here. It is an easy consequence of (3.1) that a metrizable GO space has a σ-discrete base whose members are convex open sets. But, as an elegant example due to Faber [Fa] shows, one cannot expect to obtain σ-discrete bases whose members are open intervals (i.e. convex sets with end-points in X) even when X is a metrizable LOTS.

Many of the other types of bases used in metrization theory are mutually equivalent in GO spaces. Recall that a space X is <u>quasi-developable</u> if there is a base $\mathcal{B} = \cup\{\mathcal{B}(n) : n \geq 1\}$ for X such that if p is a point of an open set V then for some $n \geq 1$, $p \in St(p, \mathcal{B}(n)) \subset V$. (It is <u>not</u> required that each $\mathcal{B}(n)$ cover X.) The following result was proved (less directly) in [Lu_3].

5.11. Theorem. Let X be a GO space. The following are equivalent:

(a) X has a σ-disjoint base;

(b) X has a σ-point finite base;

(c) X is quasi-developable.

Furthermore, a GO space satisfying any of these conditions is hereditarily paracompact.

Proof. Obviously (a) implies (b), and (b) implies (c) in the light of a theorem of Aull [Au_1]. Because no stationary set in a regular uncountable cardinal can have a quasi-development, any GO space which satisfies (c) is hereditarily paracompact. But any hereditarily paracompact quasi-developable space has a σ-disjoint base, so the proof is complete. □

The example of Faber mentioned above shows that in a quasi-developable (or even metrizable) LOTS, one cannot expect to have a σ-disjoint base of open intervals even though one can always find a σ-disjoint base of open, convex sets in such a space.

Bases which are point-countable are another familiar tool in metrization theory. While a GO space with a point-countable base must hereditarily paracompact (for the usual reason), the existence of a point-countable base does not guarantee the existence of a base having any of the stronger properties listed in (5.11). Bennett has given two examples which are relevant here.

(a) There is a LOTS which has a point-countable base and yet which is not quasi-developable [Be_2].

(b) If there is a Souslin space then there is a Souslin space having a point-countable base ([Be_2]; such a construction was also given by Ponomarev [Po]).

It is interesting to note that, unlike the situation for GO spaces having σ-discrete or σ-disjoint bases, if a GO space has a point-countable base, then it has a point-countable base whose members are open sets and intervals (note: this is not the same as an open interval). The result is due to R.W. Heath [H_2].

In closing this section let me mention two related areas in which there seem to be nice open problems.

5.12. Problem. A collection $\mathcal{C}$ is said to be minimal if $\cup\mathcal{D} \neq \cup\mathcal{C}$ whenever $\mathcal{D}$ is a proper subcollection of $\mathcal{C}$. This problem asks whether a GO space X must be metrizable given that each $Y \subset X$ has a σ-minimal base for its relative topology. It is known that any such space must be hereditarily paracompact [BeL_3]. As a first step it would be useful to know if such a space must be perfect.

5.13. Problem. There are two classes of generalized metric spaces, Hodel's γ-spaces and Ribiero's quasi-metrizable spaces, whose relationship to each other is not clear. (See [Be_3] for relevant definitions.) Every quasi-metrizable space is a γ-space, but the converse is open. This problem asks whether the two classes are equivalent if one considers only GO spaces. Recently there has been some progress: Bennett proved that the two notions are equivalent in the class of separable GO spaces (actually he proved a bit more--see [Be_3] for details), but the general case of this problem remains open.

6. Some special ordered spaces

The most famous special LOTS is a Souslin space, i.e. a non-separable LOTS having countable cellularity (every pairwise disjoint collection of open sets is countable). Whether or not such things exist depends on your set theory. Souslin spaces exist in any model of (V = L) and cannot exist in any model of (MA + $\omega_1 < c$). Two surveys of the Souslin Problem (i.e. "Do Souslin spaces exist?") are given in [Ku_1] and [R_2]. Other accessible references are [R_4], [DvJ], and [Dv]. In this section all I hope to do is to point out some general features of Souslin spaces and mention a few applications.

First of all, the fact that a Souslin space is defined to be a LOTS instead of a GO space does not seem to be important since (by densely embedding a GO space in a compact LOTS) one can see that there is a non-separable GO space which has countable cellularity iff there is a Souslin space. The general properties of Souslin spaces can be summed up as follows:

6.1. Theorem. Let X be a Souslin space. Then:

(a) X is hereditarily Lindelöf and hence is perfectly normal and paracompact [BeL_1];

(b) X has a dense set of cardinality ω_1, and $|X| \leq c$;

(c) X cannot have the Blumberg property (see below);

(d) X^2 cannot have countable cellularity [Ku_2];

(e) X^2 cannot be hereditarily normal if X is order-complete [R_8].

P. Simon [Si] sharpened (6.1(d)) by proving

6.2. Theorem. Let X be any GO space and let c and d denote cellularity and density functions [En]. Then $c(X^2) = d(X)$.

Simon's result was actually proved in case X is a LOTS. If X is only a GO space, then we densely embed X in a compact LOTS Y and we have $c(X^2) = c(Y^2) = d(Y) = d(X)$. The first equality holds because X^2 is dense in Y^2. The second is Simon's result. The third holds because for any LOTS, the density and hereditary density agree [Lu_3].

There are other properties which some, but not all, Souslin spaces have. Strictly speaking the results in the next theorem should be phrased as "If there is a Souslin space, then there is a Souslin space with additional property P."

6.3. Theorem. A Souslin space can:

(a) be compact and connected;

(b) be homogeneous [DvJ] (under V = L);

(c) have a point-countable base [Be_2], [Po];

(d) be quasi-metrizable [H_2].

Assertion (a) of (6.3) is a well-known part of the folklore and can be proved as follows. Begin with any Souslin space X.

Define a relation ~ on X by a ~ b iff the interval between a and b is separable, and collapse each equivalence class of ~ to a single point. The resulting quotient space $\hat{X}$ inherits an ordering from X; then Dedekind compactification of $\hat{X}$ is the desired compact, connected Souslin space. Assertions (c) and (d) of (6.3) can be proved by carefully selecting subspaces of a connected Souslin space.

Theorem 6.3 shows that there can be Souslin spaces with strange properties, and these spaces can be used to answer topological questions, at least up to consistency. (Example: Arhangel'skii [Ar_1] asked if a hereditarily Lindelöf space with a point-countable base must be metrizable. In the light of (6.3(c)), the answer is "Consistently, no.") In addition, Souslin spaces have provided the starting point for the construction of interesting non-Souslin spaces.

6.4. Theorem. If there is a Souslin space, then:

(a) there is a Dowker space [R_3];

(b) there is a regular hereditarily separable space which is not hereditarily Lindelöf [R_5];

(c) there is a locally compact Hausdorff space X which is neither normal nor countably paracompact and yet if μ is a σ-finite, inner and outer compact sets, measure X which is finite on compact sets, then for any $\varepsilon > 0$ and any measurable $f: X \to \mathbb{R}$ there is a continuous $g: X \to \mathbb{R}$ such that $\mu\{x | f(x) \neq g(x)\} < \varepsilon$. [Wa]

Theorem 6.4c shows that Souslin spaces can have applications outside of pure topology.

Souslin spaces--which are linearly ordered--are closely

related to Souslin trees which are certain partially ordered structures. It is often the special structures of these trees which have been used in constructions. The next theorem was given by Miller [Mi], but the idea of using tree-like structures to study Souslin spaces was certainly not new when Miller's paper was published in 1943; in particular, Đ. Kurepa had spent years studying Souslin spaces via tree-like structures (cf. [Ku_1]). Miller proved

6.5. Theorem. There is a Souslin space iff there is a Souslin tree.

The tree-to-line process has become a standard tool today. By a <u>tree</u> we mean a partially ordered set $(T,<_T)$ such that if $x \in T$, then the set $\hat{x} = \{y \in T: y <_T x\}$ is well ordered and hence isomorphic to some ordinal. For each ordinal α, let $T_\alpha = \{x: \hat{x}$ is isomorphic to $[0,\alpha)\}$. By a <u>branch</u> of T we mean an initial segment of T which is linearly ordered by $<_T$. Suppose that for each α we fix a linear ordering $<_\alpha$ of the set T_α. Now let X be the set of maximal branches of $(T,<_T)$ and linearly order X as follows. If $b \neq c$ belong to X let α be the least ordinal such that $b(\alpha) \neq c(\alpha)$, where $b(\alpha)$ is the unique point of $b \cap T_\alpha$ and $c(\alpha)$ is the unique point of $c \cap T_\alpha$. Define $b \otimes c$ to mean that $b(\alpha) <_\alpha c(\alpha)$ in T_α. Obviously, the ordering $\otimes$ depends on all of the chosen orderings $<_\alpha$ as well as on the tree $(T,<_T)$ and there is no reason to believe that different choices of $<_\alpha$ must yield the same maximal branch space.

It is a common practice to modify a given tree $(T,<_T)$ before looking at its branch space. Often one adds an initial element 0_T to T so that, as Rudin said in [R_2], T looks like a tree instead of a forest. Another useful property for T to have is that, given $x \in T$, at least two elements $y \in T$ have $\hat{y} = \{x\} \cup \hat{x}$.

By a <u>Souslin tree</u> we mean a tree $(T,<_T)$ such that T is uncountable, and has no uncountable branch and no uncountable antichain. If the tree-to-line construction is applied to a Souslin tree, one obtains a Souslin space. Details can be found in $[R_2]$, [DvJ], $[R_4]$.

There are also ways to construct trees from a linearly ordered set. Let $(X,<)$ be a linearly ordered set. If I is a convex set in X, then a point p of I is called central iff there are points $a < p < b$ with $\{a,b\} \subset I$. We construct a tree by induction. Let $C(0) = \{X\}$. If possible, choose a central point $p(0) \in X$ and let $X(0) = \{p(0)\}$. For each ordinal α, if $X(\beta)$ and $C(\beta)$ are chosen for every $\beta < \alpha$, let $C(\alpha)$ be the family of convex components of $X - \text{cl}(\cup\{X(\beta): \beta < \alpha\})$. If $I \in C(\alpha)$ has central points let $p(I)$ be one such point; otherwise let $p(I)$ be an endpoint of I. Let $X(\alpha) = \{p(I): I \in C(\alpha)\}$. This induction must terminate at some ordinal κ less than $(\exp |X|)^+$. The underlying set of our tree is $T(X) = \cup\{C(\alpha): \alpha < \kappa\}$. Observe that if $I \in C(\beta)$ and $J \in C(\alpha)$ are distinct. Then either $I \cap J = \emptyset$ or else one of I and J contains the other. Partially order $T(X)$ by the rule that $I < J$ iff $I \supset J$. Clearly, the tree $T(X)$ depends on the choices of the points $p(I)$ as well as on X itself.

Certain properties of X are mirrored in $T(X)$. For example, if X is not separable and has countable cellularity, then the construction must proceed through at least ω_1 steps. If X has countable cellularity, then no branch of $T(X)$ can contain ω_1 members of $T(X)$. For suppose b is a branch of T having ω_1 points. Write $I(\alpha) = b(\alpha)$ for each $\alpha < \omega_1$. Then each $I(\alpha)$

must have been split in half at the $(\alpha+1)$st stage of the construction, yielding two sets in $\mathcal{C}(\alpha+1)$, and one of these two must be $I(\alpha+1)$. Denote the other by $I'(\alpha+1)$. But then $\{I'(\alpha+1): \alpha < \omega_1\}$ is an uncountable pairwise disjoint collection of open sets which is impossible in a space with countable cellularity.

The tree $T(X)$ can also be used to study X. Suppose X is a Souslin space. In the above construction, let $\mathcal{C}'(\alpha) = \{I \in \mathcal{C}(\alpha): I \text{ is an infinite set}\}$. Because X is not separable, $\mathcal{C}'(\alpha) \neq \emptyset$ for each $\alpha < \omega_1$, and each $J \in \mathcal{C}'(\alpha)$ yields two members J' and J'' of $\mathcal{C}'(\alpha+1)$. Let $\mathcal{D} = \{J' \times J'': J \in \mathcal{C}'(\alpha) \text{ for some } \alpha < \omega_1\}$. Then $\mathcal{D}$ is an uncountable family of pairwise disjoint open sets in $X \times X$, showing that $X \times X$ cannot have countable cellularity. This proves (6.1d), above.

Recent work on the Blumberg problem has given us further examples of special ordered spaces. In [Bl], Blumberg proved that given any function $f: \mathbb{R} \to \mathbb{R}$, there is a dense set D in $\mathbb{R}$ such that $f|D$ is continuous. In [BrG], Blumberg's theorem was sharpened to assert that, if X is any metrizable <u>Baire space</u> (i.e. the intersection of countably many dense open sets in X is dense) and if $f: X \to \mathbb{R}$ is any function, then for some dense $D \subset X$, $f|D$ is continuous. For some years it was an open problem to determine which spaces have this <u>Blumberg property</u> (any real-valued function on the space is continuous on some dense set). In particular, it was not known whether each compact Hausdorff space had the Blumberg property. (This was called the <u>Blumberg problem</u>.)

R. Levy observed that if X is an η_1-set of cardinality $\leq c$ [Le,J] then X is a LOTS, a Baire space and does not have the Blumberg property. A second (unpublished) contribution to the Blumberg problem using ordered spaces was given by Weiss who proved that no Souslin space can have the Blumberg property. We give another proof here.

6.6. Theorem. If X is a Souslin space, then X does not have the Blumberg property.

Proof. By (6.1(b)), $|X| \leq c$ so that there is a 1-1 function f from X into $\mathbb{R}$. Suppose Y is a dense subset of X such that the function $g = f|Y$ is continuous. Then Y has a G_δ-diagonal so that, if $\mathcal{T}$ denotes the relative topology on Y, there is a metrizable topology $\mathcal{S}$ on Y having $\mathcal{S} \subset \mathcal{T}$ and such that $(Y,\mathcal{S})$ is a GO space (5.2). Because $(Y,\mathcal{T})$ is a dense subspace of X, $(Y,\mathcal{T})$ has countable cellularity; hence so does $(Y,\mathcal{S})$. But then $(Y,\mathcal{S})$ is separable. Let D be a countable subset of $(Y,\mathcal{S})$ and let I be the set of isolated points of $(Y,\mathcal{T})$. Since $(Y,\mathcal{T})$ has countable cellularlity, I is countable. Let $E = D \cup I$ and observe that, because $\mathcal{S} \subset \mathcal{T}$ are GO topologies on the same ordered set, each nonempty $U \in \mathcal{T}$ satisfies $U \subset I$ or $\mathrm{int}_{\mathcal{S}}(U) \neq \emptyset$. In either case, $U \cap E \neq \emptyset$, so that $(Y,\mathcal{T})$ is separable, which is impossible. □

Since certain Souslin spaces are compact, it follows that, consistently, some compact spaces do not have the Blumberg property. In addition, the proof of (6.6) gives us some information about the kinds of continuous real-valued functions on a Souslin space. If X is a Souslin space with no separable intervals, then every monotonic real-valued continuous function resembles the familiar Cantor ternary function--given a continuous monotonic $g: X \to \mathbb{R}$ and any open set $U \neq \emptyset$, there is an open set V with $\emptyset \neq V \subset U$ and on which g is constant.

Today we do not need to rely on Souslin spaces to solve the Blumberg problem. Weiss [We] has constructed (in ZFC) two compact spaces X and Y such that:

(a) if the Continuum Hypothesis is true, then X does not have the Blumberg property;

(b) if the Continuum Hypothesis is false, then Y does not have the Blumberg property.

Then the disjoint union $Z = X \oplus Y$ is a compact space which cannot have the Blumberg property, no matter what happens to the Continuum Hypothesis. I mention Weiss' solution in this survey because Y is a compact LOTS. (More recently, researchers in Prague have given a direct construction of a compact space Z not having the Blumberg property; their construction does not involve set theory at all.)

Weiss gave a characterization of those linearly ordered Baire spaces which have the Blumberg property. His characterization rests on the notion of an oblivious collection. A collection $\mathcal{C}$ of open subsets of a space Y is <u>oblivious</u> if there is an open subset $V \neq \emptyset$ of Y such that if $\mathcal{D} \subset \mathcal{C}$ has $V \cap (\cap\mathcal{D}) \neq \emptyset$, then $\mathrm{Int}(\cap\mathcal{D}) \neq \emptyset$.

6.7. Theorem. Let X be a LOTS and a Baire space. Then X does not have the Blumberg property iff there is a nonvoid open set $U \subset X$ such that

(a) U is the union of $\leq c$ nowhere dense sets;

(b) every countable collection of open subsets of U is oblivious.

Weiss then constructs a compact LOTS which will satisfy (a) and (b) of (6.7) provided $\omega_2 \leq c$. Weiss' space is the set of branches of a tree T (another example of the tree-to-line construction mentioned above). Weiss uses an (ω_2, c^+) tree T (i.e. T satisfies $T = \cup\{T_\alpha : \alpha < \omega_2\}$ where each T_α has cardinality $\leq c^+$) which he constructs using certain countable, strictly increasing sequences from the Dekekind completion of an η_1-set. The LOTS which Weiss constructs exists in every model of ZFC; whether it has the Blumberg property in a particular model depends upon the status of the Continuum Hypothesis in that model. It would be interesting to find an absolute example of a compact LOTS which does not have the Blumberg property.

Some recent work of Nyikos and Zenor provides other special ordered spaces. In [Ka], Katetov proved that if X is a compact Hausdorff space for which $X^3 = X \times X \times X$ is hereditarily normal, then X is metrizable. Katetov asked whether a compact space X would be metrizable given only that X^2 is hereditarily normal. A consistent negative solution to Katetov's problem has recently been given by Nyikos [Ny].

Nyikos begins with an arbitrary subset A of $I = [0,1]$ and defines $X(A) = (I \times \{0\}) \cup (A \times \{0,1\})$. Equipping X(A) with the lexicographic ordering and the associated interval topology, he obtains a compact LOTS. Special properties of the set A strongly influence the structure of X(A).

Recall that a <u>Q-set</u> is an uncountable separable metric space in which each subset is an F_σ-set; whether such spaces exist depends on your set theory. They cannot exist under the Continuum Hypothesis while under $(MA + \omega_1 < c)$ they are plentiful. Nyikos made his construction assuming $(MA + \omega_1 < c)$ to insure that every separable metric space of cardinality $< c$ must be a Q-set (and hence the square of a Q-set is again a Q-set). Recently Przymusiński [Pr] has proved that if a model of set theory has any Q-sets at all, then it has Q-sets X such that every finite power of X is also a Q-set. In the light of that result, Nyikos' theorem is:

6.8. Theorem. If there is a Q-set, then there is a compact non-metrizable LOTS whose square is hereditarily normal.

To prove (6.8), take a subset $A \subset [0,1]$ whose every finite power is a Q-set and let $X = X(A)$. Then X is a compact non-metrizable LOTS whose square is hereditarily normal. The verification that X^2 is hereditarily normal is messy.

Nyikos' theorem does not absolutely settle Katetov's question. It may be that there is an absolute example, or it may be that there are models of ZFC in which X is metrizable given that X is compact and that X^2 is hereditarily normal. However, Zenor has proved that the absolute counterexample (if it exists) cannot be a LOTS by showing:

6.9. Theorem. If there is a compact, non-metrizable LOTS whose square is hereditarily normal, then there is a Q-set.

The key steps in Zenor's proof are as follows:

(a) If X^2 is compact and hereditarily normal, then X is perfectly normal and therefore hereditarily Lindelöf so that, because the square of a Souslin space cannot be hereditarily normal ((6.1)(e)), X must be separable.

(b) Since X is separable, compact and not metrizable, X must have an uncountable family of jumps, i.e. intervals [a,b] for which $(a,b) = \emptyset$, whose endpoints are non-isolated points of X.

(c) The subspace $S = \{(x,y) \in X^2 : x < y\}$ of X^2 is open and separable, and the set $J = \{(x,y) \in S:$ x and y are non-isolated jump points of X$\}$ is a closed, discrete subspace of S. Let D be a countable subset of S - J which is dense in S.

(d) Let $Y = J \cup D$. Strengthening the topology of Y in a certain way, Zenor constructs a separable non-metrizable Moore space which is normal because Y, as a subspace of X^2, is normal. But then a result of R.W. Heath [H_1] guarantees that there is a Q-set.

Appendix: Hahn-Mazurkiewicz theory.

A well-known theorem of Hahn and Mazurkiewicz (cf. [HY]) characterizes spaces which are continuous images of the unit interval [0,1], as follows:

A1. Theorem: A Hausdorff space X is a continuous image of [0,1] if and only if X is a compact, connected, locally connected metrix space.

It is not unreasonable to ask whether connected, locally connected, compact (but possibly non-metrizable) spaces might not be characterized as being continuous images of compact connected LOTS. The earliest results on this problem were of a negative character.

A2. Example: Let X and Y be nonmetrizable compact connected LOTS. Then $X \times Y$ is compact, connected and locally connected, and yet:

(a) $X \times Y$ is not the continuous image of any compact connected LOTS [MP];

(b) $X \times Y$ is not the continuous image of any compact LOTS [Y];

(c) $X \times Y$ is not the closed continuous of any LOTS or generalized ordered space since $X \times Y$ cannot be monotonically normal [HLZ].

More recent work has yielded positive results.

A3. Theorem: A compact, connected Hausdorff space is the continuous image of some compact connected LOTS if any one of the following holds:

(a) X is tree-like, i.e. any two points of X can be separated by a third point of X[Co];

(b) X is rim-finite, i.e. there is a base $\mathcal{B}$ for the open sets of X such that each $B \in \mathcal{B}$ has finite boundary [Wd_1];

(c) X can be approximated by finite tree-like subspaces in the sense that there is a collection $\mathcal{T}$ of subsets of X satisfying

 (i) each $T \in \mathcal{T}$ is a tree-like space with only a finite number of endpoints;

 (ii) $\mathcal{T}$ is directed by inclusion and $\cup\mathcal{T}$ is a dense subspace of X;

 (iii) if $\mathcal{U}$ is an open cover then some $T_1 \in \mathcal{T}$ has the property that if $T_1 \subset T_2 \in \mathcal{T}$, then each component of $T_2 - T_1$ is contained in some member of $\mathcal{U}$ [Wd_2];

(d) X is finitely suslinian [T_n], i.e. X is metrizable and for each positive ε, X does not contain any infinite collection of pairwise disjoint subcontinua of diameter $> \varepsilon$.

To date, we do not have a characterization of spaces which are continuous images of some compact connected LOTS. Experience, and Example A2(c), suggest that conditions above and beyond connectedness, local connectness or generalizations of arcwise connectedness will play a role in such a characterization.

Besides the obvious aesthetic appeal of results such as those in Theorem A3, current research on Hahn-Mazurkiewicz theory can have applications to a general question raised by M. Maurice: which theorems known for LOTS can be proved for topological spaces whose topology is less directly tied to linear orderings? Two examples from metrization theory come to mind.

A4. Theorem: Let X be a LOTS. Then X is metrizable:

(a) if every subspace of X is a p-space [BL] (cf. section 5);

(b) if X is Eberlein compact (i.e. if X embeds as a compact subset of some Banach space with the weak topology.) [BLvW].

Maurice's general question suggests that we ask whether either of these results hold for a space Y if we know that Y is a continuous image of a compact connected LOTS. We now know that (b) does not hold for such spaces, but the situation for (a) is not yet clear. Another example which fits into the pattern suggested by Maurice appears in a paper by van Mill and Wattel [vMW] and generalizes a well-known result for LOTS [BeL_1].

A5. Theorem: Let X be a compact, connected tree-like space. Then X has countable cellularity if and only if X is hereditarily Lindelöf.

The proof of (A5) uses Theorem A2(b).

References

[A1] K. Alster, *Subparacompactness in cartesian products of generalized ordered spaces*, Fund. Math., to appear.

[Ar_1] A.V. Arhangel'skii, *Mappings and Spaces*, Russian Math. Surveys, 1 (1966), 115-162.

[Ar_2] ________, *On hereditary properties*, Gen. Top. Appl. 3 (1973), 39-46.

[Au_1] C.E. Aull, *Topological spaces with a σ-point finite base*, Proc. Amer. Math. Soc. 29 (1971), 411-417.

[Ba] B.J. Ball, *Countable paracompactness in linearly ordered spaces*, Proc. Amer. Math. Soc. 5 (1954), 190-192.

[Bn] B. Banaschewski, *Orderable spaces*, Fund. Math. 50 (1961), 2-34.

[Be_1] H.R. Bennett, *On quasi-developable spaces*, Gen. Top. Appl. 1 (1971), 253-262.

[Be_2] ________, *Point countability in ordered spaces*, Proc. Amer. Math. Soc. 28 (1971), 598-606.

[Be_3] ________, *Quasi-metrizability and the γ-space property in certain generalized ordered spaces*, Topology Proceedings, to appear.

[Be_4] ________, *A note on the metrizability of M-spaces*, Proc. Japan Acad. 45 (1969), 6-9.

[BeL_1] H.R. Bennett and D. Lutzer, *Separability, the countable chain condition and the Lindelöf property in linearly orderable spaces*, Proc. Amer. Math. Soc. 23 (1969), 664-667.

[BeL_2] ________, *A note on weak θ-refinability*, Gen. Top. Appl. 2 (1972), 49-54.

[BeL_3] ________, *Ordered spaces with σ-minimal bases*, Topology Proceedings, 2 (1977), 371-382.

[BeL_4] H.R. Bennett and D. Lutzer, *Certain hereditary properties and metrizability in generalized ordered spaces*, Fund. Math., to appear.

[BLvW] H. Bennett, D. Lutzer and J. van Wouwe, *Eberlein compact linearly ordered spaces*, Indag. Math., to appear.

[Bk] D. Burke, *On subparacompact spaces*, Proc. Amer. Math. Soc. 23 (1969), 655-663.

[BkL] D. Burke and D.J. Lutzer, *Recent advances in the theory of generalized metric spaces*, Topology, Marcel Dekker Lecture Notes in Pure and Applied Mathematics, 24, pp. 1-70.

[Bl] H. Blumberg, *New properties of all real functions*, Trans. Amer. Math. Soc. 24 (1922), 113-128.

[BrG] J.C. Bradford and C. Goffman, *Metric spaces in which Blumberg's Theorem holds*, Proc. Amer. Math. Soc. 11 (1960), 667-670.

[Bo_1] C.J.R. Borges, *On stratifiable spaces*, Pacific J. Math. 22 (1966), 1-16.

[Bo_2] ________, *On metrizability of topological spaces*, Canad. J. Math. 20 (1968), 795-804.

[C] J. Cornette, *Image of a Hausdorff arc is cyclically extensible and reducible*, Trans. Amer. Math. Soc., 199 (1974), 253-267.

[Co] H.J. Cohen, *Sur un problème de M. Dieudonné*, C.R. Acad. Sci. Paris 234 (1952), 290-292.

[Cr] G. Creede, *Concerning semistratifiable spaces*, Pacific J. Math. 32 (1970), 47-54.

[De] E. Deak, *Theory and applications of directional structures*, Colloquia Math. Soc. Janos Bolyai 8 (Topology, Keszthely, 1972), 187-211.

[Dv] K.J. Devlin, *Aspects of Constructibility*, Springer Lecture Notes in Mathematics, vol. 354, Berlin, 1973.

[DvJ] K.J. Devlin and H. Johnsbråten, *The Souslin Problem*, Lecture Notes in Mathematics, vol. 405, Springer-Verlag, Berlin, 1974.

[Do] C.H. Dowker, *On countably paracompact spaces*, Canad. J. Math. 3 (1951), 175-186.

[vD_1] E.K. van Douwen, *Simultaneous linear extensions of continuous functions*, Gen. Top. Appl. 5 (1975), 297-320.

[vD_2] ________, *Neighborhoodassignments*, manuscript.

[vDL] E.K. van Douwen and D.J. Lutzer, *On the classification of stationary sets*, Michigan Math. J., 26 (1979), 47-64.

[vDW] J. van Dalen and E. Wattel, *A topological characterization of ordered spaces*, Gen. Top. Appl. 3 (1973), 347-354.

[Ei] S. Eilenberg, *Ordered topological spaces*, Amer. J. Math. 63 (1941), 39-45.

[En] R. Engelking, *General Topology*, Polish Scientific Publishers, 1977.

[EnL] R. Engelking and D. Lutzer, *Paracompactness in ordered spaces*, Fund. Math. 94 (1976), 49-58.

[Fa] M.J. Faber, *Metrizability in generalized ordered spaces*, Math. Centre Tracts, no. 53, Amsterdam, 1974.

[Fe] V.V. Fedorčuk, *Ordered sets and the product of topological spaces*, Vestnik Moskov Univ. Ser. I Mat. Meh. 21 (1966), 66-71.

[Fl] W.M. Fleischman, *On coverings of linearly ordered spaces*, Proc. Washington State Univ. Topology Conf., March 1970, 52-55.

[GH] L. Gillman and M. Henriksen, *Concerning rings of continuous functions*, Trans. Amer. Math. Soc. 77 (1954), 340-362.

[GJ] L. Gillman and M. Jerison, *Rings of Continuous Functions*, Van Nostrand, New York, 1960.

[dGS] J. de Groot and P.S. Schnare, *A topological characterization of products of compact, totally ordered spaces*, Gen. Top. Appl. 2 (1972), 67-74.

[H_1] R.W. Heath, *Screenability, pointwise paracompactness, and metrization of Moore spaces*, Canad. J. Math. 16 (1964), 763-770.

[H_2] ________, *A construction of quasi-metric Souslin space with a point-countable base*, Set Theoretic Topology, ed. by G.M. Reed, Academic Press, New York, 1977.

[HL_1] R.W. Heath and D.J. Lutzer, *The Dugundji Extension Theorem and collectionwise normality*, Bull. Polish Acad. Sci. 22 (1974), 827-830.

[HL_2] ________, *Dugundji Extension theorems for linearly ordered spaces*, Pacific J. Math. 55 (1974), 419-425.

[HLZ] R.W. Heath, D.J. Lutzer and P.L. Zenor, *Monotonically normal spaces*, Trans. Amer. Math. Soc. 178 (1973), 481-493.

[Hr] H. Herrlich, *Ordnungsfahigkeit topologischer Raume*, Inaugural dissertation, Berlin (1962).

[HY] J. Hocking and G. Young, *Topology*, Addison-Wesley, Reading, Mass., 1961.

[Is] T. Ishii, *A new characterization of paracompactness*, Proc. Japan Acad. 35 (1959), 435-436.

[Ka] M. Katetov, *Complete normality of cartesian products*, Fund. Math. 35 (1948), 271-274.

[Kk] H. Kok, *Connected Orderable Spaces*, Math. Center Tracts, vol. 49, Amsterdam, 1973.

[Ko] H.J. Kowalsky, *Topologischer Raume*, Birkhauser-Verlag, 1961.

[Kp_1] W. Kulpa, *A factorization theorem for GO-spaces*, *Top Structures*, 2, Math. Center Tracts, Amsterdam, 1980.

[Kp_2] W. Kulpa, *On paracompactness of GO-spaces*, preprint.

[Ku_1] Đ. Kurepa, *Around the general Souslin problem,* Proc. Int. Symp. on Topology and its Appl. Herceg-Novi (1968), 239-245.

[Ku_2] ________, *La condition de Souslin et une propriete characteristique des nombres reel,* Comptes Rendu, Paris 231, 1113-1114.

[Le] R. Levy, *A totally ordered Baire space for which Blumberg's theorem fails,* Proc. Amer. Math. Soc. 41 (1973), 304.

[Lu_1] D.J. Lutzer, *A metrization theorem for linearly orderable spaces,* Proc. Amer. Math. Soc. 22 (1969), 557-558.

[Lu_2] ________, *Autorreferat of "A metrization theorem for linearly orderable spaces,"* Zentralblat fur Mat. 177 (1970), 507.

[Lu_3] ________, *On generalized ordered spaces,* Dissertationes Math., vol. 89, 1971.

[Lu_4] ________, *Ordinals and paracompactness in ordered spaces,* Proc. Second Pittsburgh Conf., December 1972, Springer Lecture Notes in Mathematics 378, pp. 258-266.

[Ly_1] I.L. Lynn, *Linearly orderable spaces,* Proc. Amer. Math. Soc. 12 (1961), 454-456.

[Ly_2] ________, *Linearly orderable spaces,* Trans. Amer. Math. Soc. 113 (1964), 189-218.

[Ma] M.J. Mansfield, *Some generalizations of full normality,* Trans. Amer. Math. Soc. 86 (1957), 489-505.

[MP] S. Mardesic and P. Papic, *Continuous images of ordered continua,* Glasnik Mat.-Fiz. Astronom. Drustro Mat. Fiz. Hevatske Ser. II 15 (1960), 171-178.

[vMW] J. van Mill and E. Wattel, *Souslin Dendrons,* Proc. Amer. Math. Soc. 72 (1978), 545-555.

[Mi] E. Miller, *A note on Souslin's problem,* Amer. J. Math. 65 (1943), 673-678.

[Mo] K. Morita, *Products of normal spaces with metric spaces,* Math. Ann. 154 (1964), 365-382.

[Na] K. Nagani, *Σ-spaces,* Fund. Math. 65 (1969), 169-192.

[Ne] S.I. Nedev, *Every linearly ordered symmetrizable space is metrizable,* Vestnik Moskov. Univ. Ser. I 24 (1969), 31-33.

[Ny] P. Nyikos, *A compact, nonmetrizable space P such that P^2 os completely normal,* preprint.

[Po] V.I. Ponomarev, *Metrizability of a finally compact p-space with a point-countable base*, Sov. Math. Dokl. 8 (1967), 765-768.

[Pr] T. Przymusiński, *The existence of Q-sets is equivalent to the existence of strong Q-sets*, Proc. Amer. Math. Soc., to appear.

[Pu] S. Purisch, *Orderability and suborderability of topological spaces*, Thesis, Carnegie-Mellon University, 1973.

[R_1] M.E. Rudin, *Internal topology in subsets of totally orderable spaces*, Trans. Amer. Math. Soc. 118 (1965), 376-389.

[R_2] ________, *Souslin's Conjecture*, Amer. Math. Monthly 76 (1969), 1113-1119.

[R_3] ________, *Countable paracompactness and Souslin's problem*, Canad. J. Math. 7 (1955), 543-547.

[R_4] ________, *Set Theoretic Topology*, CBMS Lectures, no. 23, Amer. Math. Soc. Providence, R.I., 1975.

[R_5] ________, *A normal, hereditarily separable, non-Lindelöf space*, Illinois J. Math. 16 (1972), 621-626.

[R_6] ________, *A normal space X for which $X \times I$ is not normal*, Fund. Math. 78 (1971/72), 179-186.

[R_7] ________, *A non-normal hereditarily separable space*, Illinois J. Math. 18 (1974), 481-483.

[R_8] ________, *Hereditary normality and Souslin lines*, Gen. Top. Appl. 10 (1979), 103-105.

[Sc] B. Scott, *Toward a product theory for orthocompactness*, Thesis, Univ. of Wisconsin, 1975.

[Si] P. Simon, *A note on cardinal invariants of a square*, Comm. Math. Univ. Carolinae 14 (1973), 205-213.

[St] L.A. Steen, *A direct proof that a linearly ordered space is hereditarily collectionwise normal*, Proc. Amer. Math. Soc. 24 (1970), 727-728.

[T] F. Tall, *The countable chain condition versus separability-applications of Martin's Axiom*, Gen. Top. Appl. 4 (1974), 315-340.

[Tn] E. Tymchatyn, *The Hahn-Mazurkiewicz theorem for finitely suslinian continua*, to appear.

[V] N.V. Velichko, *Ordered p-spaces*, Izvestiya Vuz. Matematika 20 (1976), 25-36.

[Wa] M. Wage, *A generalization of Luzin's theorem*, preprint.

[WW] J. Worrell and H. Wicke, *Characterizations of developable topological spaces*, Canad. J. Math. 17 (1965), 820-830.

[vW] J. van Wouwe, *GO-spaces and generalizations of metrizability*, Math. Center Tracts, 1979, to appear.

[We] W. Weiss, *Some applications of set theory to topology*, Thesis, University of Toronto, 1975.

[Wd_1] L. Ward, *The Hahn-Mazurkiewicz theorem for rim-finite continua*, Gen. Top. Appl. 6 (1976), 183-190.

[Wd_2] ________, *A generalization of the Hahn-Mazurkiewicz theorem*, Proc. Amer. Math. Soc. 58 (1976), 369-374.

[Y] G. Young, *Representations of Banach spaces*, Proc. Amer. Math. Soc. 13 (1962), 667-668.

[Z] P. Zenor, Private communication.

DIMENSION OF GENERAL TOPOLOGICAL SPACES

Kiiti Morita
Sophia University, Tokyo

As is well known, for any topological space X we can define three kinds of dimension functions: the small inductive dimension ind X, the large inductive dimension Ind X and the covering dimension dim X. The first theory of dimension is concerned with small inductive dimension and it was founded for the case of compact metrizable spaces by K. Menger and P. Urysohn independently, and then extended to the case of separable metrizable spaces by L. Tumarkin and W. Hurewicz. For the case of general (not necessarily separable) metrizable spaces the theory of dimension was established by M. Katětov and K. Morita independently; the dimension functions involved are covering dimension and large inductive dimension, and these two dimension functions are coincident here. It is the class of metrizable spaces that has a satisfactory dimension theory nowadays.

One of the fundamental theorems in dimension theory is the sum theorem. The small inductive dimension does not satisfy the finite sum theorem even in metric spaces, as observed by van Douwen [vD] and Przymusiński [Pr_1], while it was shown by Lokucievskii [L] that small and large inductive dimensions do not satisfy the finite sum theorem in compact Hausdorff spaces.

ISBN: 0-12-584960-5

As for covering dimension, the countable sum theorem was proved in the case of normal spaces by E. Čech; he is the first who defined the concept of covering dimension, although the idea goes back to H. Lebesgue. Covering dimension of normal spaces is studied by P. Alexandroff, C.H. Dowker, E. Hemmingsen, K. Morita, et al. The final form of the definition of covering dimension was given for the case of Tychonoff spaces by M. Katětov $[K_1]$ and Yu. M. Smirnov [Sm] independently; this form of definition applies equally well to arbitrary spaces. Moreover, Katětov proved the countable sum theorem for Tychonoff spaces which generalizes Čech's result.

In view of these results, it seems that covering dimension is suitable for developing dimension theory in the case of general topological spaces. However, the most important reason for our preference of covering dimension to other dimension functions lies in the fact that covering dimension is related to other topological invariants such as the Čech cohomology groups and that it is used successfully in other branches of topology such as shape theory.

The purpose of the present paper is to give a survey of our results concerning the covering dimension of general topological spaces without any separation axiom, together with some related results by other authors; for new or unpublished results of ours, their proofs will be included.

The organization of the paper is as follows. In sections 1 to 4 we shall supply the materials which are needed for developing our dimension theory. Sections 1 to 3 are a survey of the recent results and section 4 contains new results on the notion of uniformly locally finite collections of subsets in a space; the results which are of interest in their own will also be included here. Sections 5 to 8 are devoted to the dimension theory. Subset theorems, sum theorems and product theorems are discussed; in particular, the σ-uniformly locally countable sum theorem in section 7 is applied to obtain a new proof of our product theorem in section 8. In the final section 9 the usefulness of covering dimension in homotopy theory as well as in shape theory will be shown by recent results in these areas.

Throughout this paper, by a space we shall mean a topological space without any separation axiom, and by dimension we shall mean covering dimension. N will refer to the set of positive integers.

1. *The Tychonoff functor* τ

For any space X, let us construct the product space $P(X) = \Pi\{I_\phi \mid \phi \in I^X\}$, where $I_\phi = I = [0,1]$ and I^X is the set of all continuous maps from X into I. By defining $\phi_X(x)$ to be the point of P(X) whose ϕ-coordinate is $\phi(x)$, we have a continuous map $\Phi_X: X \to P(X)$. For a continuous map f: X → Y we have a continuous map P(f): P(X) → P(Y) by defining P(f)(t) to be the point of P(Y) whose ψ-coordinate is the $\psi \circ f$-coordinate of t where $t \in P(X)$, $\psi \in I^Y$. Let us put

$$\tau(X) = \text{Image of } \Phi_X,$$

$$\tau(f) = P(f)|\tau(X): \tau(X) \to \tau(Y)$$

and denote by the same letter Φ_X the map from X to τ(X) which coincides with Φ_X but has τ(X) as its range. Then τ is a covariant functor from the category of topological spaces and continuous maps to itself and $\{\Phi_X\}$ defines a natural transformation from the identity functor to τ; τ(X) is a Tychonoff space and Φ_X: $X \to \tau(X)$ is a homeomorphism if X is itself a Tychonoff space. Hence we shall call τ the Tychonoff functor. The Tychonoff functor is the reflector from the category above to its full subcategory of Tychonoff spaces; that is,

Proposition 1.1. Any continuous map f from X into a Tychonoff space R is factored through $\tau(X)$ such that $f = g \circ \Phi_X$ with some continuous map $g: \tau(X) \to R$; g is determined uniquely by f.

Let X and Y be two spaces. Then by Proposition 1.1 there exists a unique continuous map $g: \tau(X \times Y) \to \tau(X) \times \tau(Y)$ such that

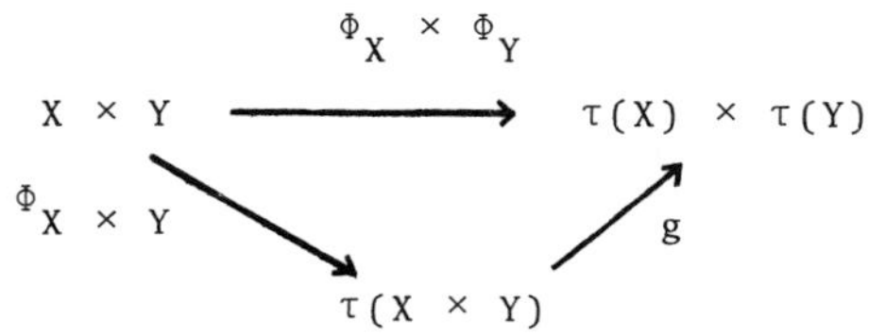

is commutative. In case g is a homeomorphism, we write $\tau(X \times Y) = \tau(X) \times \tau(Y)$. In our joint paper [HM] with T. Hoshina we have proved

Proposition 1.2. $\tau(X \times Y) = \tau(X) \times \tau(Y)$ if and only if every cozero-set of $X \times Y$ is the union of cozero-set rectangles, where by a cozero-set rectangle we mean the product of a cozero-set of X with a cozero-set of Y.

As an immediate corollary we have

Proposition 1.3 (R. Pupier [Pu]). If X is a locally compact Hausdorff space, then $\tau(X \times Y) = \tau(X) \times \tau(Y)$ for any space Y.

Indeed, let $G = \{(x,y) \mid \phi(x,y) > 0\}$ for a continuous map $\phi: X \times Y \to I$ and suppose that $\phi(x_0,y_0) > 0$; then there exists a cozero-set U of X such that $x_0 \in U$, $\mathrm{Cl}\, U \times \{y_0\} \subset G$ and $\mathrm{Cl}\, U$ is compact, and if we set $\psi(y) = \inf\{\phi(x,y) \mid x \in \mathrm{Cl}\, U\}$ the map $\psi: Y \to I$ is continuous and $(x_0, y_0) \in U \times V \subset G$ with $V = \{y \in Y \mid \psi(y) > 0\}$.

Remark. As for the problem of finding a condition for X to satisfy $\tau(X \times Y) = \tau(X) \times \tau(Y)$ for any space Y, it was solved for the case of X being Tychonoff by S. Oka [Ok] and generally by T. Ishii [Is].

2. *Normal covers*

A sequence $\{\mathcal{U}_i \mid i \in N\}$ of open covers of a space is called a normal sequence if $\mathcal{U}_{i+1}$ is a star-refinement of $\mathcal{U}_i$ for $i \in N$; that is,

$$\mathcal{U}^*_{i+1} = \{\mathrm{St}(U; \mathcal{U}_{i+1}) \mid U \in \mathcal{U}_{i+1}\} > \mathcal{U}_i$$

where ">" means "is a refinement of".

An open cover $\mathcal{U}$ of a space X is said to be normal if there is a normal sequence $\{\mathcal{U}_i \mid i \in N\}$ of open covers of X such that $\mathcal{U}_1 > \mathcal{U}$.

Then we have

Lemma 2.1. An open cover $\{G_\lambda \mid \lambda \in \Lambda\}$ of a space X is normal if and only if there exists a continuous map ϕ from X into a metric space T such that $\phi^{-1}(H_\lambda) \subset G_\lambda$, $\lambda \in \Lambda$, for some open cover $\{H_\lambda \mid \lambda \in \Lambda\}$ of T.

Hence by Proposition 1.1 we have

Lemma 2.2. An open cover $\{G_\lambda \mid \lambda \in \Lambda\}$ of X is normal if and only if there is a normal open cover $\{H_\lambda \mid \lambda \in \Lambda\}$ of $\tau(X)$ such that $\Phi_X^{-1}(H_\lambda) \subset G_\lambda$ for each $\lambda \in \Lambda$.

The following lemma is also useful.

Lemma 2.3 ([Mo_2]). An open cover of a space X is normal if and only if it is refined by a locally finite (or σ-locally finite) cozero-set cover.

Proposition 2.4 ([Mo_7]). Let X be a space and Y a compact Hausdorff space. Let $\mathcal{G} = \{G_\alpha | \alpha \in \Omega\}$ be an open cover of $X \times Y$. Then there exists an open cover $\mathcal{U} = \{U_\lambda | \lambda \in \Lambda\}$ of X satisfying conditions (a), (b) and (c) below:

(a) card $\Lambda \leq m$ or $< \aleph_0$ according as $m \geq \aleph_0$ or $m < \aleph_0$ where $m = \max(\text{card } \Omega, \text{weight of } Y)$.

(b) for a suitable collection $\{\mathcal{V}_\lambda | \lambda \in \Lambda\}$ of finite open covers of Y, the collection $\{U_\lambda \times V | V \in \mathcal{V}_\lambda, \lambda \in \Lambda\}$ is an open cover of $X \times Y$ which refines $\mathcal{G}$.

(c) $\mathcal{U}$ is normal if and only if $\mathcal{G}$ is normal.

Now, let $\Phi = \{\mathcal{U}_i | i \in N\}$ be a normal sequence of open covers of a space X. Let (X,Φ) be a space obtained from X by taking $\{St(x,\mathcal{U}_i) | i \in N\}$ as a local base at each point x of X, and let X/Φ be the quotient space of (X,Φ) obtained by identifying those points x and y for which $y \in \cap\{St(x,\mathcal{U}_i) | i \in N\}$. Let

(1) $$\phi: X \to X/\Phi$$

be the composite of the identity map from X to (X,Φ) and the quotient map from (X,Φ) onto X/Φ. Furthermore, let us set, for a subset A of X,

(2) $$\text{Int}(A;\Phi) = \{x \in X | St(x,\mathcal{U}_i) \subset A \text{ for some } i\}.$$

Then $\text{Int}(A;\Phi)$ is open in (X,Φ) and

(3) $$\phi^{-1}(\phi(\text{Int}(A;\Phi))) = \text{Int}(A;\Phi),$$

and hence $\phi(\text{Int}(A;\Phi))$ is open in X/Φ. Next, let us set

(4) $$\mathcal{V}_i = \{\text{Int}(U;\Phi) | U \in \mathcal{U}_i\} \text{ for } i \in N.$$

Then $\Psi = \{\mathcal{V}_i | i \in N\}$ is a normal sequence of open covers of X and $\mathcal{V}_{i+1} > \mathcal{U}_{i+1} > \mathcal{V}_i$ for $i \in N$, and hence $X/\Phi = X/\Psi$; let us call Ψ the normalized normal sequence associated with Φ.

Theorem 2.6 ([Mo_3]). The space X/Φ is metrizable and $\{\phi(\mathcal{V}_i) | i \in N\}$ is a normal sequence of open covers of X/Φ such that $\{St(y,\phi(\mathcal{V}_i)) | i \in N\}$ is a local base at $y \in X/\Phi$. X/Φ will be called the metrizable space associated with a normal sequence Φ of open covers of X.

Here we note

Lemma 2.7. order $\phi(\mathcal{V}_i)$ = order $\mathcal{V}_i \leq$ order $\mathcal{U}_i$.

3. *C*-embedding and z-embedding*

Let A be a subset of a space X. Then A is said to be C*-embedded (resp. C-embedded) in X if every bounded real-valued (resp. every real-valued) continuous function defined on A can be extended over X, and A is said to be P-embedded in X if every normal cover of A is refined by the restriction of a normal cover of X to the subspace A. In case we restrict covers under consideration to ones with cardinality $\leq m$ (m an infinite cardinal number), we have the notion of P^m-embedding.

Lemma 3.1. A is C*-embedded (resp. C-embedded) in X if and only if every finite (resp. countable) cozero-set cover of A is refined by the restriction of a finite (resp. countable) cozero-set cover of X to the subspace A.

In case every cozero-set of A is expressed as the intersection of a cozero-set of X with A, A is said to be z-embedded. Obviously we have the following implications.

Lemma 3.2. P-embedding $\Rightarrow$ P^m-embedding $\Rightarrow$ C-embedding $\Rightarrow$ C*-embedding $\Rightarrow$ z-embedding.

Since any cozero-set of a cozero-set of X is a cozero-set

of X, we have

Lemma 3.3. Any cozero-set of X is z-embedded in X.

Lemma 3.4. Let A be a subset of a space X. Then A is z-embedded in X if and only if for every finite cozero-set cover $\mathcal{G}$ of A there is a finite cozero-set cover $\mathcal{H}$ of some cozero-set of X containing A such that the restriction of $\mathcal{H}$ to A refines $\mathcal{G}$.

Proof is straightforward.

Lemma 3.5. Let A be a subset of a space X. If A is C*-embedded in X, so is $\Phi_X(A)$ in $\tau(X)$ and $\Phi_X(A) \cong \tau(A)$. If A is z-embedded (resp. C-embedded) in X, so is $\Phi_X(A)$ in $\tau(X)$.

As for C-embedding, we have the following theorem which contains C.H. Dowker's generalization of the Borsuk homotopy extension theorem.

Theorem 3.6 ([MoH_1]). Let A be a subset of a space X. Then the following conditions are equivalent:

(a) A is C-embedded (resp. P-embedded) in X.

(b) $X \times B \cup A \times I$ is C-embedded (resp. P-embedded) in $X \times I$ for any closed subset B of I.

(c) Any continuous map $g: X \times \{0\} \cup A \times I \to Y$ is extended to a continuous map $f: X \times I \to Y$ for any separable metric (resp. metric) Čech complete ANR space Y.

The following is also useful in our later discussions.

Proposition 3.7 ([Mo_{11}]). Let A be a compact subset of a Tychonoff space X. Then $A \times Y$ is P-embedded in $X \times Y$ for any space Y.

Proposition 3.8 ([Mo_{12}]). Let X be a space and A a subspace of X which is locally compact and paracompact Hausdorff. If A is P-embedded in X, then $A \times Y$ is P-embedded in $X \times Y$ for any space Y.

The following proposition is not used in this paper, but is of some interest in itself; it can be proved easily by a theorem of E. Michael.

Proposition 3.9. A regular Hausdorff space X, which is a countable union of closed paracompact subspaces A_i, $i \in N$, is paracompact if and only if each A_i is P-embedded in X.

This proposition, however, fails to be valid if "paracompact" is replaced by "collectionwise normal". This is seen from the fact that the Tychonoff plank $X = W(\omega_0 + 1) \times W(\omega_1 + 1) - \{(\omega_0, \omega_1)\}$ is the union of A_α, $(\alpha < \omega_0)$ and B but not normal, where $A_\alpha = \{\alpha\} \times W(\omega_1 + 1)$, $B = \{\omega_0\} \times W(\omega_1)$.

4. *Uniformly locally finite collections of subsets*

The results in this section are obtained in [Mo_{12}].

Let X be a space. Then a collection $\mathcal{K}$ of subsets of X will be called uniformly locally finite (resp. uniformly locally countable) if there is a normal cover $\mathcal{U}$ of X such that each member of $\mathcal{U}$ intersects only a finite number (resp. countable number) of the members of $\mathcal{K}$. For a uniform space X the notion of uniformly local finiteness is introduced by J.R. Isbell [I]. In case X is a Tychonoff space the notion of uniformly local finiteness in our sense is nothing else but the notion of uniformly local finiteness in the sense of Isbell with respect to X with the finest uniformity, and in the case of normal spaces it coincides with the notion of uniformly local finiteness in the sense of Katětov [K_2].

The union of a locally finite collection of zero-sets in a space X is not always a zero-set in X. In this connection, the usefulness of the notion of uniformly local finiteness is seen from the following

Proposition 4.1. Let $\{A_\lambda | \lambda \in \Lambda\}$ be a uniformly locally finite collection of subsets in a space X. If each A_λ is a zero-set in X, then the union $\cup A_\lambda$ is also a zero-set in X.

This proposition is a direct consequence of [MoH_2, Lemma 2.3] and Proposition 4.2 below.

Proposition 4.2. A collection $\{A_\lambda | \lambda \in \Lambda\}$ of subsets in a space X is uniformly locally finite if and only if there are two collections $\{F_\lambda | \lambda \in \Lambda\}$ and $\{G_\lambda | \lambda \in \Lambda\}$ of subsets of X such that

(a) $A_\lambda \subset F_\lambda \subset G_\lambda$ for each $\lambda \in \Lambda$,

(b) F_λ is a zero-set and G_λ is a cozero-set for each $\lambda \in \Lambda$, and

(c) $\{G_\lambda | \lambda \in \Lambda\}$ is locally finite (in the usual sense) in X.

Proof. Suppose that $\{A_\lambda\}$ is uniformly locally finite. Then there is a normal cover $\mathcal{U}$ of X such that each member of $\mathcal{U}$ intersects only a finite number of A_λ's. Let $\mathcal{V}$ be a normal cover of X which is a star-refinement of $\mathcal{U}$. Then $\{St(A_\lambda, \mathcal{V}) | \lambda \in \Lambda\}$ is also locally finite in X. As is easily seen, there are a zero-set F_λ and a cozero-set G_λ such that $A_\lambda \subset F_\lambda \subset G_\lambda \subset St(A_\lambda, \mathcal{V})$. This proves the "only if" part.

Conversely, suppose that there are $\{F_\lambda\}$ and $\{G_\lambda\}$ satisfying (a), (b) and (c). Let $\mathcal{W}$ be the intersection of all the binary covers $\{G_\lambda, X - F_\lambda\}$ with $\lambda \in \Lambda$. Then by [MoH$_2$, Lemma 2.3] $\mathcal{W}$ is a locally finite cozero-set cover of X and hence a normal cover of X by Lemma 2.3, and each member of $\mathcal{W}$ intersects only a finite number of A_λ's.

Proposition 4.2 was obtained also by H. Ohta [O] with a different proof independently.

The following theorem reveals the features of uniformly local finiteness.

Theorem 4.3. Let $\{\mathcal{K}_i | i \in N\}$ be a countable number of collections of subsets in a Tychonoff space X. Then the following conditions are equivalent.

(a) $\mathcal{K}_i$ is uniformly locally finite (resp. uniformly locally countable) in X for each $i \in N$.

(b) There is a paracompact space Y such that $X \subset Y \subset \beta(X)$ and $\mathcal{K}_i$ is locally finite (resp. locally countable) in Y for each $i \in N$.

Proof. (b) $\Rightarrow$ (a) is obvious. To prove (a) $\Rightarrow$ (b), suppose that for each i there exists a normal cover $\mathcal{W}_i$ of X such that each member of $\mathcal{W}_i$ intersects only a finite number (resp. countable number) of the members of $\mathcal{K}_i$. Then we can find a normal sequence $\Phi = \{\mathcal{U}_i | i \in N\}$ of open covers of X such that $\mathcal{U}_i > \mathcal{W}_i$ for each $i \in N$. Let us construct the metrizable space X/Φ associated with the normal sequence Φ; let $\phi: X \to X/\Phi$ be the canonical map defined in Section 2. Let $\{\mathcal{V}_i | i \in N\}$ be the normalized normal sequence of open covers of X which is associated with Φ; the construction of $\mathcal{V}_i$ is described in Section 2. Since $\mathcal{V}_i > \mathcal{U}_i$ and $\phi^{-1}\phi(V) = V$

for each $V \in \mathcal{V}_i$, each member of $\phi(\mathcal{V}_i)$ intersects only a finite number (resp. countable number) of the members of $\phi(\mathcal{K}_i)$. Let $\beta(\phi)\colon \beta(X) \to \beta(X/\Phi)$ be the extension of ϕ over $\beta(X)$, where $\beta(X)$ is the Stone-Čech compactification of X as usual. Let us set

$$g = \beta(\phi)|Y\colon Y \to X/\Phi, \quad Y = \beta(\phi)^{-1}(X/\Phi).$$

Then g is a perfect map and Y is a paracompact M-space. Since each member of $\phi(\mathcal{V}_i)$ intersects only a finite number (resp. countable number) of the members of $\phi(\mathcal{K}_i)$, each member of $g^{-1}(\phi(\mathcal{V}_i))$ intersects only a finite number (resp. countable number) of the members of $\mathcal{K}_i$. Thus, (b) holds.

The usefulness of the notion of uniformly local finiteness is also seen from Theorem 4.4 below, which fails to be true if the word "uniformly" is dropped; indeed, $\{(\alpha,\omega_1)\,|\,\alpha < \omega_0\}$ is locally finite in $W(\omega_0 + 1) \times W(\omega_1 + 1) - \{(\omega_0,\omega_1)\}$ but its union is not C*-embedded.

Theorem 4.4. Let $\{A_\lambda\,|\,\lambda \in \Lambda\}$ be a uniformly locally finite collection of subsets of a space X. If A_λ is C*-embedded (resp. P^m-embedded or P-embedded) in X for any $\lambda \in \Lambda$ and if $A_\lambda \cup A_\mu$ is C*-embedded in X for any λ, μ in Λ, then the union $A = \cup\{A_\lambda\,|\,\lambda \in \Lambda\}$ is also C*-embedded (resp. P^m-embedded or P-embedded) in X.

As was shown in [MoH_1], the union of two subsets A_1 and A_2 which are C*-embedded in X is not always C*-embedded in X. A necessary and sufficient condition for $A_1 \cup A_2$ to be C*-embedded in X is given in [MoH_1] and a sufficient condition given in [MoH_1] is that either A_1 or A_2 be a zero-set in X. Thus, we have

Theorem 4.5. Let $\{A_\lambda | \lambda \in \Lambda\}$ be a uniformly locally finite collection of zero-sets in a space X. If each A_λ is C*-embedded (resp. P-embedded) in X, so is $A = \cup A_\lambda$.

Proof of Theorem 4.4. Suppose that $A_\lambda \cup A_\mu$ is C*-embedded in X. By Lemmas 2.2 and 3.5 we may, and shall, assume that X is a Tychonoff space. Then by Theorem 4.3 there exists a paracompact space Y such that $X \subset Y \subset \beta(X)$ and $\{A_\lambda\}$ is locally finite also in Y. Let $f: A \to I$ be any continuous map, and let us set $f_\lambda = f|A_\lambda: A_\lambda \to I$. Since A_λ is C*-embedded in X, f_λ is extended uniquely to a continuous map $g_\lambda: \mathrm{Cl}_{\beta X}A_\lambda \to I$, and similarly $f_{\lambda,\mu} = f|A_\lambda \cup A_\mu: A_\lambda \cup A_\mu \to I$ is extended uniquely to a continuous map $g_{\lambda,\mu}: \mathrm{Cl}_{\beta X}(A_\lambda \cup A_\mu) \to I$. Therefore, we have

$$g_\lambda(z) = g_{\lambda,\mu}(z) \qquad \text{for } z \in \mathrm{Cl}_{\beta X}A_\lambda,$$
$$g_\mu(z) = g_{\lambda,\mu}(z) \qquad \text{for } z \in \mathrm{Cl}_{\beta X}A_\mu,$$

and hence

$$g_\lambda(z) = g_\mu(z) \qquad \text{for } z \in \mathrm{Cl}_{\beta X}A_\lambda \cap \mathrm{Cl}_{\beta X}A_\mu.$$

Therefore, if we set

$$h_\lambda = g_\lambda|\mathrm{Cl}_Y A_\lambda: \mathrm{Cl}_Y A_\lambda \to I,$$

then we have

$$h_\lambda(y) = h_\mu(y) \qquad \text{for } y \in \mathrm{Cl}_Y A_\lambda \cap \mathrm{Cl}_Y A_\mu.$$

Let us define a map $h: B \to I$, where $B = \cup\{\mathrm{Cl}_Y A_\lambda | \lambda \in \Lambda\}$, by

$$h(y) = h_\lambda(y) \qquad \text{if } y \in \mathrm{Cl}_Y A_\lambda;$$

then h is single-valued and $h_\lambda = h|\mathrm{Cl}_Y A_\lambda$ is continuous. Since $\{A_\lambda | \lambda \in \Lambda\}$ is locally finite in Y, so is also $\{\mathrm{Cl}_Y A_\lambda | \lambda \in \Lambda\}$ in Y. Hence $h: B \to I$ is a continuous map. Since B is closed in Y and Y is normal, h is extended to a continuous map $k: Y \to I$. Thus, $k|X: X \to I$ is an extension of $f: A \to I$ and hence the theorem is proved for the case of C*-embedding.

To treat the case of P^m-embedding, let K be any compact Hausdorff space of weight $\leq m$. Then by [MoH_1, Theorem 2.9] and [MoH_2, Theorem 1.3] $A_\lambda \times K \cup A_\mu \times K$ is C*-embedded in $X \times K$ for any λ, μ in Λ. Hence it follows from the theorem for the case of C*-embedding that $A \times K$ is C*-embedded in $X \times K$. Applying [MoH_2, Theorem 1.3] again, we see that A is P^m-embedded in X. This completes the proof of Theorem 4.4.

A collection $\mathcal{K}$ of subsets in a space is called σ-uniformly locally finite (resp. countable) if $\mathcal{K}$ is a countable union of uniformly locally finite (resp. countable) subcollections $\mathcal{K}_i$, $i \in N$.

Theorem 4.6. A Tychonoff space X admits a σ-uniformly locally finite cover by compact sets if and only if X is paracompact and a countable union of locally compact closed subspaces.

Proof. The "if" part is obvious. To prove the "only if" part, assume that $\mathcal{K}_i$ is a uniformly locally finite collection of compact subsets of X for $i \in N$ and that $\cup\mathcal{K}_i$ is a cover of X. Then by Theorem 4.3 there is a paracompact space Y such that $X \subset Y$ and each $\mathcal{K}_i$ is locally finite still in Y. Thus, X is an F_σ-subset of Y and hence X is paracompact. This proves the "only if" part.

Remark 4.7. Theorem 4.6 remains true if "locally finite" is replaced by "locally countable."

A space X will be said to have property (U) if the notion of local finiteness coincides with the notion of uniformly local finiteness for collections of subsets of X. Then by a theorem of Katětov [K_2] a normal space has property (U) if and only if it is strongly normal (i.e. countably paracompact and collectionwise normal). Hence the pre-image of a strongly normal space under a quasi-perfect map has property (U). In particular, every M-space has property (U). However, an M*-space or an M'-space, which are both generalizations of M-spaces, does not have property (U), as is observed by T. Hoshina. Thus, property (U) is different from expandability in the sense of L.L. Krajewski.

Problem 4.7. Find a property which, combined with expandability, is equivalent to property (U).

5. *Definition and basic properties of dimension*

Let us begin with the definition of dimension.

Definition 5.1. The (covering) dimension of a space X, dim X in notation, is the least integer n such that any finite normal cover of X is refined by a normal cover of X of order $\leq n+1$; in case there is no such integer n, the dimension of X is infinite and we write dim X = ∞.

In this paper we are concerned exclusively with the finite dimensional spaces.

Since a space is normal if and only if any finite open cover is normal, we have

Proposition 5.2. If X is normal, then dim X is the least integer n such that any finite open cover of X is refined by a finite open cover of X of order $\leq n+1$.

Thus, the above definition of dimension is equivalent to the definition given by E. Čech for the case of normal spaces, and it was given for the case of Tychonoff spaces by M. Katětov and Yu. M. Smirnov.

Theorem 5.3 ([Mo_7]). $\dim X = \dim \tau(X)$ for any space X.

This follows directly from Lemma 2.2.

By virtue of Theorem 5.3, together with Lemmas 2.2 and 3.5, many of the theorems on the dimension of Tychonoff spaces are extended to the case of general spaces. The following are such a kind of theorems. Here by $\dot{I}^{n+1}$ we mean the boundary of I^{n+1}, where I^{n+1} is the product of n + 1 copies of $I = [0,1]$.

Theorem 5.4 ([Mo_7]). If A is C*-embedded in a space X, then $\dim A \leq \dim X$. (Cf. Theorem 5.16 below)

Theorem 5.5 ([Mo_7]). For a space X, $\dim X \leq n$ if and only if every normal cover of X is refined by a normal cover of X of order $\leq n+1$.

Theorem 5.6 ([GJ],[Mo_7]). For a space X, $\dim X \leq n$ if and only if for any continuous map $f: X \to I^{n+1}$ there is a continuous map $g: X \to \dot{I}^{n+1}$ such that $g(x) = f(x)$ for $x \in f^{-1}(\dot{I}^{n+1})$.

Indeed, in case X is a Tychonoff space, Theorems 5.4, 5.5 and 5.6 were proved by M. Katětov [K_1], B. Pasynkov [P_1] and Yu. M. Smirnov [Sm] respectively; for another proof of Theorem 5.5 in this case cf. Morita [Mo_4].

The following theorem is proved by J.R. Gard and R.D. Johnson [GJ], together with Theorem 5.6 above.

Theorem 5.7 ([GJ]). For a space X the following are equivalent:

(a) $\dim X \leq n$.

(b) For each continuous map $f: X \to I^{n+1}$ any point y of I^{n+1} is an unstable value of f (i.e. for each $\varepsilon > 0$ there is a continuous map $g: X \to I^{n+1}$ such that $|f - g| < \varepsilon$ and $y \notin g(X)$).

(c) For any n + 1 pairs (A_i, B_i), $i=1,\ldots,n+1$ of disjoint zero-sets there exist n + 1 zero-sets C_i, $i=1,\ldots,n+1$, such that A_i and B_i are separated in $X - C_i$ and $\cap\{C_i | i=1,\ldots,n+1\} = \emptyset$.

Another proof of (a) $\Leftrightarrow$ (c) in Tychonoff spaces is found in Holsztyński [H].

From these results one might think that it would be sufficient to develop the dimension theory within the framework of Tychonoff spaces. However, it is not the case as will be seen below. Indeed, the space X/A, which is obtained as the quotient space of X by contracting A to a point (this point will be denoted by q_A in this paper), is not necessarily a Tychonoff space even if X is Tychonoff and A is closed (in this paper, in case $A = \emptyset$, we mean by X/A the disjoint union of X and a single point q_A), and consideration about the dimensional behavior of X/A will be seen to play an important role in our theory of dimension.

Theorem 5.8 ([Mo_7]). $\dim X/A \leq \dim X$ for any subset A of a space X.

Theorem 5.9 ([Mo_9]). If A is C*-embedded in a space X, we have $\dim X = \max(\dim X/A, \dim A)$.

On the other hand, it is easy to prove

Proposition 5.10. Let A be a closed set of a normal space X. Then X/A is normal and $\dim X/A \leq n$ if and only if $\dim F \leq n$ for any closed set F of X such that $F \cap A = \emptyset$; if, in addition, A is a zero-set, then $\dim X/A = \dim(X - A)$.

Thus, Theorem 5.9 may be considered as a generalization of Proposition 5.11 below which is well known.

Proposition 5.11. In case A is a closed set in a normal space X, $\dim X \leq n$ if and only if $\dim A \leq n$ and $\dim F \leq n$ for any closed set F of X such that $F \cap A = \emptyset$.

For a cover $\mathcal{V} = \{V_\lambda | \lambda \in \Lambda\}$ of a space X, the nerve $N(\mathcal{V})$ of $\mathcal{V}$ is defined to be the simplicial complex over Λ whose simplexes are the finite subsets Γ of Λ such that $\cap\{V_\lambda | \lambda \in \Gamma\} \neq \emptyset$; if $A \subset X$, we denote by $\mathcal{V} \cap A$ the cover of the subspace A which consists of $V \cap A$ with $V \in \mathcal{V}$. The geometric realization of the nerve $N(\mathcal{V})$ with the weak topology is denoted also by the same letter $N(\mathcal{V})$. Thus, $N(\mathcal{V})$ is a perfectly normal, paracompact, Hausdorff space and $N(\mathcal{V} \cap A)$ is closed in $N(\mathcal{V})$ and hence $\dim N(\mathcal{V})/N(\mathcal{V} \cap A) = \dim (N(\mathcal{V}) - N(\mathcal{V} \cap A))$ by Proposition 5.10.

Theorem 5.12 ([Mo_9]). If A is a subset of a space X, then the following conditions are equivalent:

(a) $\dim X/A \leq n$.

(b) For any finite (resp. locally finite) normal cover $\mathcal{U}$ of X there exists a finite (resp. locally finite) normal cover $\mathcal{V}$ of X such that $\mathcal{V} > \mathcal{U}$ and $\dim N(\mathcal{V})/N(\mathcal{V} \cap A) \leq n$.

(c) For any countable collection $\{\mathcal{U}_i | i \in N\}$ of normal covers of X there exists a continuous map ψ from X to a metrizable space T such that $\psi: X \to T$ is a $\mathcal{U}_i$-map for each $i \in N$ and $\dim T/\mathrm{Cl}\,\psi(A) \leq n$.

Here $\psi: X \to T$ is a $\mathcal{U}$-map for an open cover $\mathcal{U}$ of X if $\mathcal{U}$ is refined by $\psi^{-1}(\mathcal{V})$ with some open cover $\mathcal{V}$ of T. In case G is open in X, then $G \cap A = \emptyset$ if and only if $G \cap \mathrm{Cl}\, A = \emptyset$. Hence

(b) of Theorem 5.12 implies

Lemma 5.13. dim X/A = dim X/Cl A for a subset A in a space X.

In view of Lemma 5.13 the following well-known theorem may be considered as a corollary to Theorem 5.9.

Theorem 5.14. If X a is Tychonoff space, then dim X = dim $\beta(X)$.

Indeed, since the proof of Theorem 5.8 in [Mo_7] as well as the proof of the inequality dim X $\leq$ max(dim X/A, dim A) in [Mo_9] does not appeal to Theorem 5.14, we have only to show that Theorem 5.4 can be proved without using Theorem 5.14, but such a proof can be obtained from the proof of the equivalence (a) $\Leftrightarrow$ (b) in Theorem 5.15 below, which may be compared with Theorem 5.12.

Theorem 5.15. In case A is C*-embedded in a space X, the following conditions are equivalent:

(a) dim A $\leq$ n.

(b) For any finite normal cover $\mathcal{U}$ of X there is a finite normal cover $\mathcal{V}$ of X such that $\mathcal{V} > \mathcal{U}$ and dim N($\mathcal{V} \cap$ A) $\leq$ n.

(c) For any countable collection $\{\mathcal{U}_i \mid i \in N\}$ of finite normal covers of X there exists a continuous map ψ from X to a separable metric space T such that ψ is a $\mathcal{U}_i$-map for each $i \in N$ and dim Clψ(A) $\leq$ n.

Proof. (a) $\Leftrightarrow$ (b) follows immediately from Lemma 3.1. Since (c) $\Rightarrow$ (b) is also easy to see, we have only to prove (b) $\Rightarrow$ (c). Assume (b). Then there is a normal sequence $\Psi = \{\mathcal{W}_i \mid i \in N\}$ of finite normal covers of X such that

$$\mathcal{W}_i > \mathcal{U}_i \text{ and dim } N(\mathcal{W}_i \cap A) \leq n \text{ for } i \in N.$$

Then the continuous map $\psi: X \to X/\Psi$ is the desired map with $T = X/\Psi$.

Now, we shall give a generalization of Theorem 5.4.

Theorem 5.16. If A is z-embedded in X, then dim A $\leq$ dim X.

Proof. Let $\{G_1, \ldots, G_s\}$ be a finite cozero-set cover of A and let dim X $\leq$ n. Then for each i there is a cozero-set H_i of $\tau(X)$ such that $G_i = A \cap \Phi_X^{-1}(H_i)$; this is obvious by Proposition 1.1. Moreover, for each i there is a cozero-set $\tilde{H}_i$ of $\beta(\tau(X))$ such that $H_i = \tau(X) \cap \tilde{H}_i$. Since $\cup\tilde{H}_i$ is an F_σ-subset of $\beta(\tau(X))$, we have dim $\cup \tilde{H}_i \leq$ dim $\beta(\tau(X))$ = dim X $\leq$ n. Hence there exists a finite cozero-set cover $L = \{L_i \mid i=1,\ldots,s\}$ of $\cup\tilde{H}_i$ such that $L_i \subset \tilde{H}_i$ for i=1,...,s and order $L \leq n+1$. Since each L_i is also a cozero-set of $\beta(\tau(X))$, $K_i = L_i \cap \tau(X)$ is a cozero-set of $\tau(X)$. Hence, if we set $P_i = \Phi_X^{-1}(K_i) \cap A$ for i=1,...,s, each P_i is a cozero-set of A and $P_i \subset G_i$ for each i. Moreover, $\{P_i \mid i=1,..,s\}$ is a finite cozero-set cover of A of order $\leq$ n+1. This prove dim A $\leq$ n.

The formula dim X = max(dim X/A, dim A) in Theorem 5.9 does not hold in general for the case of A being z-embedded in X, as is seen from Example 5.17 below.

Example 5.17. In [Po] E. Pol constructed a Tychonoff space X satisfying the conditions below:

(a) dim X > 0;

(b) $X = F_1 \cup F_2$, where F_i is a zero-set and dim $F_i = 0$ for i=1,2;

(c) $F_i = G_i \cup F$, where G_i are cozero-sets and F is discrete;

(d) $G_1 \cup G_2$ is a countable dense subset of X.

If we set $A = G_1 \cup G_2$, then A is a cozero-set of X and $\dim A = 0$, $\dim X/A = \dim X/\mathrm{Cl}\, A = 0$.

The following is of a peculiar character in this paper; it fails to be valid if "strongly" is removed.

Proposition 5.18 ([Mo_1]). If a subspace A of a Tychonoff space X is strongly paracompact, then $\dim A \leq \dim X$.

Recently E. Pol and R. Pol have succeeded in obtaining a hereditarily normal space X_1 such that $\dim X_1 = 0$ but $\dim A_1 > 0$ for some subset A_1 of X_1 and a hereditarily normal space X_2 such that $\dim X_2 = 0$ but $\dim B_n = n$ for each natural number n and for some subspace B_n of X_2 in [PoP_1] and [PoP_2] respectively. Thus, the subset theorem does not hold in general even in hereditarily normal spaces.

Finally, we conclude this section by posing a problem. In a previous paper we proved the following theorem.

Theorem 5.19 ([Mo_7]). If X is a Tychonoff space with $\dim X \leq n$ and topologically complete, then X can be expressed as the inverse limit of an inverse system of at most n-dimensional metric spaces.

In another paper [Mo_8] we have proved

Theorem 5.20 ([Mo_8]). A Tychonoff space X is expressed as the inverse limit of an inverse system of polyhedra if and only if X is topologically complete.

It is not possible to improve Theorem 5.20 so that if $\dim X \leq n$ then each polyhedron in the inverse system can be chosen to be at most n-dimensional, even in the case of X being compact Hausdorff; this fact was proved by S. Mardešić $[M_1]$.

Problem 5.21. Find a necessary and sufficient condition for a space X with dim X $\leq$ n to be the inverse limit of an inverse system of polyhedra of dimension $\leq$n.

6. *The Hopf extension theorem for general spaces and a cohomological characterization of dimension*

Let A be a subset of a space X. For each integer r $\geq$ 0 and any additive group G, let $H^r(X,A;G)$ be the r-th Čech cohomology group with coefficients in G which is based on all the locally finite normal covers of X.

A continuous map f: $(X,A) \to (I^n, \dot{I}^n)$ is called essential if any continuous map g: $(X,A) \to (I^n, \dot{I}^n)$ with $g|A = f|A$ satisfies $g(X) = I^n$; otherwise f is called inessential. Hence f is inessential if and only if there is a continuous map g: $X \to \dot{I}^n$ which is an extension of $f|A: A \to \dot{I}^n$.

Thus, the following theorem may be viewed as a generalization of the Hopf extension theorem.

Theorem 6.1 ($[Mo_9]$). Let A be a subset of a space X such that dim X/A $\leq$ n. Then a continuous map f: $(X,A) \to (I^n, \dot{I}^n)$ is inessential if and only if the induced homomorphism $H^n(f)$: $H^n(I^n, \dot{I}^n; \mathbb{Z}) \to H^n(X,A;\mathbb{Z})$ is zero, where n $\geq$ 2 and $\mathbb{Z}$ is the additive group of integers.

In case A is C-embedded in X, we obtain the same formulation as the usual Hopf extension theorem on CW-complexes or paracompact spaces; the latter case is due to C.H. Dowker.

Theorem 6.2 ($[Mo_9]$). Let X be a space and A a subset of X which is C-embedded in X. Let dim X/A $\leq$ n and n $\geq$ 2. Then a continuous map $g: A \to \dot{I}^n$ can be extended to a continuous map from X to $\dot{I}^n$ if and only if Image $H^{n-1}(g) \subset$ Image $H^{n-1}(i)$, where $H^{n-1}(g)$: $H^{n-1}(\dot{I}^n; \mathbb{Z}) \to H^{n-1}(A; \mathbb{Z})$ and $H^{n-1}(i)$: $H^{n-1}(X; \mathbb{Z}) \to H^{n-1}(A; \mathbb{Z})$ are the induced homomorphisms by g and i respectively, and $i: A \to X$ is the inclusion map.

As a corollary to Theorem 6.1 we have, by Lemma 6.4 below,

Theorem 6.3. If $\dim X/A \leq n$, then every continuous map $f: (X,A) \to (I^{n+1}, \dot{I}^{n+1})$ is inessential.

Lemma 6.4 ([Mo_7]). $H^r(X,A;G) \cong H^r(X/A,q_A;G)$. In particular, $H^r(X,A;G) \cong H^r(X/A;G)$ for $r \geq 1$, and $\dim X/A \leq n$ implies $H^r(X,A;G) = 0$ for $r > n$.

The following theorem is a cohomological characterization of dimension; it has been proved hitherto for the case of paracompact Hausdorff spaces by C.H. Dowker, and such an approach to dimension theory of compact metric spaces from the point of view of homology theory was originated by P. Alexandroff.

Theorem 6.5. Let $\dim X < \infty$. Then $\dim X \leq n$ if and only if $H^{n+1}(X,A;\mathbb{Z}) = 0$ for any subset A of X.

Proof is obvious from Theorems 5.6 and 6.1.

If Y is a compact Hausdorff space and B is its closed subset, then we have the Künneth formula, which was proved in [Mo_7]:

$$H^n((X,A) \times (Y,B);G) \cong \bigoplus_{q=0}^{n} H^q(X,A;H^{n-q}(Y,B;G))$$

where $\oplus$ means the operation of taking the direct sum of additive groups.

Applying this formula to the case where $Y = I$, $B = \dot{I}$ and noting that $H^1(I,\dot{I};\mathbb{Z}) \cong \mathbb{Z}$, we see that $H^n(X,A;\mathbb{Z}) \neq 0$ implies

$H^{n+1}((X,A) \times (I,\dot{I});\mathbb{Z}) \neq 0$. Thus, we obtain from Theorem 6.5,

Theorem 6.6 ([Mo_7]). $\dim (X \times I) \geq \dim X + 1$ for any space X. (The equality will be proved in section 8 below.)

Furthermore, we can prove Theorem 6.7 below by using the cohomological method.

Theorem 6.7 ([Mo_7], [Mo_9]). Let A be a subset of a space X such that $\dim X/A = n$, and let $f: (X,A) \to (I^n,\dot{I}^n)$ be a continuous map. If f is essential, so is the product map $f \times 1: (X,A) \times (I,\dot{I}) \to (I^n,\dot{I}^n) \times (I,\dot{I}) = (I^{n+1},\dot{I}^{n+1})$, where $f \times 1$ is defined by $(f \times 1)(x, t) = (f(x), t)$ for $x \in X$, $t \in I$.

This theorem implies also Theorem 6.6.

Quite recently, Ščepin [Sc] has obtained essentially the same result as Theorm 6.7 in the case of compact Hausdorff spaces also with the cohomological method and applied it to prove an interesting theorem asserting that a finite-dimensional compact absolute neighborhood retract is metrizable.

Problem 6.8. Prove Theorem 6.7 without using the cohomological method.

7. *Sum theorems*

The results in this section are taken from [Mo_{12}]; most of the results in [Mo_{12}] are described in Sections 4 and 7.

The following countable sum theorem is obtained for the case of Tychonoff spaces by M. Katětov [K_1].

Theorem 7.1 ([Mo_7]). Let $\{A_i | i \in N\}$ be a countable cover of a space X. If each A_i is C*-embedded in X and $\dim A_i \leq n$ for each $i \in N$, then $\dim X \leq n$.

On the other hand, we have proved another sum theorem.

Theorem 7.2 ([Mo_9]). Let $\{G_\lambda | \lambda \in \Lambda\}$ be a normal cover of a space X. If $\dim \text{Cl } G_\lambda/\text{Bd } G_\lambda \leq n$ for each $\lambda \in \Lambda$, then $\dim X \leq n$, and conversely.

As a corollary to Theorem 7.2 we have

Theorem 7.3. Let $\mathcal{G}$ be a σ-locally finite cozero-set cover of a space X. Then $\dim X \leq n$ if and only if $\dim G \leq n$ for each $G \in \mathcal{G}$.

Indeed, the "only if" part is obvious from Lemma 3.3 and Theorem 5.16. To prove the "if" part, let $\mathcal{H}$ be a normal cover of X which is a star-refinement of $\mathcal{G}$; such a cover $\mathcal{H}$ exists since $\mathcal{G}$ is normal by Lemma 2.3. Then for each $H \in \mathcal{H}$ there is $G \in \mathcal{G}$ such that $\text{St}(H,\mathcal{H}) \subset G$. Then $\text{Cl } H \subset G$ and hence $\text{Cl } H/\text{Bd } H \cong G/(G - H)$. Therefore, we have

$$\dim \text{Cl } H/\text{Bd } H = \dim G/(G - H) \leq \dim G \leq n.$$

Thus, Theorem 7.2 can be applied to $\mathcal{H}$, and the "if" part is proved.

As an application of Theorem 7.3 we have

Proposition 7.4. If $\{A_\lambda | \lambda \in \Lambda\}$ is a uniformly locally countable cover of X such that A_λ is C*-embedded in X and $\dim A_\lambda \leq n$ for each $\lambda \in \Lambda$, then $\dim X \leq n$.

Proof. Let $\mathcal{G}$ be a σ-locally finite cozero-set cover of X such that each member of $\mathcal{G}$ intersects only a countable number of A_λ's. Hence for each $G \in \mathcal{G}$ there is a countable subset Γ of Λ such that

$$G \subset \cup\{A_\lambda | \lambda \in \Gamma\} = A.$$

By Theorem 7.1 we have $\dim A \leq n$ and hence $\dim G \leq \dim A \leq n$ by Lemma 3.3 and Theorem 5.16. Hence Proposition 7.4 follows readily from Theorem 7.3.

Next, we shall establish a more general theorem, which contains Theorem 7.1 and Proposition 7.4 simultaneously.

Theorem 7.5 (The σ-uniformly locally countable sum theorem). Let $\mathcal{A} = \{A_\lambda | \lambda \in \Lambda\}$ be a σ-uniformly locally countable cover of a space X such that each A_λ is C*-embedded in X. If $\dim A_\lambda \leq n$ for each $\lambda \in \Lambda$, then $\dim X \leq n$.

Proof. Suppose that $\mathcal{A}$ is decomposed into the countable union of subcollections $\mathcal{A}_i$, $i \in N$, such that each $\mathcal{A}_i$ is uniformly locally countable. By Lemmas 2.2 and 3.5 and by Theorem 5.3, we may, and shall, assume that X is a Tychonoff space. Then by Theorem 4.3 there exists a paracompact space Y such that

(5) $X \subset Y \subset \beta(X)$,

(6) each $\mathcal{A}_i$ is locally countable in Y.

Let $\mathcal{U}_i$ be a locally finite open cover of Y such that each member of $\mathcal{U}_i$ intersects only a countable number of the members of $\mathcal{A}_i$ and let $\mathcal{V}_i$ be a locally finite open cover of Y which is a star-refinement of $\mathcal{U}_i$. Let $\mathcal{A}_i = \{A_{i\lambda} | \lambda \in \Lambda_i\}$. Then for each $V \in \mathcal{V}_i$ there is a countable subset Γ_V of Λ_i such that $\mathrm{Cl}_Y V \cap \mathrm{Cl}_Y A_{i\lambda} = \emptyset$ for $\lambda \in \Lambda_i - \Gamma_V$. Let us put

$$F_V = \cup\{\mathrm{Cl}_Y V \cap \mathrm{Cl}_Y A_{i\lambda} | \lambda \in \Gamma_V\}.$$

Since each $A_{i\lambda}$ is C*-embedded in X, we have

$$\dim A_{i\lambda} = \dim \beta(A_{i\lambda}) = \dim \mathrm{Cl}_{\beta X} A_{i\lambda} = \dim \mathrm{Cl}_Y A_{i\lambda},$$

and consequently

$$\dim F_V \leq n$$

by the countable sum theorem in paracompact Hausdorff spaces.

Let us set

$$F_i = \cup\{F_V | V \in \mathcal{V}_i\}.$$

Since each F_V is an F_σ-subset of Y, F_i is also an F_σ-subset of Y and we have

$$\dim F_i \leq n$$

by virtue of the countable and the locally finite sum theorems in paracompact spaces. Applying again the countable sum theorem in paracompact spaces, we have

$$\dim \cup F_i \leq n.$$

Since F_i contains the union of all the members in $\mathcal{A}_i$, we have

$$\cup F_i = \cup\{Cl_Y A_{i\lambda} \mid \lambda \in \Lambda_i,\ i \in N\}.$$

Therefore, $X \subset \cup F_i \subset Y \subset \beta(X)$, and consequently

$$\dim X = \dim \cup F_i \leq n.$$

This proves Theorem 7.5.

Theorem 7.5 fails to be valid if "σ-uniformly locally countable" is replaced by "σ-locally finite," or "locally countable". Indeed, in Example 5.17, the space X is the union of a countable dense cozero-set and a discrete zero-set, but $\dim X > 0$.

Thus, the σ-locally finite sum theorem as well as the locally countable sum theorem does not hold in general. However, as is proved by us some thirty years ago, the locally finite sum theorem holds in normal spaces.

Problem 7.6. Improve Theorem 7.5 so that it may contain the locally finite sum theorem in normal spaces as its special case.

Likewise, in the case of normal spaces Theorem 7.3 remains true if "σ-locally finite" is replaced by "point-finite", as is seen from a theorem of A.V. Zarelua [Z].

Problem 7.7. Improve Theorem 7.3 so that it may contain the "point-finite" case mentioned above.

8. Product Theorems

A theorem asserting the validity of the inequality

$$\dim (X \times Y) \leq \dim X + \dim Y$$

under a certain condition is called a product theorem (on dimension); we shall say also that the product theorem holds under this condition. Here X and Y are assumed to be non-empty.

Recently M. Wage [W] has succeeded in proving under CH (= the Continuum Hypothesis) that the product theorem does not hold in general even if $X \times Y$ is locally compact and normal, $\dim X = \dim Y = 0$, and $X = Y$. T. Przymusiński [Pr_2] pointed out that CH can be avoided and proved also that the product theorem does not hold in general even if X, Y are Lindelöf, $X \times Y$ is normal and $\dim X = \dim Y = 0$.

Thus, it becomes an important problem to find a product theorem which has wide applications. The following is one of such a kind of theorems.

Theorem 8.1 ([Mo_{11}]). Let X be a paracompact Hausdorff space which is a countable union of locally compact closed subspaces. Then we have

$$\dim (X \times Y) \leq \dim X + \dim Y$$

for any space Y.

In [Mo_{11}] we have proved Theorem 8.1 by using Proposition 3.8. Here we shall give another proof by virtue of the σ-uniformly locally finite sum theorem in Section 7, combined with Lemma 8.2 below.

Lemma 8.2. If X is a compact Hausdorff space, then $\dim (X \times Y) \leq \dim X + \dim Y$ for any space Y.

Lemma 8.2 was proved first in [Mo_4] under a weaker assumption that X is locally compact and paracompact Hausdorff. Here we point out that Lemma 8.2 can be proved more simply if we utilize Proposition 2.4. Indeed, let G be any (finite) normal cover of $X \times Y$; then by Proposition 2.4 there is a normal cover $\mathcal{V} = \{V_\lambda | \lambda \in \Lambda\}$ of Y such that $\{U \times V_\lambda | U \in \mathcal{U}_\lambda, \lambda \in \Lambda\}$ refines G, where $\{\mathcal{U}_\lambda | \lambda \in \Lambda\}$ is a family of finite open covers of X. Suppose that $\dim Y \leq n$. Then, by Theorem 5.12 for the case $A = \emptyset$, there is a continuous $\mathcal{V}$-map $\psi: Y \to T$ such that T is a metrizable space with $\dim T \leq n$. By the product theorem which was proved in our previous paper [Mo_1] we have $\dim (X \times T) \leq \dim X + \dim T$. Since $1 \times \psi: X \times Y \to X \times T$ is a G-map, we can find a normal cover $\mathcal{W}$ of $X \times Y$ which refines G and has order $\leq \dim X + \dim T + 1$. This proves Lemma 8.2.

Now. let us proceed to the proof of Theorem 8.1.

Proof of Theorem 8.1. Let $\{A_\lambda | \lambda \in \Lambda\}$ be a σ-uniformly locally finite cover of X by compact subsets; such a cover exists certainly by Theorem 4.6. Then $\{A_\lambda \times Y | \lambda \in \Lambda\}$ is also a σ-uniformly locally finite cover of $X \times Y$ and by Proposition 3.7 $A_\lambda \times Y$ is P-embedded in $X \times Y$ for each λ. Since by Lemma 8.2

$$\dim (A_\lambda \times Y) \leq \dim A_\lambda + \dim Y \leq \dim X + \dim Y$$

for each $\lambda \in \Lambda$, it follows from Theorem 7.5 that $\dim (X \times Y) \leq \dim X + \dim Y$. This proves Theorem 8.1.

As an example of a space X with the property stated in Theorem 8.1 we can mention any CW-complex; the paracompactness of CW-complexes was proved first by us as was remarked in [Mo_8, footnote on p. 69] and the fact that a CW-complex is a countable union of locally compact, closed (metrizable) subspaces is easy to see. Thus, Theorem 6.6, together with Theorem 8.1, implies

Theorem 8.3 ([Mo_{11}]). If X is a CW-complex, then $\dim (X \times Y) = \dim X + \dim Y$ for any space Y.

As another case where the equality holds, we have

Theorem 8.4 ([Mo_{11}]). If X is a paracompact Hausdorff space which is a countable union of locally compact, closed subspaces and if Y is any space with $\dim Y = 1$, then $\dim (X \times Y) = \dim X + \dim Y$.

Therefore, all of the product theorems in [Mo_1] have been generalized herewith except Theorem 8.5 below.

Theorem 8.5 ([Mo_1]). If the product $X \times Y$ of Hausdorff spaces is strongly paracompact (in other words, has the star-finite property), then $\dim (X \times Y) \leq \dim X + \dim Y$.

The proof of Theorem 8.5 is based on Proposition 5.18, which does not hold if "strongly" is deleted. The following problem seems to be still open:

Problem 8.6. If the product space $X \times Y$ is paracompact Hausdorff, is it true that $\dim (X \times Y) \leq \dim X + \dim Y$?

The product theorems 8.1 and 8.3 have many applications because one of the factor spaces in these theorems is allowed to be quite arbitrary.

On the other hand, B.A. Pasynkov [P_2] has announced a beautiful product theorem by introducing the notion of rectangular products. A product space $X \times Y$ is called rectangular if every finite normal cover of $X \times Y$ is refined by a σ-locally finite cover of $X \times Y$ which consists of cozero-set rectangles. This notion is a generalization of the notion of F-products in the sense of J. Nagata [Na] for the case of normal product spaces; the special case where "σ-locally finite" is replaced by "countable" is studied by R.L. Blair and A.W. Hager [BH]. Pasynkov states Theorem 8.7 below for the case of Tychonoff spaces.

Theorem 8.7. If the product $X \times Y$ is rectangular, then $\dim (X \times Y) \leq \dim X + \dim Y$.

The general case is reduced to the special case of Tychonoff spaces by Lemma 8.8 below.

Lemma 8.8 ([HM]). A product space $X \times Y$ is rectangular if and only if $\tau(X \times Y) = \tau(X) \times \tau(Y)$ and $\tau(X) \times \tau(Y)$ is rectangular (for τ cf. section 1).

The product theorem of Pasynkov is remarkable in that many of the product theorems can be derived from it. Indeed, as cases in which the product $X \times Y$ of Tychonoff spaces is rectangular Pasynkov mentioned the following ones:

1) X is locally compact and paracompact.

2) The projection $p: X \times Y \to X$ is a closed map.

3) X is metrizable and there is a perfect map $g: Y \to Y_0$ such that $X \times Y_0$ is normal and countably paracompact.

4) X is a paracompact M-space, Y is normal and there is a perfect map $g: Y \to Y_0$ such that $X \times Y_0$ is normal and countably paracompact.

5) X is a paracompact Σ-space in the sense of Nagami [N] and Y is a paracompact P-space in the sense of Morita [Mo_2].

Filippov's product theorem [F] deals with cases 2) and 4) with $Y = Y_0$. Here we note that case 2) is contained in case 2)' below, since by [No] $X \times Y$ is C*-embedded in $X \times \beta Y$ in case 2) and that by [HM] case 4) is contained in case 4)' below, a special case of which is case 6) below.

2)' $X \times Y$ is z-embedded in $X \times \beta Y$. (This is the case if and only if every finite normal cover of $X \times Y$ is refined by an open cover $\{U_\lambda \times V_\lambda | \lambda \in \Lambda\}$ of $X \times Y$ consisting of cozero-set rectangles such that $\{U_\lambda | \lambda \in \Lambda\}$ is σ-locally finite in X.) Cf. Theorems 5.16 and 8.1.

4)' X is a paracompact M-space which admits a perfect map onto a metric space T, Y is a normal space which admits a quasi-perfect map onto Y_0, and $T \times Y_0$ is normal.

6) X is a paracompact M-space and Y is an M-space or a normal P-space.

One might think that all the product theorems could be derived from Pasykov's theorem. However, it is not the case. Indeed, it was shown in our joint paper [HM] with T. Hoshina that a Tychonoff space X is locally compact and paracompact if and only if $X \times Y$ is rectangular for any Tychonoff space Y. Thus, Theorem 8.1 is not contained in Pasynkov's theorem.

9. *Applications*

The following theorems are direct applications of the theorems obtained in previous sections. Here, by "$\simeq$" we mean "is homotopic to" as usual.

Theorem 9.1. Let A be a subset of a space X with $\dim X/A \leq n-1$ and let $f, g: X \to \dot{I}^{n+1}$ be continuous maps. If (i) $f|A = g|A$ or (ii) $f|A \simeq g|A$ and A is C-embedded in X, then $f \simeq g$.

Proof. Case (i). In the product space $X \times I$ let us set

$$K = X \times \{0,1\} \cup A \times I$$

and define ϕ: $X \times I \to I^{n+1}$ by

$$\phi(x,t) = (1-t)f(x) + tg(x) \text{ for } x \in X,\ t \in I.$$

Then ϕ: $(X \times I, K) \to (I^{n+1}, \dot{I}^{n+1})$ is continuous and we have

$$\dim (X \times I)/K \leq \dim (X \times I)/(A \times I) \leq \dim (X/A) \times I \leq n-1+1=n$$

by Theorems 5.8 and 8.1. Hence ϕ is inessential by Theorem 6.3.

Case (ii). Let h:$A \times I \to \dot{I}^{n+1}$ be a continuous map such that $h(x,0) = f(x)$, $h(x,1) = g(x)$ for $x \in A$. Let ψ: $K \to \dot{I}^{n+1}$ be a map defined by

$$\psi(x,0) = f(x),\ \psi(x,1) = g(x) \text{ for } x \in X,$$

$$\psi(x,t) = h(x,t) \text{ for } x \in A,\ t \in I.$$

Since $A \times I$ and $X \times \{0,1\}$ are C-embedded in $X \times I$ and the latter is a zero-set, we have by [MoH$_1$, Theorem 2.10 and Remark 2.7]

$$\text{Cl}(A \times I \cap X \times \{0,1\}) = \text{Cl}(A \times I) \cap \text{Cl}(X \times \{0,1\})$$

and K is C-embedded in $X \times I$. Thus ψ is continuous, and ψ is extended to a continuous map from $X \times I$ to $\dot{I}^{n+1}$ by Theorem 6.3 since $\dim (X \times I)/K \leq n$ as was proved in case (i) above.

Theorem 9.2. Let A and B be closed subsets of a space X such that $X = A \cup B$ and $A \cap B$ is C-embedded in A as well as in B, let D be a subset of $A \cap B$ such that $\dim (A \cap B)/D \leq n-1$, and let f: $A \to \dot{I}^{n+1}$, g: $B \to \dot{I}^{n+1}$ be continuous maps. If (i) $f|D = g|D$ or (ii) $f|D \simeq g|D$ and D is C-embedded in $A \cap B$, then f and g are extended to continuous maps f': $X \to \dot{I}^{n+1}$ and g': $X \to \dot{I}^{n+1}$ so that $f' \simeq g'$.

Proof. By Theorem 9.1 there is a continuous map h: $(A \cap B) \times I \to \dot{I}^{n+1}$ such that $h(x,0) = f(x)$, $h(x,1) = g(x)$ for $x \in A \cap B$. By Theorem 3.6 there are continuous maps

$$\phi: A \times I \to \dot{I}^{n+1}, \quad \psi: B \times I \to \dot{I}^{n+1}$$

such that

$$\phi(x,0) = f(x),\ x \in A; \quad \psi(x,1) = g(x),\ x \in B;$$

$$\phi(x,t) = h(x,t), \text{ for } (x,t) \in (A \cap B) \times I.$$

The maps ϕ and ψ are united to a continuous map $\chi: X \times I \to \dot{I}^{n+1}$ and this proves Theorem 9.2.

There are several results in homotopy theory of CW-complexes which require the dimensional restriction. In discussing their generalizations to arbitrary spaces, covering dimension, which we are considering, is most suitable. We shall state here some of these generalizations.

Theorem 9.3 (The Hopf classification theorem; [Mo_7]). Let (X,A) be a pair of spaces such that $\dim X/A \leq n$ and $A \neq \emptyset$. Then $[X,A;\ S^n,\ s_0] \simeq H^n(X,A;\ \mathbb{Z})$.

Here, $[X,A;\ Y,\ B]$ denotes the set of all the homotopy classes of continuous maps from (X,A) to (Y,B), and (S^n,s_0) is a pointed n-sphere.

Theorem 9.4 ([Mo_7]). Let (X,A) be a pair of spaces such that $\dim X/A < 2n-1$ and $A \neq \emptyset$. Then

$$\pi^n(X,A) = [X,A;\ S^n,s_0]$$

has an abelian group structure and is called the n-th cohomotopy group of (X,A).

$\pi^n(X,A)$ is introduced by K. Borsuk for the case of compact metric spaces and studied in detail by E.H. Spanier for the case of compact Hausdorff spaces. The following is also a generalization of his result.

Theorem 9.5 ([Mo_7]). Let X be a space with $\dim X < 2n-1$ and A a C-embedded subset of X with $x_0 \in A$. Then the cohomotopy sequence of (X,A)

$$\pi^n(X,A) \to \pi^n(X,x_0) \to \pi^n(A,x_0) \to \pi^{n+1}(X,A) \to \cdots$$

is exact.

In case dim X < 2n-1 and dim X/A $\leq$ n-1, we have $\pi^n(X,A) = 0$ and $\pi^{n+1}(X,A) = 0$, and hence $\pi^n(X,x_0) \simeq \pi^n(A,x_0)$ by Theorem 9.5; Theorem 9.1 generalizes this fact.

In 1968 shape theory was founded by K. Borsuk for the case of compact metric spaces, and then extended to the case of compact Hausdorff spaces and metric spaces by S. Mardešić - J. Segal and R.H. Fox respectively, and finally to the case of general spaces by S. Mardešić. Our approach [Mo_6] to shape theory is based on the Čech construction of forming the nerves of all the locally finite normal covers of a space, which yields an inverse system in the homotopy category of CW-complexes. Thus, our approach is useful in cases where the dimensional restrictions are involved. For example, by this approach we have obtained a shape version of the Whitehead theorem in homotopy theory of CW-complexes.

Theorem 9.6 ([Mo_5]). Let $f: (X,x_0) \to (Y,y_0)$ be a shape morphism of pointed connected spaces such that $n_0 = \max(1 + \dim X, \dim Y) < \infty$. If the morphism

$$\text{pro-}\pi_k(f)\colon \text{pro-}\pi_k(X,x_0) \to \text{pro-}\pi_k(Y,y_0)$$

of homotopy pro-groups is an isomorphism for $1 \leq k < n_0$ and an epimorphism for $k = n_0$, then f is a shape equivalence.

It was shown by J. Draper and J. Keesling [DK] that the finite-dimensionality of spaces is crucial in this theorem. Theorem 9.6 was proved first by M. Moszyńska [Mos] for the case of compact metric spaces with

$n_0 = 1 + \max(1 + \dim X, \dim Y)$. In his generalization of her result to the case where f is induced by a continuous map and X, Y are arbitrary spaces, S. Mardešić [M_2] used the mapping cylinder similarly as in the case of CW-complexes by proving Lemma 9.7 below. Here we give another proof.

Lemma 9.7 ([M_2]). Let M_f be the mapping cylinder of a continuous map $f: (X, x_0) \to (Y, y_0)$. Then $\dim M_f = \max(1 + \dim X, \dim Y)$.

Proof. Since Y is regarded as a retract of M_f, Y is P-embedded in M_f and hence $\dim M_f = \max(\dim Y, \dim M_f/Y)$ by Theorem 5.9. Since $\dim \Sigma X \leq \dim M_f/Y \leq 1 + \dim X$, we have the lemma by Lemma 9.8 below.

Lemma 9.8 ([Mo_7]). Let ΣX be the reduced suspension of a pointed space (X, x_0). Then $\dim \Sigma X = 1 + \dim X$.

The following is a generalization of a theorem of Mardešić [M_2]; the assumption "$\dim X \leq n-1$" in his theorem is weakened to "$\dim X/A \leq n$" here.

Theorem 9.9 ([Mo_{13}]). Let (X, A, x_0) be a pair of pointed connected spaces such that A is P-embedded in X. If $\dim X/A \leq n$ and pro-$\pi_k(X, A, x_0) = 0$ for $1 \leq k \leq n$, then the inclusion map $i: (A, x_0) \to (X, x_0)$ induces a shape equivalence.

The operation of taking suspension is one of the important tools in homotopy theory of CW-complexes and likewise in shape theory. Let $S_0[(X, x_0), (Y, y_0)]$ be the set of all shape morphisms from (X, x_0) to (Y, y_0). Then we have

Theorem 9.10 ([Mo_{10}]). Let (X, x_0) be a pointed space with dim $X < 2n-1$ and Y a shape $(n-1)$-connected space ($n \leq 2$). Then the suspension map

$$\Sigma: S_0[(X, x_0), (Y, y_0)] \to S_0[(\Sigma X, x_0), (\Sigma Y, y_0)]$$

is bijective.

It is to be noted that Theorem 9.4 follows readily from Theorem 9.10, since any shape morphism to a CW-complex (Y, y_0) is induced by a continuous map and determined uniquely by its homotopy class (that is, $S_0[(X, x_0), (Y, y_0)] = [X, x_0; Y, y_0]$) and $S_0[(\Sigma X, x_0), (Y, y_0)]$ has a group structure for any pointed spaces (X, x_0) and (Y, y_0); for more details cf. [Mo_{10}]).

References

[BH] R.L. Blair and A.W. Hager, *z-embedding in $\beta X \times \beta Y$*, Set-Theoretic Topology, Academic Press, 1977, 47-72.

[vD] E.K. van Douwen, *The small inductive dimension can be raised by the adjunction of a single point*, Indag. Math. 35 (1973), 434-442.

[DK] J. Draper and J. Keesling, *An example concerning the Whitehead Theorem in shape theory*, Fund. Math. 92 (1976), 255-259.

[F] V.V. Filippov, *On the dimension of normal spaces*, Soviet Math. Dokl. 14 (1973), 547-550.

[GJ] J.R. Gard and R.D. Johnson, *Four dimension equivalences*, Canad. J. Math. 20 (1968), 48-50.

[H] W. Holsztyński, *Topological dimension of lattices*, Bull. Acad. Pol. Sci. Ser. Math. 14 (1966), 63-69.

[HM] T. Hoshina and K. Morita, *On rectangular products of topological spaces*, to appear.

[I] J.R. Isbell, *Uniform spaces*, Providence, 1964.

[Is] T. Ishii, *On the Tychonoff functor and w-compactness*, preprint, 1978.

[K_1] M. Katětov, *A theorem on the Lebesgue dimension,* Casopis Pest. Mat. Fys. 75 (1950), 79-87.

[K_2] ________, *Extensions of locally finite covers,* Colloq. Math. 6 (1958), 145-151.

[L] O.V. Lokucievskii, *On the dimension of bicompacta,* Dokl. Akad. Nauk. SSSR 67 (1949), 217-219.

[M_1] S. Mardešić, *On covering dimension and inverse limits of compact spaces,* Ill. J. Math. 4 (1960), 278-291.

[M_2] ________, *On the Whitehead theorem in shape theory I,* Fund. Math. 91 (1976), 51-64.

[Mo_1] K. Morita, *On the dimension of product spaces,* Amer. J. Math. 75 (1953), 205-223.

[Mo_2] ________, *Products of normal spaces with metric spaces,* Math. Ann. 154 (1964), 365-382.

[Mo_3] ________, *Topological completions and M-spaces,* Sci. Rep. Tokyo Kyoiku Daigaku 10 (1970), 271-288.

[Mo_4] ________, *On the dimension of the product of Tychonoff spaces,* Gen. Topology Appl. 3 (1973), 125-133.

[Mo_5] ________, *The Hurewicz and the Whitehead theorems in shape theory,* Sci. Rep. Tokyo Kyoiku Daigaku 12 (1974), 246-258.

[Mo_6] ________, *On shapes of topological spaces,* Fund. Math. 86 (1975), 251-259.

[Mo_7] ________, *Čech cohomology and covering dimension for topological spaces,* Fund. Math. 87 (1975), 31-52.

[Mo_8] ________, *On expansions of Tychonoff spaces into inverse systems of polyhedra,* Sci. Rep. Tokyo Kyoiku Daigaku 13 (1975), 66-74.

[Mo_9] ________, *The Hopf extension theorem for topological spaces,* Houston J. Math. 1 (1975), 121-129.

[Mo_{10}] ________, *The suspension theorem in shape theory,* Math. Jap. 20 (1975), 179-183.

[Mo_{11}] ________, *On the dimension of the product of topological spaces,* Tsukuba J. Math. 1 (1977), 1-6.

[Mo_{12}] ________, *Uniformly local finiteness and sum theorems on covering dimension,* unpublished, 1977.

[Mo_{13}] ________, *The Whitehead theorems in shape theory,* Proc. Japan Acad. 50 (1974), 458-461.

[MoH_1] K. Morita and T. Hoshina, *C-embedding and the homotopy extension property*, Gen. Topology Appl. 5 (1975), 69-81.

[MoH_2] ________, *P-embedding and product spaces*, Fund. Math. 93 (1976), 71-80.

[Mos] M. Moszyńska, *The Whitehead Theorem in the theory of shapes*, Fund. Math. 80 (1973), 221-263.

[N] K. Nagami, *Σ-spaces*, Fund. Math. 65 (1969), 169-192.

[Na] J. Nagata, *Product theorems in dimension theory I*, Bull. Acad. Pol. Sci. Sér. Math. 15 (1967), 439-448.

[No] N. Noble, *Products with closed projections*, Trans. AMS 140 (1969), 381-391.

[O] H. Ohta, *Topologically complete spaces and perfect maps*, Tsukuba J. Math. 1 (1977), 77-90.

[Ok] S. Oka, *Tychonoff functor and product spaces*, Proc. Japan Acad. 54 (1978), 97-100.

[P_1] B.A. Pasynkov, *On the spectral decomposition of topological spaces*, AMS Transl. Ser. 2, 73 (1968), 87-134.

[P_2] ________, *On the dimension of rectangular products*, Soviet Math. Dokl. 16 (1975), 344-347.

[Po] E. Pol, *Some examples in the dimension theory of Tychonoff spaces*, Fund. Math. 102 (1979), 29-43.

[PoP_1] E. Pol and R. Pol, *A hereditarily normal strongly zero-dimensional space with a subspace of positive dimension and an N-compact space of positive dimension*, Fund. Math. 97 (1977), 43-50.

[PoP_2] ________, *A hereditarily normal strongly zero-dimensional space containing subspaces of arbitrarily large dimension*, Fund. Math. 102 (1979), 137-142.

[Pr_1] T. Przymusiński, *A note on dimension theory of metric spaces*, Fund. Math. 85 (1974), 277-284.

[Pr_2] ________, *On the dimension of product spaces and an example of M. Wage*, to appear.

[Pu] R. Pupier, *La completion universelle d'un produit d'espaces*, Publ. Dept. Math. Lyon 6 (2) (1969), 75-84.

[Sc] E.V. Šc̆epin, *A finite dimensional compact absolute neighborhood retract is metrizable*, Soviet Math. Dokl. 18 (1977), 402-406.

[Sm] Yu M. Smirnov, *On the dimension of proximity spaces*, Mat. Sb. 38 (1956), 283-302.

[W] M. Wage, *The dimension of product spaces*, preprint 1977; Proc. Nat. Acad. Sci. U.S.A. 75 (1978), 4671-4672.

[Z] A.V. Zarelua, *On a theorem of Hurewicz*, Mat. Sb. 60 (1963), 17-28.

COMBINATORIAL TECHNIQUES IN FUNCTIONAL ANALYSIS

S. Negrepontis
Athens University, Greece

In this paper I describe how some infinitary combinatorial techniques, long familiar and useful in set-theoretic topology, have recently been employed for the treatment and solution of fundamental problems in functional analysis, and in particular in the isomorphic theory of Banach spaces. In this description the interrelation that these applications have with questions in set-theoretic topology will be apparent. I will roughly follow the historical development of these ideas, as they grew during the last three years in the research group of Mathematical Analysis at Athens University. Essentially all the results described here deal with the possibility of isomorphically embedding the (generalized) sequence spaces ℓ^1_α of all absolutely summable real functions on (a set of cardinality) α into various classes of Banach spaces. The combinatorial techniques associated with such questions bear strong relations with questions about calibers of various class of compact spaces.

The organization of the paper is as follows: In A. we consider the class of dyadic spaces, and the larger class of Γ-spaces; here the Erdös-Rado theorem on quasi-disjoint (finite) sets and its extension to singular cardinals are used. In B. we consider the class of $L^\infty(\mu)$-spaces, for σ-finite measure spaces (X,S,μ); in effect, we deal with compact

ISBN: 0-12-584960-5

hyper-Stonian spaces. Here the Erdös-Rado theorem on quasi-disjoint (countable) sets and its extension to singular cardinals are used. In C. we consider compact spaces Ω as a function of their Souslin number; here tree arguments of a more fundamental type are called for, for regular and for singular cardinals. In D. we consider for general Banach spaces Pełczynski's conjecture (and Hagler-Stezall's conjecture); here, Hajnal's theorem on free sets is mostly used. In E. we describe Rosenthal's criterion for embedding ℓ^1 into any Banach space, some of its applications, and some of the difficulties in extending it to higher cardinalities. Here, the Nash-Williams-Galvin-Prikry infinite type version of Ramsay's theorem is used.

Some of these results have been discussed, at an earlier stage of their development, in [Ne_1], [Ne_2].

A. Dyadic Spaces and Generalizations

In set-theoretic topology the dyadic spaces (i.e. the continuous images of the generalized Cantor space $\{0,1\}^I$ for any set I) have been intensively studied in [Ef], [EfE], [EP], [E], [Hay_1], [S_1], [S_2], [S_3], etc.

In particular, Šanin proved for the first time in [S_1] what is now usually referred to as the Erdös-Rado theorem on quasi-disjoint sets, and used it in [S_2] to prove that dyadic spaces have as caliber every uncountable regular cardinal. By definition, a space X has *caliber* a cardinal α if for every family $\{U_\xi\colon \xi < \alpha\}$ of non-empty open subsets of X there is $A \subset \alpha$, with $|A| = \alpha$, such that

$$\cap_{\xi \in A} U_\xi \neq \emptyset .$$

A related universal property is for X to have *pre-caliber* α; here, from the same datum, we conclude that the family $\{U_\xi : \xi \in A\}$ has merely the finite intersection property. Of course, a compact Hausdorff space X the two notions coincide.

A crucial step now is to consider families of *pairs* of non-empty open sets, instead of families of non-empty open sets. In fact, let $\{(U_\xi, V_\xi): \xi < \alpha\}$ be a family of pairs of non-empty open sets, with $U_\xi \cap V_\xi = \emptyset$ for $\xi < \alpha$. The suitable generalization of the notion of family with the finite intersection property is an independent family. Indeed, $\{(U_\xi, V_\xi): \xi < \alpha\}$ is *independent* if for every $\xi_1 < \ldots < \xi_n < \alpha$ and $\varepsilon_1, \ldots, \varepsilon_n \in \{-1,+1\}$, we have that

$$\bigcap_{k=1}^{n} \varepsilon_k U_{\xi_k} \neq \emptyset$$

(where $(+1)U_\xi = U_\xi$, $(-1)U_\xi = V_\xi$, by convention).

It is not unreasonable to expect to find, under mild conditions, independent families in dyadic spaces, using combinatorial techniques, in much the same way that Šanin established the caliber properties of such spaces. Before doing this, however, one might wonder why go into the trouble of doing so. Here, comes a crucial observation due to Haskell Rosenthal [R_2], who gave a criterion for a set of functions to be isomorphically equivalent to the canonical basis of the (generalized) sequence space ℓ^1_α using the notion of independence, and thus to a large extent transforming a Banach space problem to a set-theoretic one, and opening the way for its treatment with combinatorial techniques. We first give some definitions. For a cardinal α, we set

$$\ell^1_\alpha = \{f: \alpha \to \mathbb{R} \mid \sum_{\xi<\alpha} |f(\xi)| < \infty\},$$

with $\|f\| = \sum_{\xi<\alpha} |f(\xi)|$; and,

$$\ell_\alpha^\infty = \{f\colon \alpha \to \mathbb{R} \mid f \text{ bounded}\},$$

with $\|f\| = \sup_{\xi<\alpha} |f(\xi)|$.

By ℓ^1, ℓ^∞ we denote $\ell_\omega^1, \ell_\omega^\infty$, respectively.

If X and Y are Banach spaces, a linear bounded operator $T\colon X \to Y$ is a *(Banach space) isomorphism* if T is one-to-one, onto, and T^{-1} is also bounded. It follows immediately, by the open mapping theorem for Banach spaces, that T is an isomorphism if and only if there are positive constants m,M such that

$$m\|x\| \leq \|T(x)\| \leq M\|x\| \qquad \text{for } x \in X.$$

We say that T is a *(Banach space) embedding* if T is a Banach space isomorphism from X to T(X). Finally, a set $\{f_\xi\colon \xi < \alpha\}$ in X will be *(isomorphically) equivalent to the canonical basis of* ℓ_α^1 if the correspondence $e_\xi \to f_\xi$ (for $\xi < \alpha$) (where e_ξ is the element of ℓ_α^1 defined by $e_\xi(\zeta) = 1$ if $\xi = \zeta$, $= 0$ if $\xi \neq \zeta$) can be extended (in the unique way) to an embedding of ℓ_α^1 into X.

We now state (for any cardinal α)

Rosenthal's criterion [R_2]. Let α be an infinite cardinal, γ and δ real numbers, $\delta > 0$, and $\{f_\xi\colon \xi < \alpha\} \subset \ell_\kappa^\infty$ (where κ is some cardinal) such that

$$\|f_\xi\| = 1 \qquad \text{for } \xi < \alpha,$$

and setting

$$A_\xi = \{\zeta\colon f_\xi(\zeta) \geq \gamma + \delta\}$$

$$B_\xi = \{\zeta\colon f_\xi(\zeta) \leq \gamma\}$$

for $\xi < \alpha$, we have that the family

$$\{(A_\xi, B_\xi)\colon \xi < \alpha\} \text{ is independent.}$$

Then $\{f_\xi\colon \xi < \alpha\}$ is $\frac{\delta}{2}$-equivalent to the canonical basis of ℓ_α^1, i.e.,

$$\frac{\delta}{2}\sum_{k=1}^{n}|t_k| \le \|\sum_{k=1}^{n} t_i f_{\xi_k}\| \le \sum_{k=1}^{n}|t_k|$$

for $\xi_1 < \ldots < \xi_n < \alpha$, and $t_1,\ldots,t_n \in \mathbb{R}$.

The preceding remarks and associations make the next result expressing a universal property axiom to caliber, due to J. Hagher [H] for regular cardinal α, appear plausible. We need some definitions. For a compact Hausdorff space X, we denote by C(X) the Banach space of all continuous real-valued functions f on X, with $\|f\| = \sup_{x\in X}|f(x)|$, the *supremum norm* (or, *norm of uniform convergence*).

The *dimension* dim X of a Banach space X is the smallest cardinality α, such that there is a set $A \subset X$, with $|A| = \alpha$, and the set of linear combinations of A is dense in X. (If dim X is infinite, then in fact dim X = density character of X). We note that if X is a compact Hausdorff space then dim C(X) = weight of X.

Theorem (on dyadic spaces). Let α be an infinite cardinal with $cf(\alpha) > \omega$, X a dyadic space, and E a closed linear subspace of C(X), with dim $X \geq \alpha$. Then X contains isomorphically the space ℓ^1_α.

We remark that, by a simple example of Pełczynski [P_2] (Example 7), the cardinality restriction on α (i.e., that $cf(\alpha) > \omega$) cannot be removed.

The proof of this theorem and of the next one make use of the Erdos-Rado theorem on quasi-disjoint sets, and its extension to singular cardinals that we now state. Some notation is needed. By $P_\kappa(\alpha)$ we denote the set of all subsets A of α with $|A| < \kappa$; thus, $P_\omega(\alpha)$ denotes the set of finite subsets of α. We set $P^*_\kappa(\alpha) = P_\kappa(\alpha)\setminus\{\emptyset\}$. We write $\kappa << \alpha$ if $\kappa < \alpha$ and in addition for every $\beta < \alpha$ and $\lambda < \kappa$ we have that $\beta^\lambda < \alpha$.

For example: $\alpha^+ << (2^\alpha)^+$ for any infinite cardinal α.

An ordinal ξ is identified with the set of all ordinals smaller than ξ, and a cardinal is an ordinal not in a one-to-one correspondence with any smaller ordinal. The cofinality of a cardinal α is denoted $cf(\alpha)$.

Erdös-Rado theorem on quasi-disjoint sets (Erdös-Rado [ErR]). Let α be a regular infinite cardinal, $\kappa << \alpha$, and let $\{N_\xi : \xi < \alpha\} \subset P_\kappa(\alpha)$. Then there are a set $A \subset \alpha$, with $|A| = \alpha$ and a set N such that

$$N_\xi \cap N_\zeta = N \text{ for } \xi, \zeta \in A,\ \xi \neq \zeta.$$

Extension of the Erdös-Rado theorem to singular cardinals. Let α be singular with $cf(\alpha) > \omega$, $\kappa << \alpha$ and $\kappa << cf(\alpha)$, and let $\{N_\xi : \xi < \alpha\} \subset P_\kappa(\alpha)$. Then there are $\Gamma, N, \{A_\sigma : \sigma \in \Gamma\}$, and $\{N_{(\sigma)} : \sigma \in \Gamma\}$ such that

(i) $\Gamma \subset cf(\alpha)$, $|\Gamma| = cf(\alpha)$,

(ii) $|\cup_{\sigma\in\Gamma} A_\sigma| = \alpha$,

(iii) $A_\sigma \cap A_{\sigma'} = \emptyset$ for $\sigma', \sigma \in \Gamma$, $\sigma' \neq \sigma$, and

(iv) $N_{\xi'} \cap N_\xi = N_{(\sigma)}$ for $\xi', \xi \in A_\sigma$, $\xi' \neq \xi$
$= N$ for $\xi' \in A_{\sigma'}$, $\xi \in A_\sigma$, $\sigma, \sigma' \in \Gamma$, $\sigma = \sigma'$.

A proof of this extension can be found in Argyros $[A_1]$, and Argyros-Negrepontis $[AN_2]$ (Theorem 2.5), where some historical comments are also given. For a large number of applications of this theorem to product theorems, the reader should consult the forthcoming monograph by Comfort and Negrepontis [CN].

We now outline the proof of the theorem on dyadic spaces. Let θ be a real number $0 < \theta < 1$. From an application of Hahn-Banach's theorem, there is a family $\{f_\xi : \xi < \alpha\}$ such that

$$\|f_\xi\| = 1 \text{ for } \xi < \alpha,$$
$$\|f_\xi - f_\zeta\| > \theta \text{ for } \xi < \zeta < \alpha, \text{ and}$$
$$f_\xi \in E \text{ for } \xi < \alpha.$$

Also, without loss of generality we may assume that $X = \{0,1\}^I$ for some set I. By the Stone-Weierstrass theorem, the simple continuous functions with rational values are uniformly dense in C(X). Thus there is a family $\{g_\xi\colon \xi < \alpha\}$ such that

g_ξ is continuous, simple, with rational values for $\xi < \alpha$, and

$$\|g_\xi - g_\zeta\| > \theta \text{ for } \xi < \zeta < \alpha.$$

(Note that we do not require that $g_\xi \in E$). Thus

$$g_\xi = z_{\xi_1}\chi_{A_{\xi,1}} + \dots + z_{\xi_k}\chi_{A_{\xi,k}} \qquad \text{for } \xi < \alpha,$$

for some rational numbers $z_{\xi_1},\dots,z_{\xi_k}$, and $A_{\xi,1},\dots, A_{\xi,k}$ open-and-closed subsets of X. Since $\mathrm{cf}(\alpha) > \omega$, it follows that there are $\Gamma \subset \alpha$, with $|\Gamma| = \alpha$, and rational numbers $z_1,\dots,z_k$ such that

$$g_\xi = z_1\chi_{A_{\xi,1}} + \dots + z_k\chi_{A_{\xi,k}} \qquad \text{for } \xi \in \Gamma.$$

By the compactness of X, g_ξ depends on a finite set N_ξ of coordinates for $\xi \in \Gamma$. We now consider the family

$$\{N_\xi\colon \xi \in \Gamma\}$$

and we make a crucial application of the Erdös-Rado combinatorial theorem. (For this outline, we assume that α is regular; if α is singular, then we will apply the extension of the Erdös-Rado theorem to singular cardinals for $k = \omega$). Thus, there are $\Gamma' \subset \Gamma$, with $|\Gamma'| = \alpha$, and a set N such that

$$N_\xi \cap N_\xi = N \text{ for } \xi,\zeta \in \Gamma',\ \xi \neq \zeta.$$

For the remainder of the proof we must consider separately the cases: $N = \emptyset$; $N \neq \emptyset$. The former case is simpler, and thus we look at the latter. We set $X_N = \{0,1\}^N$. It is clear that there open-and-closed subsets $A_1,\dots,A_k$ in X_N, and $\Delta \subset \Gamma'$, with $|A| = \alpha$, such that

$$\mathrm{proj}_N(A_{\xi,m}) = A_m \text{ for } m=1,2,\ldots,k \text{ and } \xi \in \Delta$$

(where proj_N is the projection function from X onto X_N). It now follows without difficulty that there are indices k, ℓ, with $1 \leq k \leq \ell \leq m$, such that $|z_k - z_\ell| > \theta$ and $A_k \cap A_\ell \neq \emptyset$. We set

$$w_0 = z_k,\ w_1 = z_\ell,$$
$$A_\xi = g_\xi^{-1}(\{w_0\}) = A_{\xi,k},$$
$$B_\xi = g_\xi^{-1}(\{w_1\}) = A_{\xi,\ell}$$

for $\xi \in \Delta$, and it is not difficult to prove that the family

$$\{(A_\xi, B_\xi):\ \xi \in \Delta\}$$

is independent. It then follows, from the nearness of each f_ξ to g_ξ, for $\xi \in \Delta$, setting

$$\tilde{A}_\xi = f_\xi^{-1}(\{w_0\}),$$
$$\tilde{B}_\xi = f_\xi^{-1}(\{w_1\}),$$

that the family

$$\{(\tilde{A}_\xi, \tilde{B}_\xi):\ \xi \in \Delta\}$$

is also independent. Rosenthal's criterion mentioned earlier now gives us that $\{f_\xi:\ \xi \in \Delta\}$ is equivalent to the canonical basis $\{e_\xi:\ \xi \in \Delta\}$ of ℓ_α^1.

The theorem on dyadic spaces admits a rather straightforward generalization that we now state. A compact Hausdorff space X is called a Γ-*space* if there is a family of compact spaces $\{X_i:\ i \in I\}$, with $w(X_i) < w(X)$ for all $i \in I$ (where $w(X)$ denotes the topological weight of X), and a continuous onto function $\phi: \pi_{i \in I} X_i \to X$. The *secondary weight* $\Delta w(X)$ of a Γ-space is the smallest infinite cardinal α, such that the above family of compact spaces $\{X_i:\ i \in I\}$ can be chosen so that $w(X_i) < \alpha$ for all $i \in I$. We then have:

(I) Let X be a compact Γ-space, and α a cardinal such that $w < cf(\alpha)$ and $\Delta w(X) \leq cf(\alpha)$. If E is a closed linear subspace of C(X), with dim $E \geq \alpha$, then there is an isomorphic embedding of ℓ^1_α into E.

(II) Let X be a compact Γ-space with $w(X) = \alpha$, such that $w < cf(\alpha)$ and $\Delta w(X) \leq cf(\alpha)$. Then the compact space $\{0,1\}^\alpha$ is homeomorphic to a subspace of X.

The proofs of (I) and (II) appear in Argyros-Negrepontis [AN_2] (Theorems 3.11 and 3.13). Hagler [H] proved (I) and (II) for dyadic spaces and for α regular uncountable. These results were also announced by the present author at the Proceedings of the Fourth Prague Topological Symposium (August 26, 1976); at the same symposium J. Gerlits announced related results [Ge_1], that include a proof of (II) for dyadic spaces and for α a cardinal of uncountable cofinality (his method of proof is different from ours, employing a simple form of Hajnal's [Ha] theorem on free sets. Results related to (II) can also be found in a subsequent paper of Gerlits [Ge_2].

A very general form of the universal ℓ^1_α-embedding into closed linear subspaces E, with dim $E \geq \alpha$, for suitable cardinals α, of $C(\beta(\pi_{i\in I}X_i))$, for suitable (generally non-compact) spaces $\{X_i: i \in I\}$ has been proved by Zachariades [Z].

B. $L^\infty(\mu)$-spaces

How far can these very strong results (for calibers of dyadic spaces, or for universal ℓ^1_α-embeddings into closed linear subspaces of C (dyadic space)) be generalized to more extensive classes of Banach spaces? The next result is very crucial for this direction, for as we shall see it is the connecting link between the very restrictive class of dyadic spaces, and--in a sense--the class of all compact spaces. Some more definitions are needed at this point.

Let (X,S,μ) be a σ-finite measure space (i.e. X is a set, S is a σ-algebra of subsets of X, and μ is a non-negative σ-additive measure defined on S, so that $X = \cup_{n<\omega} A_n$, with $\mu(A_n) < \infty$ for $n < \omega$). By $L^\infty(\mu)$ (or sometimes by $L^\infty(X,S,\mu)$, $L^\infty(X)$) we denote the Banach space of all μ-essentially bounded real-valued functions f on X, with $\|f\| = \mu\text{-ess.}\sup_{x\in X}|f(x)|$. Now, two basic facts are known about the basic structure of the Banach space $L^\infty(\mu)$, that we will find useful.

(a) The first, and quite deep one, is Rosenthal's classification theorem of $L^\infty(\mu)$ (appearing in Rosenthal $[R_1]$ (Theorem 3.5)) based on Mahavam's [M] fundamental classification theorem of measure spaces. It states that for every σ-finite measure space (X,S,μ) there is a cardinal α, such that $L^\infty(\mu)$ is Banach isomorphic to $L^\infty(\{0,1\}^\alpha)$, where $\{0,1\}^\alpha$ is equipped with the Borel regular Haar probability measure λ (or, equivalently, with the product measure $\lambda = \Pi_{\xi<\alpha}\mu_\xi$, where $\mu_\xi(\{0\}) = \mu_\xi(\{1\}) = 1/2$ for $\xi < \alpha$).

(b) The second fact about $L^\infty(\mu)$ is Dixmier's [D] (Théorème 1): There is a compact Hausdorff extremally disconnected space Ω and a finite regular Borel positive measure ν on Ω such that

$C(\Omega)$ is (isometrically) isomorphic to $L^\infty(\mu)$, and

$\nu(U) = \nu(\overline{U}) > 0$ for every non-empty open U of Ω.

We may ask now what, if anything, can be proved about universal embeddings in $L^\infty(\mu)$ and about calibers of its related Dixmier compact space Ω? To begin with, there are certain negative indications. If we consider the simplest such space, namely $L^\infty(\lambda)$, where λ is Lebesque measure on $[0,1]$, then we observe that

(i) There is a non-separable closed linear subspace of $L^{\infty}(\lambda)$ that does not contain (even) ℓ^1. This is a consequence of a (difficult) example, due to R.C. James [J], of a Banach space X which is separable (and therefore its dual space X* of bounded linear functionals can be embedded isomorphically into $L^{\infty}(\lambda)$, but such that (X and) X* do not contain isomorphically ℓ^1, and X* is non-separable; and,

(ii) Assuming the continuum hypothesis, the space Ω (corresponding by (b) to $L^{\infty}(\lambda)$), according to an (unpublished) result of P. Erdös, does not have caliber ω^+.

Nevertheless, let us look more carefully into the situation. Fact (a) provides for $L^{\infty}(\mu)$ a product structure for the underlying space (in fact a generalized Cantor set), and for its measure, and thus opens the way for an application of the Erdös-Rado combinatorial result on quasi-disjoint sets, and its extension to singular cardinals. Of course, on this space we deal with measurable, and not with continuous functions; what is the difference? A crucial one is this: A continuous function on a generalized Cantor set can be approximated uniformly by a simple function depending on a *finite* set of coordinates (as follows from the Stone-Weierstrass theorem, and as in fact we need to know for the proof of the previous theorem on dyadic spaces). There is no such approximation for measurable functions, and the best thing we can hope for is a corresponding approximation by a simple function depending on a *countable* set of coordinates. The size of the set of coordinates on which these simple functions depend should determine the conditions imposed upon us on the use of the Erdös-Rado theorem: for dyadic spaces, we deal with finite sets of coordinates and there the cardinalities α should only obey the restrictions $cf(\alpha) > \omega$; for $L^\infty(\mu)$, we deal with countable sets of coordinates and here the cardinalities α should obey the stronger restriction $\alpha >> \omega^+$ and $cf(\alpha) >> \omega^+$.

These heuristic remarks explain the difficulties with (i) and (ii), and suggest the correct statement of the theorem for $L^\infty(\mu)$, and to a large extent the correct procedure for its proof. There is an additional difficulty with the proof that must be faced. It is not enough to find a large independent family for an application of Rosenthal's lemma of pairs of open subsets of the generalized Cantor set, but a large family of pairs $\{(A_\xi, B_\xi): \xi < \alpha\}$ of measurable subsets of the generalized Cantor set, that is μ-*essentially independent* (i.e. for $\xi_1 < \ldots < \xi_n < \alpha$ and $\varepsilon_1, \ldots, \varepsilon_n \in \{-1,1\}$ we have that

$$\mu\left(\bigcap_{k=1}^{n} \varepsilon_k A_{\xi_k}\right) > 0).$$

A basic new tool in our search for such a μ-essentially independent family will be Fubini's theorem (for the product Haar measure). We can now state the

$L^\infty(\mu)$-*Theorem*. Let (X, S, μ) be a σ-finite measure space, and E a closed linear subspace of $L^\infty(\mu)$, such that $\dim E = \alpha >> \omega^+$ and $cf(\alpha) >> \omega^+$. Then the space ℓ^1_α can be isomorphically embedded in E.

The proof of this theorem is quite technical. The details can be found in Argyros $[A_1]$, Argyros-Negrepontis $[AN_2]$ (cf. also $[A_2]$, $[AN_1]$).

It is interesting that with a related argument Talagrand [T] has proved that there is a compact Hausdorff space X (in fact, the set of all closed subsets of $\{0,1\}^{\aleph_2}$ with the

Hausdorff topology), and a Borel measure μ on X with support all of X, such that there is no strong Borel linear lifting of $L^\infty(\mu)$.

For the case $L^\infty\{0,1\}^{\omega^+}$, not covered by the $L^\infty(\mu)$-theorem (since $\omega^+ >> \omega^+$ fails!), there is an interesting result by S. Argyros and Th. Zachariades:

Assume Martin's axiom together with the denial of the continuum hypothesis, and let $\{f_\xi : \xi < \omega^+\} \subset L^\infty\{0,1\}^{\omega^+}$, with $\|f_\xi\|_\infty \leq 1$ for $\xi < \omega^+$, and $\|f_\xi - f_\zeta\|_1 = \int_{\{0,1\}^{\omega^+}} |f_\xi - f_\zeta| d\mu \geq \varepsilon > 0$ for $\xi < \zeta < \omega^+$ (for some $\varepsilon > 0$) (where μ denotes the Haar (product) probability measure on $\{0,1\}^{\omega^+}$). Then there is $A \subset \omega^+$, with $|A| = \omega^+$, such that the family $\{f_\xi : \xi \in A\}$ is equivalent to the canonical basis of $\ell^1_{\omega^+}$ (in L^∞-norm).

Furthermore, if we assume Martin's axiom and that $\omega^+ < 2^\omega < \aleph_\omega$ then the above statement holds with any cardinal α, replacing ω^+.

C. A compact space as a function of its Souslin number

The product structure of the underlying space and measure (described in fact (a) above) was the essential reason that made possible the combinatorial proof of the universal ℓ^1_α-embedding property of $L^\infty(\mu)$, but it is its Dixmier structure as a Banach space of continuous functions on the compact space Ω, carrying a *strictly positive* measure ν (i.e. a measure ν such that $\nu(U) > 0$ for every non-empty open subset U of Ω) (described in fact (b) above) that points the direction to the most fruitful and extensive generalization. Since Ω has a strictly positive measure, it is clear that it also has the *countable chain condition* (c.c.c.) (i.e. every family of pairwise disjoint non-empty subsets of Ω is countable). We introduce the following definition. For a topological space X, its *Souslin number* $S(X)$ is the smallest cardinal α such that there is no family of pairwise disjoint non-empty subsets of X of cardinality α. Thus X has the c.c.c. if and only if $S(X) \leq \omega^+$. It turns out that in fact the reason for the validity of the $L^\infty(\mu)$-theorem is simply that $L^\infty(\mu)$ is isomorphic to $C(\Omega)$, for some compact Hausdorff Ω, satisfying c.c.c. This is an immediate consequence of the following powerful result.

$C(\Omega)$-*Theorem.* Let Ω be a compact Hausdorff space and let $S(\Omega) = \kappa$. Let E be a closed linear subspace of $C(\Omega)$ with $\dim E = \alpha$, and assume that

$$\kappa \ll \alpha \text{ and } \kappa \ll cf(\alpha).$$

Then there is an isomorphic embedding of the space ℓ^1_α into E.

There is also a corresponding result for calibers as follows:

Caliber Theorem. Let Ω be a compact Hausdorff space, let $S(\Omega) = \kappa$ and let α be a cardinal such that

$$\kappa \ll \alpha \text{ and } \kappa \ll cf(\alpha).$$

Then Ω has caliber α.

The proofs of these theorems are due to Argyros for regular cardinals α, and to Tsarpalias for singular cardinals α (cf. [A_1], [A_3], [A_5], [AT_1], [AT_2], [AT_3], [Ts_1], [Ts_2], [Ts_3]). Both are (despite subsequent partial simplifications) proofs of considerable ingenuity and complexity. Although they are of a combinatorial nature, the quite arbitrary nature of the space X, and its lack of any cartesian product structure, precludes the usefulness of the Erdös-Rado theorem on quasi-disjoint sets in favor of more basic tree arguments intertwined with the topological or functional analytic nature of our assumptions.

By way of illustration we will now describe the proof for the theorem on calibers for the regular cardinal α case.

The following lemma (given in Argyros [A_1]) is a convenient form for the combinatorial basis of the proof.

Lemma. Let α be an infinite regular cardinal, and $\kappa << \alpha$. We assume that for every $A \subset \alpha$, with $|A| = \alpha$, there is a partition $\mathcal{P}_A$ of A, such that $|\mathcal{P}_A| < \alpha$. Then there is a family $\{A_\eta : \eta < \kappa\}$ of subsets of α such that

$$|A_\eta| = \alpha \text{ for } \eta < \kappa,$$
$$A_{\eta+1} \in \mathcal{P}_{A_\eta} \text{ for } \eta < \kappa,$$
$$A_{\eta'} \subset A_\eta \text{ for } \eta < \eta' < \kappa, \text{ and}$$
$$\cap\{A_\eta : \eta < \kappa\} \neq \emptyset.$$

We now proceed with the *proof*. Suppose that $\alpha >> \kappa$, α regular, and α not a caliber for Ω. Then there is a family $\{U_\xi : \xi < \alpha\}$ of non-empty open subsets of Ω such that

if $B \subset \alpha$ and $\{U_\xi : \xi \in B\}$ has the finite intersection property then $|B| < \alpha$.

For $A \subset \alpha$ with $|A| = \alpha$ we set

$\mathcal{B}_A = \{B \subset A: \{U_\xi : \xi \in B\}$ has the finite intersection property$\}$.

The family $\mathcal{B}_A$, partially ordered by set inclusion, is inductive; hence there is a maximal element $B_A \in \mathcal{B}_A$. It is clear that $B_A \neq \emptyset$ and that $|B_A| < \alpha$. We define

$$\phi_A : A \setminus B_A \to \mathcal{P}^*(B_A)$$

as follows. Let $\eta \in A \setminus B_A$. The maximality of B_A implies that there is a finite non-empty subset F_η of B_A such that

$$\cap_{\xi \in F_\eta} U_\xi \cap U_\eta = \emptyset;$$

we set $\phi_A(\eta) = F_\eta$ for $\eta \in A \setminus B_A$. We set

$$\mathcal{P}_A = \{B_A\} \cup \{\phi_A^{-1}(F): F \in \mathcal{P}_\omega^*(B_A)\}$$

and we note that

$$\mathcal{P}_A \text{ is a partition of } A, \text{ with } |\mathcal{P}_A| < \alpha.$$

From the above lemma, it follows that there is a family $\{A_i: i < \kappa\}$ such that

$$|A_i| = \alpha \text{ for } i < \kappa,$$

$$A_{i'} \subset A_i \text{ for } i < i' < \kappa, \text{ and}$$

$$A_{i+1} \in \mathcal{P}_{A_i} \text{ for } i < \kappa.$$

Thus, since $|A_i| = \alpha$ for $i < \kappa$, we have

$$A_{i+1} = \phi_{A_i}^{-1}(\{F_i\})$$

for some $F_i \in \mathcal{P}_\omega^*(B_{A_i})$, $i < \kappa$. We set

$$V_i = \cap_{\xi \in F_i} U_\xi \text{ for } i < \kappa,$$

and we proved that the family $\{V_i: i < \kappa\}$ consists of pairwise disjoint non-empty subsets of Ω. Indeed, since $F_i \subset B_{A_i}$, it follows that V_i is open and non-empty. If $i < i' < \kappa$, then $F_{i'} \subset A_{i+1} = \phi_{A_i}^{-1}(\{F_i\})$; hence, if $\xi' \in F'$, we have that $V_i \cap U_{\xi'} = \emptyset$, and hence $V_i \cap V_{i'} = \emptyset$. This is a contradiction to $S(\Omega) = \kappa$, proving our claim.

The proof of the $C(\Omega)$-theorem is a more refined and difficult version of the proof of the caliber theorem, since in the former case we must produce (starting from certain data) a large independent family, while in the latter case we must only produce (starting from the same data) a large family with the finite intersection property. However, the combinatorial principles involved are the same. The following result expresses clearly a part of this interrelation.

(A) Let Ω be a compact Hausdorff space, and α a regular cardinal with $\alpha >> \omega^+$, and such that every closed linear subspace E of $C(\Omega)$, with dim $E \geq \alpha$, contains isomorphically ℓ^1_α. Then Ω has caliber α. (Argyros-Negrepontis [AN_1] (Theorem 4.28)).

However, the analogue of (A) for $\alpha = \omega^+$, does not hold, at least assuming CH. S. Argyros, N. Kalamidas, and Th. Zachariades have given an ingenious example, with CH, of a compact totally disconnected space Ω, that does not have caliber ω^+ and does not have any independent family $\{(A_\xi, B_\xi): \xi < \omega^+\}$ of closed subsets, while every closed linear subspace E of $C(\Omega)$ of dimension ω^+ (and generated by characteristic functions of open-and-closed sets of Ω) contains isomorphically $\ell^1_{\omega^+}$. Furthermore, $|\Omega| = \omega^+$, while the dimension of the dual space of $C(\Omega)$ is 2^{ω^+}.

There is a useful complement to the caliber theorem that, at least with GCH, essentially completes the description of calibers of a compact space. The surprising result is that the Souslin number of a compact space essentially determines the calibers of the space. In fact, let X be a compact Hausdorff space, with $S(X) = \kappa$, and assume GCH and that α is a regular cardinal (as we may assume on account of the caliber theorem). Then:

a) if $\alpha > \kappa$ and α either a limit cardinal or $\alpha = \beta^+$, with $cf(\beta) \geq \kappa$, then X has caliber α;

b) if $\alpha > \kappa$ and $\alpha = \beta^+$, with $cf(\beta) < \kappa$ then X has *caliber* (α, β) (i.e. for every family $\{U_\xi: \xi < \alpha\}$ of non-empty open subsets of X, there is $B \subset \alpha$, with $|B| = \beta$, such that $\bigcap_{\xi \in B} U_\xi \neq \emptyset$); and,

c) if $\alpha < \kappa$, then X does not have even cal$(\alpha,2)$. Furthermore, by the example of Erdös mentioned earlier, and by some more general related examples, case b) cannot be improved. (The proofs of these results appear in Argyros-Tsarpalias [AT_3]).

Incidentially, the following is the simplest case of the main open problem on calibers of compact spaces (G.C.H. is assumed). Is there a compact Hausdorff space X, with c.c.c., such that X does not have caliber α, for some $\alpha > \omega^+$ with $\alpha = \beta^+$, $cf(\beta) = \omega$ (e.g. $\alpha = (\aleph_\omega)^+$), and X has caliber ω^+?

The $C(\Omega)$-theorem has some interesting immediate topological consequences: If Ω is a compact totally disconnected Hausdorff space with topological weight α, and Souslin number κ, and $\kappa \ll \alpha$, $\kappa \ll cf(\alpha)$, then there is a continuous function from Ω onto $\{0,1\}^\alpha$ (and, hence Ω has a discrete subset of (maximal) cardinality α). For Ω compact Hausdorff and extremally disconnected (i.e. such that the closure of any open set is open) a recent result of Balcar (generalizing a result of Solevay and, independently, of Ketonen [Ke] for $S(\Omega) = \omega^+$) shows that there is always a continuous function from Ω onto $\{0,1\}^\alpha$ (where α = weight of Ω). An interesting result in the same direction, for continuous function of any compact Hausdorff space Ω onto $[0,1]^{\omega(\Omega)}$ is given by Šapirouskiĭ [Sa].

The Balcar result has a significant application towards a conjecture by Rosenthal [R_1], concerning the structure of injective Banach spaces. A Banach space B is *injective* if for every Banach space X, every linear subspace Y of X, and every bounded linear operator T: Y → B there is a bounded linear operator S: X → B such that $S|Y = T$; and B is a P_1-*space* if in addition, we can choose S, so that $||S|| = ||T||$. The structure of P_1-spaces is completely known, from work of Nachbin [N], Goodner [Go], and Kelley [K]: a Banach space B is a P_1-*space* if and only if B is isomorphic to $C(\Omega)$ for some compact Hausdorff extremally disconnected space Ω. On the other hand, the structure of the injective Banach spaces is not known; it is not even known if there is an injective, not P_1-space. Rosenthal's conjecture is the following: If X is an injective Banach space of dimension α, then ℓ^1_α can be embedded in X, and the dual space X* is isomorphic to the space

$$\underbrace{(L^1\{0,1\}^\alpha \oplus \ldots \oplus L^1\{0,1\}^\alpha \oplus \ldots)}_{2^\alpha \text{ times}}{}_1$$

(which by definition is the set of all $(x_\xi)_{\xi<2^\alpha}$, with $x_\xi \in L^1\{0,1\}^\alpha$ for $\xi < 2^\alpha$, and $\Sigma_{\xi<2^\alpha} ||x|| < \infty$, and norm given by $||(x_\xi)_{\xi<2^\alpha}|| = \Sigma_{\xi<2^\alpha} ||x_\xi||$).

The Balcar result (and the $C(\Omega)$-theorem for certain spaces Ω) proves Rosenthal's conjecture for P_1-spaces (since a continuous function from Ω onto $\{0,1\}^{\omega(\Omega)}$ implies the embedding of $\ell^1_{\omega(\Omega)}$ into $C(\Omega)$, and this in turn implies Rosenthal's second statement of the conjecture concerning the dual space).

The $C(\Omega)$-theorem and the caliber theorem can be generalized, using essentially the same techniques, in the following way.

1) If Ω is a compact Hausdorff space, with $S(\Omega) = \kappa$ and $\kappa << \alpha$, $\kappa << cf(\alpha)$, $\lambda < \kappa$, then, denoting by Ω_{λ^+} the set Ω, with the λ^+-completion topology, i.e. the topology generated by the family of all sets that are intersections of at most λ open subsets of Ω, we have that

1a) α is a caliber for Ω_{λ^+}; and,

1b) if E is a closed linear subspace of $C(\beta(\Omega_{\lambda^+}))$ (where

βX denotes the Stone-Čech compactification of X) with $\dim E \geq \alpha$, then the space ℓ^1_α can be isomorphically embedded in E.

2) If Ω_i are compact Hausdorff spaces, with $S(\Omega_i) \leq \kappa$ for $i \in I$, κ a regular cardinal, $\kappa << \alpha$, $\kappa << cf(\alpha)$, $\omega \leq \lambda < \kappa$, then, denoting by

$(\pi_{i\in I}\Omega_i)_{(\lambda+)}$ the set $\pi_{i\in I}\Omega_i$, with the λ^+-box topology, i.e. the topology generated by the family of all sets $\pi_{i\in I}U_i$, where V_i is open in Ω_i for $i \in I$ and $|\{i \in I: V_i \neq \Omega_i\}| \leq \lambda$, we have that

2a) α is a precaliber for $(\pi_{i\in I}\Omega_i)_{\lambda^+}$; and,

2b) if E is a closed linear subspace of $C(\beta(\pi_{i\in I}\Omega_i)_{\lambda^+})$, with $\dim E \geq \alpha$, then the space ℓ^1_α can be isomorphically embedded in E.

These results were proved by N. Kalamidas, S. Negrepontis, and Th. Zachariades; the topological results 1a) and 2a) appear in the Comfort-Negrepontis [CN] monograph. These topological results strengthen results by Kurepa [Ku], and Juhász [Ju].

D. Pełczynski's conjecture

The $L^\infty(\mu)$-theorem has as a consequence a partial solution to a problem posed by Pełczynski [P_1] in 1968. This problem, which we will now describe, has subsequently been solved completely by Argyros [A_4], [A_6], in a strikingly simple way, using again heavily infinitary combinatorial methods.

Pełczynski [P_1], using the facts that every Banach space of dimension at most α is the linear bounded image of ℓ^1_α and that the double dual $C^{**}(\Omega)$ for any compact space Ω is a P_1 (injective) Banach space, proved that if ℓ^1_α can be isomorphically embedded into a Banach space X, then the Banach space $L^1(\{0,1\}^\alpha)$ of λ-equivalence classes of real-valued (absolutely) integrable with respect to Haar (product) probability measure λ on $\{0,1\}^\alpha$ can be isomorphically embedded in the dual space X*. *Pełczynski conjectured* that the converse statement is also true; and, he was able to prove it for $\alpha = \omega$. R. Haydon [Hay_3] proved Pełczynski's conjecture for α regular and $\alpha >> \omega^+$, using also directly the Erdös-Rado theorem on quasi-disjoint sets, while Argyros and Negrepontis [AN_2] proved this conjecture for $\alpha >> \omega^+$ and $cf(\alpha) >> \omega^+$, as an application of the $L^\infty(\mu)$-result. Next, Haydon [Hay_4], assuming the continuum hypothesis, found with an interesting argument a compact space Ω such that

$\ell^1_{\omega^+}$ cannot be embedded isomorphically in $C(\Omega)$, and

$L^1\{0,1\}^{\omega^+}$ is isomorphically embedded in $C^*(\Omega)$,

thus obtaining a counterexample to Pełczynski's conjecture for $\alpha = \omega^+$ (This example will be described below.). Subsequently, in March 1979, Fleissner and Negrepontis [FlN] proved that there is a model of set theory in which $\omega^+ < 2^\omega < 2^{\omega^+}$ and there is a sequence $\{N_\xi: \xi < \omega^+\}$ of null sets of the Cantor set such that every compact null set is contained in N_ξ for some $\xi < \omega^+$. There follows the existince of a space Ω with the same properties, in a model set theory in which the continuum hypothesis fails Argyros [A_6] proved that for $\alpha > \omega^+$, Pełczynski's conjecture is true. His extremely ingenious proof uses Hajnal's [Ha] strong combinatorial theorem on the existence of free sets: if α, κ are cardinals, $\kappa < \alpha$, and $f: \alpha \to P_\kappa(\alpha)$ is any function, then there is $A \subset \alpha$ with $|A| = \alpha$, such that $\zeta \notin f(\xi)$ for any $\xi, \zeta \in A$, $\xi \neq \zeta$. Hajnal's theorem gets used in the following way: If X is a Banach space,

$\alpha > \omega^+$, and there is an isomorphic embedding

$$L^1\{0,1\}^\alpha \hookrightarrow X^*,$$

then there is a ω^*-continuous, onto (and, hence open) operator $T: X^{**} \twoheadrightarrow L^\infty\{0,1\}^\alpha$. We now consider the diagram

$$X \overset{J}{\hookrightarrow} X^{**} \overset{T}{\twoheadrightarrow} L^\infty\{0,1\}^\alpha \overset{P_\xi^*}{\to} L^\infty\{0,1\}^{\{\xi\}}$$

for $\xi < \alpha$, where J is the canonical embedding of X into its second dual (and, by Goldstine's theorem $J(X)$ is ω^*-dense in X^{**}), and P_ξ^* is the canonical projection for $\xi < \alpha$. Since T is open, there is $\kappa > 0$, such that $T \circ J(S_X^\kappa)$ is ω^*-dense $S^1_{L^\infty}$, where $S_X^\varepsilon = \{x \in X: \|x\| < \varepsilon\}$. For $\varepsilon > 0$ "suitably small", we can then choose $x_\xi \in S_X^{\kappa+\varepsilon}$, such that

$$P_\xi^* \circ T \circ J(x_\xi) = \text{identity function from } \{0,1\}^{\{\xi\}} \text{ to } \mathbb{R}.$$

The element $T \circ J(x_\xi) \in L^\infty\{0,1\}^\alpha$, and hence it depends on a countable set $N_\xi \subset \alpha$ of coordinates for $\xi < \alpha$. We apply Hajnal's theorem for the function

$$f: \alpha \to \mathcal{P}_{\omega^+}(\alpha)$$

given by $f(\xi) = N_\xi$. There is $A \subset \alpha$, with $|A| = \alpha$, such that $\zeta \notin N_\xi$ for $\xi, \zeta \in A$, $\xi \neq \zeta$. The rest of the Argyros' proof is devoted in showing that $\{x_\xi: \xi \in A\}$ is equivalent to the canonical basis of ℓ^1_α.

Furthermore, Argyros [A_6] has proved that if we assume Martin's axiom together with the denial of the continuum hypothesis (MA + ¬CH) then Pełczynski's conjecture is true for $\alpha = \omega^+$. As a consequence, Pełczynski's conjecture is true for all $\alpha \neq \omega^+$, while its truth depends on our model of set theory for $\alpha = \omega^+$.

The methods of Argyros for proving Pełczynski's conjecture for $\alpha > \omega^+$, and for $\alpha = \omega^+$ under MA + ¬CH work also to prove a conjecture by Hagler and Stegall [HS]: If $L^1\{0,1\}^\alpha$ is isomorphically embedded into the dual X^* of a Banach space X as a complementary subspace, then the space Z_α can be embedded into X, where

$$Z_\alpha = \underbrace{(Z \oplus \ldots \oplus Z \oplus \ldots)}_{\alpha\text{-times}}{}_1,$$

where $Z = (\ell_1^\infty \oplus \ell_2^\infty \oplus \ldots \oplus \ell_n^\infty \oplus \ldots)_1$.
For $\alpha = \omega$ this had been proved by Hagler and Stegall [HS], while for $\alpha = \omega^+$ Haydon's [Hay_4] example is a counterexample to this conjecture under CH. The proofs for $\alpha > \omega^+$, and for $\alpha = \omega^+$ with MA + ¬CH are given in Argyros [A_6].

We finally give a description of

Haydon's space (assuming CH). Let Ω_0 be the Cantor set, with μ_0 its Haar Borel probability measure. Let $\xi < \omega^+$, and suppose that Ω_ζ, μ_ζ have been defined for $\zeta < \xi$, such that

Ω_ζ is a compact, metric, totally disconnected space,

μ_ζ is a probability regular Borel measure on Ω_ζ,

$p_{\zeta',\zeta}\colon \Omega_{\zeta'} \to \Omega_\zeta$ has been defined for $\zeta < \zeta' < \xi$, such that

$p_{\zeta',\zeta}$ is continuous, onto, and

$p_{\zeta',\zeta}{}^*(\mu_{\zeta'}) = \mu_\zeta$,

and let $\{N_{\zeta,\eta}\colon \eta < \omega^+\}$ be a well-ordering (assuming CH) of all closed μ_ζ-null subsets of Ω_ζ for $\zeta < \xi$. We now define Ω_ξ, μ_ξ, $p_{\xi,\zeta}$ for $\zeta < \xi$, and $\{N_{\xi,\eta}\colon \eta < \omega^+\}$. If ξ is a limit ordinal, we set

$$\Omega_\xi = \varprojlim_{\xi<\zeta} \Omega_\zeta,$$

$$\mu_\xi = \varprojlim_{\xi<\zeta} \mu_\zeta,$$

let $p_{\xi,\zeta}\colon \Omega_\xi \to \Omega_\zeta$ be the projective limit natural projections for $\zeta < \xi$, and let $\{N_{\xi,\eta}\colon \eta < \omega^+\}$ be a well-ordering (assuming CH) of all closed μ_ξ-null subsets of Ω_ξ.

Let now $\xi = \rho + 1$; set

$$E = \bigcup_{\zeta,\eta\leq\rho} p^{-1}_{\rho,\zeta}(N_{\zeta,\eta}),$$

and note that E is a μ_ρ-null set (as a countable union of such sets), and hence by the regularlity of μ_ρ, there is a sequence $\{K_{\rho,n}: n < \omega\}$ of closed pairwise-disjoint subsets of Ω_ρ, such that $\mu_\rho(\cup_{n<\omega}K_{\rho,n}) = 1$. We let

$$\Omega_\xi = \Omega_{\rho+1} = (\Omega_\rho \times \{0\}) \cup (\cup_{n<\omega}(K_{\rho,n} \times \{\tfrac{1}{2^n}\}))$$

(considered as a subset of $\Omega_\rho \times [0,1]$), and we note that clearly Ω_ξ is a compact, metric, totally disconnected space. We set

$$\begin{aligned} \mu_\xi &= \tfrac{1}{2}\mu_\rho \text{ on } \Omega_\rho \times \{0\} \\ &= \tfrac{1}{2}(\mu_\rho | K_{\rho,n}) \text{ on } K_{\rho,n} \times \{\tfrac{1}{2^n}\} \text{ for } n < \omega, \end{aligned}$$

and we note that clearly μ_ξ is a probability regular Borel measure on Ω_ξ. The function $p_{\xi,\rho}: \Omega_\xi \to \Omega_\rho$ is the obvious (2-1) one, and $\{N_{\xi,\eta}: \eta < \omega^+\}$ is a well-ordering (again assuming CH) of all closed μ_ξ-null subsets of Ω_ξ.

We finally let $\Omega = \varprojlim_{\xi<\omega^+} \Omega_\xi$, $\mu = \varprojlim_{\xi<\omega^+} \mu_\xi$.

E. *Rosenthal's criterion for the embedding of ℓ^1*

Rosenthal $[R_2]$ proved the following quite satisfying and general non-compactness criterion on the embedding of ℓ^1.

Rosenthal's theorem. Let X be a Banach space. The space ℓ^1 is isomorphically embedded in X if and only if there is a sequence $(x_n)_{n<\omega}$ in X such that $\|x_n\| \leq M$ (for some M and $n < \omega$), and such that it has no weakly Cauchy subsequence.

(A sequence $(y_n)_{n<\omega}$ in a Banach space is *weakly Cauchy* if $(f(y_n))_{n<\omega}$ converges for every element $f \in X^*$.)

Rosenthal's original proof was quite ingenious and elementary. A simple proof was subsequently discovered by Farahat [F], (see also Lindenstrauss-Tzafriri [LT]) using the following Nash-Williams, Galvin, Prikry [GP], Silver infinite combinatorial version of Ramsey's theorem:

Denote by $\omega^{[\omega]}$ the subset of Baire's space ω^{ω} consisting of all infinite subsets of ω, and let A be an analytic subset (i.e. a continuous image of a Borel subset of Baire's space) of $\omega^{[\omega]}$, then there is an infinite subset M of ω, such that either $M^{[\omega]} \subset A$ or $M^{[\omega]} \subset \omega^{[\omega]} \setminus A$.

Rosenthal's embedding theorem, and the ideas connected with its proof, had interesting applications. One of them is the following:

Theorem (Odell-Rosenthal [OR]). A separable Banach space X contains a subspace isomorphic to ℓ^1 if and only if there is an element of the double dual X** of X, which is not a weak*-limit of a sequence of elements of X (i.e. there is an element of X** that is not Baire-1 as a function on the unit ball of X* with the weak*-topology) if and only if $|X^{**}| > 2^{\omega}$.

There is a version of this theorem for general (and not necessarily separable) Banach spaces due to Haydon [Hay_2]: A Banach space X contains a subspace isomorphic to ℓ^1 if and only if there is an element of the double dual X** that is not universally measurable (as a function on the unit ball of X* with the weak*-topology).

These ideas have also led to some remarkable theorems on the compactness properties of the set of Baire-1 functions, or of the set of Borel functions, equipped with the pointwise topology, on a Polish space. These results are due to Rosenthal [R_3], Fremlin [Fr], Bourgain-Fremlin-Talagrand [BFT].

It would be highly desirable to have a satisfactory generalization of Rosenthal's embedding theorem for ℓ^1_α, with uncountable cardinal. However, there are some rather strong limitations on what we may be able to prove, because of the existence of two examples that we now describe. One, by Haydon [Hay_4], states that there is a compact Hausdorff space Ω, and a family $F = \{f_\xi : \xi < \omega^+\}$ in $C(\Omega)$, with $\|f_\xi\| \leq 1$ for $\xi < \omega^+$, such that if $\xi_1 < \ldots < \xi_n < \ldots < \omega^+$, then $\{f_{\xi_n} : n < \omega\}$ is not a weakly Cauchy sequence, and such that $L^1\{0,1\}^{\omega^+}$ cannot be embedded in X* (and hence by Pełczynski's result mentioned in D, $\ell^1_{\omega^+}$ cannot be embedded in X). The other, a recent example by C. Gryllakis, states that in $\ell^\infty_{\omega^+}$ there is a family $F = \{f_\xi : \xi < \omega^+\}$, with $\|f_\xi\| \leq 1$ for $\xi < \omega^+$, such that if $\xi_1 < \ldots < \xi_n < \ldots < \omega^+$, then $\{f_{\xi_n} : n < \omega\}$ is not a weakly Cauchy sequence, and such that if $A \subset \omega^+$ and $\{f_\xi : \xi \in A\}$ is equivalent to the canonical base of ℓ^1, then the order type of A, as a subset of ω^+, is less than the ordinal product $\omega \circ \omega$.

The author wishes to thank Mr. Theodosios Zachariades for valuable help in the preparation of this paper.

References

[A_1] S. Argyros, *Isomorphic embeddings of $\ell^1(\Gamma)$ in classical Banach spaces*, Doctoral dissertation, Athens University, 1977. [In Greek.]

[A_2] ________, *Applications of isomorphic embeddings of $\ell^1(\Gamma)$*, Notices Amer. Math. Soc. 24 (1977), A-539.

[A_3] ________, *Isomorphic embeddings of ℓ^1_m into m-dimensional subspace of $C(\Omega)$: the regular cardinal case*, Notices Amer. Math. Soc. 25 (1978), A-234.

[A_4] ________, *On some isomorphic embeddings of ℓ^1_m and Z_m*, Notices Amer. Math. Soc. 25 (1978), A-588.

[A_5] ________, *Some isomorphic embeddings of ℓ^1_α and Z_α in Banach spaces*, Colloquia 1978, Greek Mathematical Society, Athens 1978, pp. 123-137. [In Greek.]

[A_6] ________, *On non-separable Banach spaces*, to appear.

[AN_1] S. Argyros and S. Negrepontis, *Isomorphic embedding of $\ell^1(\Gamma)$*, Notices Amer. Math. Soc. 24 (1977), A-538.

[AN_2] S. Argyros and S. Negrepontis, *Universal embeddings of ℓ_α^∞ into* C(X) *and* $L^\infty(\mu)$, Proceedings of Colloquium on Topology, Budapest, 1978, to appear.

[AT_1] S. Argyros and A. Tsarpalias, *Calibers of compact spaces,* Notices Amer. Math. Soc. 25 (1978), A-234.

[AT_2] ________, *Calibers of compact spaces,* Abstracts of Colloquium on Topology, Budapest, 1978.

[AT_3] ________, *Calibers of compact spaces,* to appear.

[BFT] J. Bourgain, D.H. Fremlin, and M. Talagrand, *Pointwise compact sets of Baire-measurable functions,* Amer. J. Math. 100 (1978), 845-886.

[CN] W.W. Comfort and S. Negrepontis, *Chain conditions in topology,* Cambridge Tracts in Mathematics, Cambridge University Press, to appear.

[D] J. Dixmier, *Sur un théorème de Banach,* Duke Math. J. 15 (1948), 1057-1071.

[ErR] P. Erdös and R. Rado, *Intersection theorems for systems of sets,* J. London Math. Soc. 35 (1960), 85-90; (II) J. London Math. Soc. 44 (1969), 467-479.

[Ef] B.V. Efimov, *Subspaces of dyadic bicompacta,* Dokl. Akad. Nauk SSSR 185 (1969), 987-990.

[EfE] B.V. Efimov and R. Engelking, *Remarks on dyadic spaces, II,* Colloq. Math. 13 (1964/65), 181-197.

[EP] R. Engelking and A. Pełczynski, *Remarks on dyadic spaces,* Colloq. Math. 11 (1963), 55-63.

[E] R. Engelking, *Cartesian products and dyadic spaces,* Fund. Math. 57 (1965), 287-304.

[F] J. Farahat, *Espaces de Banach contenant* ℓ^1, d'opres H.P. Rosenthal, Seminaire Maurey-Schwartz 1973-1974, no. 26.

[FlN] W. Fleissner and S. Negrepontis, *Haydon's counter-example with not CH,* Notices Amer. Math. Soc. 26 (1979), A-381.

[Fr] D.H. Fremlin, *Pointwise compact sets of measurable functions,* Mansucripta Math. 15 (1975), 219-242.

[GP] F. Galvin and K. Prikry, *Borel sets and Ramsey's theorem,* J. Symbolic Logic 38 (1974), 193-198.

[Ge_1] J. Gerlits, *On subspaces of dyadic compacta,* Studia Math. Hung., to appear.

[Ge_2] ________, *Continuous functions on products of topological spaces,* to appear.

[Go] D.A. Goodner, *Projections in normed linear spaces*, Trans. Amer. Math. Soc. 69 (1950), 89-108.

[H] J. Hagler, *On the structure of S and C(S) for S dyadic*, Trans. Amer. Math. Soc. 214 (1975), 415-428.

[HS] J. Hagler and G. Stegall, *Banach spaces whose duals contain complemented subspaces isomorphic to* $C^*[0,1]$, J. Functional Analysis 13 (1973), 233-251.

[Ha] A. Hajnal, *Proof of a conjecture of S. Ruziewicz*, Fund. Math. 50 (1961), 123-128.

[Hay_1] R. Haydon, *Embedding* D^τ *in Dugundji spaces, with an application to linear topological classification of spaces of continuous functions*, Studia Math. 56 (1976), 229-242.

[Hay_2] ________, *Some more characterizations of Banach spaces containing* ℓ^1, Math. Proc. Cambridge Philos. Soc. 80 (1976), 269-276.

[Hay_3] ________, *On Banach spaces which contain* $\ell^1(\tau)$ *and types of measures on compact spaces*, Israel J. Math. 28 (1977), 313-323.

[Hay_4] ________, *On dual* L^1*-spaces and injective bidual Banach spaces*, Israel J. Math. 31 (1978), 142-152.

[J] R.C. James, *A separable somewhat reflexive Banach space with non-separable dual*, Bull. Amer. Math. Soc. 80 (1974), 738-743.

[Ju] I. Juhász, *Cardinal functions in topology*, Math. Centre Tracts 34, Mathematisch Centrum, Amsterdam, 1971.

[K] J.L. Kelley, *Banach spaces with the extension property*, Trans. Amer. Math. Soc. 72 (1954), 323-326.

[Ke] J. Ketonen, *On non-regular ultrafilters*, J. Symbolic Logic 37 (1972), 71-74.

[Ku] G. Kurepa, *The cartesian multiplication and the cellularity number*, Publ. Inst. Math. (Beograd) (N.S.) 2 (1962), 121-139.

[LT] J. Lindenstrauss and L. Tzafriri, *Classical Banach Spaces I, Sequence Spaces*, Ergebnisse Series vol. 92, Springer-Verlag, 1977.

[M] D. Maharam, *On homogeneous measure algebras*, Proc. Nat. Acad. Sci. U.S.A. 28 (1942), 108-111.

[N] L. Nachbin, *A theorem of Hahn-Banach type for linear transformations*, Trans. Amer. Math. Soc. 68 (1950), 28-46.

[Ne_1] S. Negrepontis, *Combinatorial methods in the isomorphic theory of Banach spaces*, Colloquia 1977, Greek Mathematical Society, Athens 1977, pp. 24-26. [In Greek.]

[Ne_2] ________, *Infinitary combinatorics in general topology and functional analysis*, Abstracts of Colloquium on Topology, Budapest, 1978.

[OR] E. Odell and H.P. Rosenthal, *A double-dual characterization of separable Banach spaces containing* ℓ^1, Israel J. Math. 20 (1975), 375-384.

[P_1] A. Pełczynski, *On Banach spaces containing* $L^1(\mu)$, Studia Math. 30 (1968), 231-246.

[P_2] ________, *Linear extensions, linear averagingsm and their applications to linear topological classification of spaces of continuous functions*, Dissertationes Mat. Roszpawy Mat. 58 (1968).

[R_1] H.P. Rosenthal, *On injective Banach spaces and the spaces* $L^\infty(\mu)$ *for finite measures* μ, Acta Mathematica 124 (1970), 205-248.

[R_2] ________, *A characterization of Banach spaces containing* ℓ^1, Proc. Nat. Acad. Sci. U.S.A. 71 (1974), 2411-2413.

[R_3] ________, *Pointwise compact subsets of the first Baire class*, Amer. J. Math. 99 (1977), 362-378.

[S_1] N.A. Šanin, *A theorem from the general theory of sets*, C.R. (Doklady) Acad. Sci. URSS 53 (1946), 399-400.

[S_2] ________, *On intersection of open subsets in the product of topological spaces*, C.R. (Doklady) Acad. Sci. URSS 53 (1946), 499-501.

[S_3] ________, *On dyadic bicompacta*, C.R. (Doklady) Acad. Sci. URSS 53 (1946), 777-779.

[Sa] B.E. Šapirovskiĭ, *On decomposition of perfect mappings into irreducible one and a retraction*, Abstracts of Colloquium on Topology, Budapest, 1978.

[T] M. Talagrand, *En général il n'existe pas de relèvement linéaire borélien fort*, C.R. Acad. Sc. Paris 287 (1978), Series A - 633-634.

[Ts_1] A. Tsarpalias, *Isomorphic embedding of* ℓ^1_m *into m-dimensional subspaces of* $C(\Omega)$*: the singular case*, Notices Amer. Math. Soc. 25 (1978), A-234.

[Ts_2] ________, *Methods of infinitary combinatorics and weak compactness in the theory of Banach spaces*, Doctoral dissertation, Athens University, Athens, 1978. [In Greek.]

[Ts_3] A. Tsarpalias, *Isomorphic embeddings of* $\ell^1(\Gamma)$ *in the space* $C(\Omega)$, Colloquia 1978, Greek Mathematical Society, Athens 1978, pp. 155-165. [In Greek.]

[Z] Th. Zachariades, *Isomorphic embeddings of* ℓ_m^1 *into subspaces of* C(X) *for arbitrary product spaces* X, Notices Amer. Math. Soc. 25 (1978), A-587.

ORDER-THEORETIC BASE AXIOMS

Peter J. Nyikos
Institute for Medicine and Mathematics
Ohio University

The study of base axioms has expanded considerably in the last few years, as attested to by C. Aull's survey [A_6] and the references he cites (especially the Burke-Lutzer paper [BL]). In this paper I will confine myself to those base properties which are defined with the help of partial order relationships, either between the members of a base, or between the points of the space, or both. Even so, some topics had to be left out due to lack of space. In particular, the theory of generalized ordered spaces comes under this heading, but (fortunately!) there is a paper by Lutzer elsewhere in this volume, so it is only mentioned where it is relevant to other concepts which are treated at length. I have also (reluctantly!) omitted the topic of bases that are Boolean algebras.

This paper can be read as a sequel to [Ny_1] and [Ny_3], but to make it self-contained I have recapitulated some of the main results announced in those papers in a slightly different format. As in the case of those papers, this one is also an announcement of many results which I and others have not had the time to publish yet.

Section 1 will be on the properties of local bases at points. After that, the emphasis will be on the properties of (global) bases. Section 2 will be on non-archimedean

ISBN: 0-12-584960-5

spaces. The next two sections will be on the various kinds of rank (large, ordinary, small), the first being a general survey of ranks and large ranks, and the other focusing on cardinal invariants, especially the S and L space situation. Section 5 will be on co-scattered spaces, especially Pixley-Roy hyperspaces. Section 6 has to do with the preservation of metacompactness in finite and countable products.

"Space" will always mean "T_1-space" unless preceded by "topological." And "compact," unless followed by "T_1," is always assumed to include "Hausdorff."

Section 1: Lobs, blobs, and globs

The concepts in this section are all more general than first countability, so one cannot expect too much in the way of structure theory; nevertheless, we do have a number of interesting results.

The first generalization is that of a space in which each point has a linearly ordered base ("lob"). Whenever a point p in any topological space X has a lob, the cofinality of that base (with respect to reverse inclusion) is a regular cardinal, and equals the character $\chi(p,X)$ of that space at p. Any other base at p will contain a subset order-isomorphic to $\chi(p,X)$. It is easy to show that the tightness of p, the pseudo-character at p, and (if p is not isolated) the least cardinal of a set which has p as an accumulation point, are all equal to $\chi(p,X)$. [The tightness at p, designated $t(p,X)$, is the smallest infinite cardinal κ such that if $A \subset X$ and $p \in \overline{A}$, then there is a set $C \subset A$ with $|C| \leq \kappa$ and $p \in \overline{C}$; the pseudo-character $\psi(p,X)$ at p is the smallest cardinal of a collection of open subsets of X whose intersection is p.]

Moreover, if Y is a subspace of X in which p is not isolated, these numbers will be the same for Y. And if p is in the closure of a subset A of X, there is a well-ordered net from A converging to p. It follows that in a space X where every point has a lob (called a "lob-space") the cardinal invariants of tightness, character, and pseudo-character of X (defined to be the supremum, over all points of X, of the local versions) are all equal.

Sheldon Davis has proven or collected a number of theorems about covering properties on lob-spaces in [D_2]. Some of them have to do with pre-paracomapctness, which involves a kind of "non-heterogeneity" condition on open refinements:

1.1. Definition. A Hausdorff space is *preparacompact* [*resp.* *$\aleph$-preparacompact*] if each open cover of X has an open refinement $\{H_\alpha : \alpha \in A\}$ such that, if $B \subset A$ is infinite [resp. uncountable] and if, for each $\beta \in B$, one chooses p_β and q_β in H_β with $p_\alpha \neq p_\beta$ and $q_\alpha \neq q_\beta$ for $\alpha \neq \beta$, then the set $Q = \{q_\beta : \beta \in B\}$ has a limit point whenever $P = \{p_\beta : \beta \in B\}$ has one.

1.2. Theorem [D_2]. If X is a regular lob-space, the following are equivalent:

(1) X is paracompact,

(2) X is irreducible and $\aleph$-preparacompact,

(3) X is $\delta\theta$-refinable and $\aleph$-preparacompact.

1.3. Theorem [D_2]. Every regular preparacompact lob-space is collectionwise Hausdorff; every normal preparacompact lob-space is collectionwise normal.

One disadvantage of lob-spaces is that the product of even two of them is not a lob-space, unless all nonisolated points in both spaces have the same local character. To overcome this, B.M. Scott has studied the class of what he calls "globular" spaces: spaces in which each point has a local base which is isomorphic as a partially ordered set to some finite product of linear orders.

1.4. Definition. Let Ω be a finite (possibly empty) set of infinite regular cardinals. Let $P = \pi\Omega$, with the convention that $\pi\phi = 1$. A local base $\mathcal{B}$ at a point p is called an *Ω-generalized linearly ordered base* (*Ω-glob*)[1] at p if $\langle \mathcal{B}, \supseteq \rangle$ is isomorphic to $\langle P, \leq \rangle$. A point p of a space X is *globular* is there is a glob at p in X, and X is *globular* if every point of X is globular. An Ω-glob such that $|\Omega| = 2$ is a *bilinearly ordered base* (*blob*). A glob $\mathcal{B}$ at p is *obese* if for all $a \in P$, there exists $x(a) \in B(a)$ [where $B: P \to \mathcal{B}$ is an isomorphism] such that for all $b \in P$, $x(a) \in B(b)$ only if $b \leq a$.

Of course, the class of globular spaces is closed under finite products, and the Ω involved in the product at each point will be the *union* of the Ω's of the factors. If the same cardinal appears in several factors, a subset of the product-base can be obtained by taking finite intersections so that each cardinal appears in the expression for P once. The Pixley-Roy hyperspace (see Section 5) of a globular space is globular, and a similar analysis works for it.

We do not yet know how much of S. Davis's work on covering properties generalizes to globular spaces, and to what extent.

---

[1]B. Scott takes sole responsibility for these expressions.

Many of the earlier cardinal-invariant results do not go through without change, either:

1.5. Example. (Originally described in [V].) Let X be ω_1+1 with the following topology: all points except ω_1 are isolated, while ω_1 has a local base of sets of the form $B_\alpha \cap B_n$ (α countably infinite, n finite), where $B_\alpha = (\alpha,\omega_1]$ and B_n consists of ω_1 together with all ordinals of the form $\lambda+m$ where λ is a limit ordinal and $m \geq n$. This is an obese $\{\omega_0,\omega_1\}$-blob. Now the character and tightness at ω_1 is $\aleph_1$, while the pseudo-character at ω_1 is $\aleph_0$ (ω_1 is a G_δ). There is no nontrivial well-ordered set converging to ω_1, and ω_1 is not the accumulation point of any countable subset of X.

However, B. Scott has found some interesting invariants with the help of his brother, David W. Scott. One of the more unexpected ones is the "glob-character."

1.6. Lemma [Scott and Scott]. If p has an Ω-glob and an Ω'-glob, then $\Omega = \Omega'$. If $p \in Y \in X$, then p has an Ω'-glob for some $\Omega' \subset \Omega$.

1.7. Definition. The *glob-character* of a globular point p is that unique Ω for which p has an Ω-glob.

1.8. Lemma. The following relationships hold for a point p which is globular in a space X.

$\chi(p,X) = \max \Omega$; $t(p,X) = \max \Omega$; $\psi(p,X) \in \Omega$.

Example 1.5, together with the Tychonoff plank, shows that the last result is the best possible for $\Omega = \{\omega_0,\omega_1\}$; similar examples can be constructed for any other Ω.

In the case of blobs, there is more:

1.9. Definition. Let $\mathcal{B}$ be a $\{\kappa,\lambda\}$-blob at p, and let $B: \kappa \times \lambda \to \mathcal{B}$ be an isomorphism. For $\alpha \in \kappa$, let $V_\alpha = \cap\{B(\gamma,0): \gamma \geq \alpha\}$, and for $\beta \in \lambda$ let $H_\beta = \cap\{B(0,\gamma): \gamma \geq \beta$. Then $\mathcal{B}$ is *ectomorphic* if $B(\alpha,\beta) = V_\alpha \cup H_\beta$ for each $\langle\alpha,\beta\rangle \in \kappa \times \lambda$.

1.10. Theorem [Scott and Scott]. If there is a glob at p, there is either an obese blob or an ecotmorphic blob at p; but no point can have both an obese and an ectomorphic blob associated with it.

The possibilities become more numerous with higher products, and no exhaustive classification has been verified yet. B. Scott has conjectured that there is one class for each abstract simplicial complex on Ω which contains all the 1-simplexes. But the details are so tedious that the only case that has been verified is:

1.11. Lemma [Scott and Scott]. If there is an obese Ω-glob at p, then every Ω-glob at p is obese.

I will close this section with some remarks about a class of spaces with a more "uniform" structure, those in which each nonisolated point has a κ-lob for a fixed regular cardinal κ. I have called these spaces *of characteristic* κ, in analogy with fields. Spaces with the same characteristics mix well together (for example, the κ-box product of κ or fewer spaces of characteristic κ is again of characteristic κ) but not with spaces of another characteristic. Among these spaces we have a highly successful analogue of metric spaces, the κ-metrizable spaces, which are the spaces with a compatible uniformity with a totally ordered base of cofinality

κ [NR_2]. We also have analogues of several classes of generalized metric spaces, the best known of which are the linearly stratifiable spaces of Vaughan and Tamano [V]--but all the "generalized metric spaces" of Hodel [H_2], [H_3] have similar analogues among the spaces of characteristic κ. That much more has not been written about them can be ascribed to simply inertia: it is unexciting to publish proofs which so closely parallel the case $\kappa = \omega$. What little has appeared, has mostly to do with theorems and proofs which are markedly different from the countable case. Two examples are the analogue of the Nagata general metrization theorem with its short proof for $\kappa > \omega$ [NR_2] and the following, which deviates from the countable case:

1.12. Theorem. Let κ be an uncountable regular cardinal. The following are equivalent:

(1) X is non-archimedianly κ-quasimetrizable,

(2) X is κ-quasimetrizable,

(3) X is a κ-γ-space.

That (1) and (2) are equivalent was proven in [R], which contains some other material on spaces of characteristic κ. Combining ideas from this proof and Theorem 3 of [NR_2] gives a proof that (2) and (3) are equivalent.

On the other hand, Kofner [K] gave an example of a space which is quasimetrizable, but not non-archimedianly, and it is a major unsolved problem whether every γ-space is quasimetrizable (see the Problem Section in vol. 2 of Topology Proceedings).

Section 2: Non-archimedean spaces

The concept of a non-archimedean space can be thought of as "globalizing" the definition of a lob-space.

2.1. Definition. A base $\mathcal{B}$ is *of rank* 1 if for each point p the set $\mathcal{B}(p)$ of all members of $\mathcal{B}$ containing p is totally ordered. A space is *non-archimedean* if it has a rank 1 base.

But the increase in strength as one goes from "lob" to "rank 1" is enormous, even more so than the increase in going from "first countable" to "point-countable base." This shows up most quickly if we use the alternative definition [it takes some effort to show the equivalence] where "totally ordered" is replaced by "well-ordered." Then one quickly obtains some strong structure theorems, for example, that every non-archimedean space is paracompact, and strongly zero-dimensional, and suborderable. [These were shown in a more roundabout way in [NR_1].] It also points the way toward one easy way of constructing non-archimedean spaces, that of taking a properly branching tree T and letting the points of our space X_T be the branches of T, while the members of the base are of the form B(t) where t is a node of T and B(t) is the set of all branches running through t [Ny_2]. It can be shown that every non-archimedean space is embeddable in such a space X_T. This can get us tangled in set-theoretic consistency problems: the existence of a non-archimedean space which is hereditarily Lindelöf but not hereditarily separable (hence metrizable) is equivalent to the existence of a Souslin tree [AF], [Ny_2]. And Juhász and Weiss showed [JW] that the existence of an ω_1-metrizable Lindelöf space of cardinality $>\omega_1$ is equivalent to the existence of a Kurepa tree with no

Aronszajn subtree, settling a problem raised in 1950 by Sikorski. (If κ is uncountable, every κ-metrizable space is non-archimedean [NR_1].)

The most important unsolved problem concerning non-archimedean spaces is still:

2.2. Problem. Does there exist a "real" perfectly normal non-archimedean space which is not metrizable?

No progress has been made on this problem since I raised it in [Ny_1], when it was already known that if T is a Souslin tree, X_T is a perfectly normal non-metrizable non-archimedean space with a point-countable base. We do not even know what happens if we interpolate the existence of a point-countable base: whether every perfectly normal non-archimedean space must have one (even consistency results are lacking here) or whether, if it has one, it is consistent that it be metrizable (a problem open for perfectly normal spaces in general!).

The other main area where little is known about non-archimedean spaces has to do with their finite and countable products: under what conditions are they normal, or strongly zero-dimensional? The first question includes the special problem of characterizing non-archimedean Morita P-spaces; also, E. Michael's still-open problem of whether a countable product of paracompact Σ# spaces is paracompact is unsolved for non-archimedean spaces. All that we know about the second question is contained in an early paper by J. Teresawa [T].

Other than that, the theory of non-archimedean spaces has been one big success story, already told mostly in [Ny_1], [Ny_2], and [Ny_3]. I thought it would be helpful here to draw up a list of the main distinct classes of non-archimedean spaces. Some of the classes are numbered in increasing order of generality. The conditions in each heading are equivalent to each other, and (except for the ZFC-independent 2*) we have real examples to show that headings with different integers are distinct.

1. [A_1], [NR_1] compact metrizable, compact, sequentially compact, countably compact, pseudocompact

2. [AF], [NR_1] separable metrizable, hereditarily separable, separable

3. metrizable, developable, semimetrizable, symmetrizable, stratifiable, semi-stratifiable, M-space, p-space, wΔ-space, quasi-complete, Σ-space, Σ*-space, first countable Σ#-space, first countable β-space

4. submetrizable, G_δ-diagonal

5. tree-base of countable height α

6. σ-disjoint base, σ-point-finite base, quasi-developable, primitive base, quasi-G_δ-diagonal

7. point-countable base, point-countable separating open cover, tree-base of height $\leq\omega_1$ with no uncountable branch

8. δθ-base

9. first countable, countable tightness, points G_δ.

The following is more general than 2, less general than 7, and incomparable with 3 through 6 (and exists separately from 2 iff there is a Souslin tree)

2*. hereditarily Lindelöf, countable spread, ccc

The following is more general than 6, less general than 9, and incomparable with 7 and 8:

A. [G_1] σ-Q-base, quasi-metrizable, γ-space, θ-space, wγ-space, wθ-space.

Another that is more general than 3 but incomparable with all the others is:

B. Σ#-spaces, β-spaces

It is easy to see where the following two classes fit in: C. locally compact (metrizable) and D. completely metrizable, Cech complete. And the class of Lindelöf non-archimedean spaces is more general than 2* and incomparable with all the others.

The location of spaces with tree-bases of various kinds is the outcome of a correspondence between J. Roitman and myself. Roitman also showed that every non-archimedean space in class A has a σ-disjoint π-base.

Examples to show the distinctness of the various classes are: the Michael line (see description in [Ny_2]) and its Lindelöf modification [M_1], Mishchenko's space [M_2], the spaces of Aull and Bennett and their extensions beyond ω_1 [A_5], [B], the space of infinite repetitions [G_1], spaces obtainable from Souslin trees and special Aronszajn trees, and the spaces D^*_μ of regular cardinality $\omega\mu$ with a single non-isolated point p whose neighborhoods are all sets whose complements are of cardinal $<\omega\mu$.

New tidbits of information on non-archimedean spaces keep cropping up intermittently. The following was obtained well after writing [Ny_3]:

2.3. Theorem. A scattered space is non-archimedean if, and only if, it is a hereditarily paracompact generalized ordered lob-space.

Closely related is the following pretty "cross-theorem," which takes a basic result about non-archimedean spaces (they are hereditarily paracompact) and a trivial fact about subspaces of well-ordered spaces, and switches them:

2.4. Theorem. The following are equivalent for a space X:

1. X is non-archimedean and scattered.

2. X is embeddable in a well-ordered space and heredi-

Some problems still remain in the area of orderability. One of the most interesting is:

2.5. Problem. If κ is an uncountable regular cardinal, is every κ-metrizable space a LOTS?

2.5. Theorem. Let κ be an uncountable regular cardinal. Every κ-metrizable space is a LOTS.

Section 3: Large rank and rank

The various kinds of rank were introduced by Nagata [N] and Arhangel'skii [A_1]. They all agree in the case of rank 1, the case of non-archimedean spaces. The strongest variety is large rank:

3.1. Definition. A collection Γ of subsets of a set X is of *large rank* n if it is the union of n collections of rank 1 [n may be either finite or infinite]. A space is *of countable large basis dimension* if it has a base which is the union of $\aleph_0$ collections of rank 1, such that each point has a local base belonging to one of the collections.

3.2. Definition. A collection Γ of subsets of a set X is *of rank* $\leq$n *at an element* x if every incomparable (with respect to containment) subcollection containing x is of cardinality $\leq$n; of *rank* $\leq$n if it is of rank $\leq$n at each element x; and *of point-finite rank* if for each x ∈ X there exists n such that Γ is of rank $\leq$n at x.

3.3. Definition. A collection Γ of subsets of a set X is *Noetherian* if every ascending sequence from Γ (with respect to containment) is finite, and *of subinfinite rank* if every incomparable subcollection with nonempty intersection is finite. A base for a space X is *well-ranked* if it is the union of countably many Noetherian collections of subinfinite rank.

These are real "nuts and bolts" concepts, the sort that could be taught to a student shortly after the notion of the base of a topological space, the sort that one might use in constructing a space "from scratch," without any heavy topological or set-theoretic machinery. Which is why I kept being surprised by how useful they are [Ny_1], [Ny_3].

Besides non-archimedean spaces, the spaces with bases of finite large rank include all finite-dimensional metric spaces [N], [A_1], and *some* finite products of non-archimedean spaces. For example, (Michael line) × (irrationals) has a base of large rank 2, and (Michael line)^2 has a base of large rank 3. On the other hand, the Dieudonne plank, $D_1^* \times D_0^*$, does not have a base of any finite large rank; but if one removes the corner point, what remains has a base of large rank 2.

Spaces of countable large basis dimension include all spaces with bases of finite large rank [AF], all countable-dimensional metric spaces, σ-discrete collectionwise Hausdorff lob-spaces, and *some* subspaces of σ-products and σ-box products of other spaces of countable large basis dimension.

Spaces with bases of finite rank include all spaces with bases of finite large rank and all "truncated Pixley-Roy spaces" of the form $\{p \in F[x]: |p| \leq 2\}$ for a lob-space X. For example, Heath's tangent V space [H_1] is the case $X = \mathbb{R}$. They also include all spaces of the form $\{p \in F[x]: |p| \leq n\}$ for some fixed n if X is a space with some characteristic κ. *Every* Pixley-Roy space on a space of characteristic κ has a base of point-finite rank (see Section 5) and so does every hereditarily metacompact σ-scattered lob-space (see Section 5).

So far, the classes do not even include all metric spaces. The Hilbert cube, for example, does not have a base satisfying any of the above conditions. But this is no deficiency, since the concepts were first introduced in connection with dimension theory [N], [A_1]. Anyway, the third definition is more general. Every metric space (in fact, every metacompact Moore space) has a Noetherian base of subinfinite rank, in fact, a uniform base [H_1]. (A base is *uniform* if for each point p and each neighborhood U of p, there are only finitely many members of the base which contain p and are not contained in U.) So does every finite product of non-archimedean spaces, since a tree-base is Noetherian of rank 1, and spaces with Noetherian bases of subinfinite rank are closed under finite products [GN]. In a way, "Noetherian of subinfinite rank," does for "rank 1" what "glob" did for "lob." The analogy is not absolutely precise, but there are many parallels; compare the following with Theorem 2.3:

3.4. Theorem. Every σ-scattered, hereditarily metacompact globular space has a Noetherian base of subinfinite rank. (See Section 5.)

It can also be shown that every space with a base of finite large rank has a Noetherian base of finite large rank, hence of subinfinite rank. All this is important because:

3.5. Theorem [GN]. Every space with a base of point-finite rank, and every space with a Noetherian base of subinfinite rank is (hereditarily) metacompact.

The proof is much harder for point-finite rank, and we do not know whether the finite product of spaces with bases of point-finite (or even finite) rank is metacompact, so it would be nice to know:

3.6. Problem. Does every space with a base of point-finite rank have a Noetherian base of subinfinite (equivalently, point-finite) rank?

There has been no significant progress on this in the last four years, nor on:

3.7. Problem. Is every space with a base of subinfinite rank metacompact?

The classification of these various spaces has not progressed as far as that of non-archimedean spaces. Most of the work has been done in metrization theory; foremost in the theorem of Gruenhage:

3.8. Theorem [GN]. Every compact space with a base of finite rank is metrizable.

I believe this is as difficult and important a theorem as any that Gruenhage has proven, and that includes his justly celebrated theorem that stratifiable spaces are M_2. The proof takes up two full pages of GTA, and relies upon such high-powered theorems as the Dilworth theorem on partially ordered sets [D_1], Mishchenko's lemma [M_2], and my earlier result in the same article, that every space with a base of point-finite rank is metacompact, itself requiring a full page for its proof. No one has significantly improved on either the theorem or the proof. [I extended it to point-finite rank and locally compact spaces in the same article, but these were trivial modifications.] That the result has not received the recognition it deserves, is probably due to two publishing ironies. The article in which the problem was raised by Arhangel'skii in 1963 was printed in Russian in Fundamenta Mathematicae, which is not regularly translated into English. And Gruenhage's proof appeared over three years after he obtained it, and over two years after the final referee's report on [GN].[2]

The proof of the following theorem (also due to Gruenhage) was almost as long, and required entirely different techniques.

3.9. Theorem [GN]. Every locally compact space with a well-ranked base is metrizable.

We have analogous, but much easier [except for "point-finite" as distinct from "finite"] results for regular spaces of caliber $\aleph_1$ (in particular, separable regular spaces).

3.10. Theorem [GN], [G_2]. Every regular space which is locally of caliber $\aleph_1$, and has either a well-ranked base or a base of point-finite rank, is metrizable.

Part of this comes from:

---

[2]The notation "Received 30 July 1976" in [GN] should read "Received 30 July 1975."

3.11. Theorem [GN]. Every ccc regular space with a well-ranked base or a base of finite rank has a point-countable base.

I do not know whether "finite" can be strengthened to "point-finite" here. It can in one instance:

3.12. Theorem [G_2]. Every regular hereditarily Lindelöf space with a base of point-finite rank has a point-countable base.

Szentmiklóssy has showed that every first countable regular hereditarily Lindelof space is hereditarily separable under MA + ¬CH. It is therefore ZFC-independent whether every hereditarily Lindelöf space with a well-ranked base or a base of point-finite rank is metrizable. (Yes if MA + ¬CH, no if there is a Souslin tree.)

We have stronger theorems for countable large basis dimension:

3.13. Theorem [GZ]. Every normal Σ-space and every normal $w\Delta$-space of countable large basis dimension is metrizable.

3.14. Theorem [Ny_1]. Every normal semi-stratifiable space of countable large basis dimension is metrizable.

The normality condition is annoying, but there is no help for it: the non-normal, zero-dimensional screenable Moore space introduced in [H_1] has a base of large rank 2.

Subinfinite rank, without the Noetherian condition, is a horse of a different color. Two results, both difficult, are worth mentioning here.

3.15. Theorem [G_2]. Every countably compact space with a base of subinfinite rank is compact.

3.16. Theorem (M. Ismail [IN]). Every compact space with a base of subinfinite rank is first countable.

The lexicographically ordered unit square is a striking example of a compact nonmetrizable space with a base of subinfinite rank.

Section 4: Small rank and cardinal invariants

The concept of small rank is even more general than that of rank, and it too was introduced originally as an adjunct to dimension theory [A_1]. The expressions "small rank n," "point-finite small rank," and "subinfinite small rank" are defined as in rank, but with "irreducible" in place of "incomparable."

4.1. Definition. A collection Γ of subsets of a set X is *irreducible* if each contains an element not in the union of the others.

In addition to the examples already mentioned for the stronger concepts, we note that every generalized ordered space has a base of small rank 2. But products can be very badly behaved: every base for the Sorgenfrey plane is of small rank $\underline{c}$.

Small rank goes well together with the concept of spread. The spread of a space is the supremum of the cardinals of its discrete subspaces, and (equivalently) of its irreducible collections of open sets. So it is easy to see:

4.2. Theorem [Ny_4]. The following are equivalent for a space X.

1. X is of countable spread
2. X has a base in which every irreducible collection

is countable.

3. Every base for the topology on X is of countable small rank.

The theorem easily generalizes to arbitrary cardinals. The concept of "point-additively $\aleph_0$-Noetherian" was introduced in the same article and shown to have the same affinity for hereditarily Lindelöf spaces.

Of course, having a base of countable small rank is much more general than every base being of countable small rank, but there is one important class of spaces for which these are equivalent.

4.3. Theorem [Ny_4]. Let X be a space of caliber $\aleph_1$. Then X has countable spread if, and only if, it has a base of countable small rank.

As a corollary, every separable space with a base of countable small rank is of countable spread.

Countable spread is intimately connected with the famous S and L problem. The S-space problem is equivalent to whether every regular space of countable spread is Lindelöf; the L-space problem, to whether every regular space of countable spread is hereditarily separable. (See the paper by M.E. Rudin.) At about the same time (in 1974) that I obtained the above results, I noticed that every ccc space with a base of finite small rank is of countable spread, and that every space of countable spread with a base of point-finite rank is hereditarily Lindelöf [GN]. Muhammad Ismail started the ball rolling again in 1976 with:

4.4. Theorem [IN]. If MA + $\neg$CH, then every regular space of countable spread with a base of point-finite rank is hereditarily separable (hence metrizable).

In rapid succession we obtained four theorems:

4.5. Theorem [G_2]. If a space X has spread κ, then every open cover of X which is of subinfinite rank has a subcollection of cardinal $\leq\kappa$ whose closures cover X.

A corollary is that in every regular space with a base of subinfinite rank, the spread equals the hereditary Lindelöf degree; the proof of 4.5 can be extracted from the proof of this corollary in [G_2].

4.6. Theorem [IN]. If a space X has spread κ, then every open cover of X which is of point-finite small rank has a subcollection of cardinal $\leq\kappa$ whose closures cover X.

. . . and a similar corollary follows from this theorem. In the next theorem, the notation ($\downarrow$) refers to the axiom that there is an uncountable Noetherian collection of subsets of ω which has no uncountable incomparable subcollection. Kunen showed CH $\rightarrow$ ($\downarrow$) and MA + $\neg$CH $\rightarrow$ $\neg$($\downarrow$) [vDK].

4.7. Theorem [IN]. If $\neg$($\downarrow$), then every separable regular space with a base of countable rank is hereditarily Lindelöf.

4.8. Theorem [IN]. Every separable regular space with a base of subinfinite small rank is hereditarily Lindelöf.

By 4.2 and 4.3, "countable spread" or "caliber $\aleph_1$" may be substituted for "separable" in these last two theorems.

We now have the illusion of closing in on the S-space problem from two directions: insert "small" in 4.7 or change "subinfinite" to "countable" in 4.8, and you've got it made. Of course, any result along these lines must be a consistency result, in view of the many examples of S-spaces constructed using various axioms. In particular, Theorem 4.7 is definitive: assuming (↓), van Douwen constructed a hereditarily separable, non-Lindelöf space with a base of countable rank [vDK]. So we now have two equivalences in the area of rank: (↓) is equivalent to there being an S-space with a base of countable rank, while "there is a Souslin tree" is equivalent to there being an L-space with a base of rank one. An interesting sidelight on these two axioms (which can be cast into the same mold, with "subset of $P(\omega)$" in one place and "tree" in the other) is that while there is considerable overlap, neither implies the other [vDK].

The axiom ¬(↓) figures in another result.

4.9. Theorem [IN]. If ¬(↓), then every separable regular space with a Noetherian base of countable rank is first countable.

Assuming CH, Kunen constructed a countable space with a single nonisolated point which violated the conclusion of this theorem [vDK].

4.10. Problem [GN]. Is it consistent that every separable regular space with a Noetherian base of countable rank is metrizable?

Section 5: Co-scattered spaces

In this section the partial orders are on the spaces rather than the bases, but the orders give us coverings which can often be used to define nice bases. And our main focus of attention, the Pixley-Roy spaces, have a base defined by an ordering.

For the sake of convenience, I will call a partial order $\leq$ on a space X "c-good" if for each $x \in X$, $\{y: y \leq x\}$ is closed, and "o-good" if $\{y: y \leq x\}$ is open. The term "acceptable" was used in [BFL] for the latter concept.

Right-separations and left-separations are examples of o-good and c-good partial orders, respectively, and their reverses are c-good and o-good, respectively.

5.1. *Definition.* Let X be a space with a well-ordering $\leq$. Then $\langle X, \leq \rangle$ is *right-separated* if $\leq$ is o-good, and *left-separated* if $\geq$ is o-good.

Some use "right-separated" to refer to the underlying space itself, but since a space can be right-separated if and only if it scattered (that is, contains no dense-in-itself subspace), I will reserve "right [left] separated" for the space together with a specific well-ordering, and employ the following terminology.

5.2. *Definition.* A space X is *co-scattered* if it admits a well-ordering $\leq$ such that $\langle X, \leq \rangle$ is left-separated. A space X is *σ-scattered* if it is the countable union of closed scattered subspaces.

There has been a recent surge of interest in co-scattered spaces. The first half of the following threorem was proven independently by Arhangel'skii, by Gerlits and Juhász, and by Ismail and myself. And while there was considerable overlap, one could put together two disjoint proofs from the material.

5.3. Theorem [A_2]. Every countably compact co-scattered space is scattered, as is every Cech complete co-scattered space.

One drawback of co-scattered spaces is that they have no canonical division into levels the way scattered spaces do. The zeroth level of any space consists of its isolated points; and if all levels $<\alpha$ have been defined, the αth level is the set of isolated points of the subspace remaining when all lower levels have been subtracted off. Scattered spaces are those in which each point belongs to the αth level for some α. This is an example of a right layering:

5.4. Definition. A right [resp. left] layering of a space is a well-ordered sequence of subspaces (*layers*) $\{X_\alpha: \alpha < \tau\}$ such that each X_α is discrete and $\cup\{X_\alpha: \alpha \leq \beta\}$ is open [resp. $\cup\{X_\alpha: \alpha \geq \beta\}$ is open] for all $\beta < \tau$.

Every right layering gives an o-good partial order $\leq$ if one defines $x \leq y$ if either $x = y$ or x is in a lower layer than y. This can be converted to a right-separation by arbitrarily well-ordering the individual layers. (Of course, every right separation "is" a right layering into singletons.) One can do the same, *mutatis mutandis*, for left layering.

Despite the lack of a general way of left-layering a co-scattered space, it can be done in many special cases. A σ-discrete space obviously admits a left layering, and so do some more general kinds of spaces: spaces in which the non-isolated points form a σ-discrete subspace (such as the Michael line, and scattered spaces without level 2 points) and more generally, spaces (scattered or otherwise) in which one is left with a σ-discrete subspace after finitely many levels have been subtracted off. One simply lists the countably many closed discrete subspaces first, then the levels in reverse order. Also, if a space is a countable union of closed co-scattered subspaces, it is co-scattered, and a left-layering can be obtained by listing the subspaces in a sequence and taking advantage of the individual left-layerings.

As one studies these layerings, it becomes natural to ask, as Gerlits and Juhász did, whether every space which is scattered and co-scattered is a countable union of discrete subspaces. J. Roitman found a regular counterexample using CH, then posed the following question, for which we have no consistency results:

5.5. Problem. Is every compact co-scattered space the countable union of discrete subspaces? [It is scattered by 5.3.]

Pixley-Roy spaces come equipped with a natural left layering. Given any space X, its *Pixley-Roy hyperspace*, denoted F[X], is the space of all finite nonempty subsets of X, with the following topology. For each finite subset F of X and each open subset U containing F, let $[F,U] = \{G: G$ is a finite subset of X such that $F \subset G \subset U\}$. Then the sets of the form [F,U] form a base for the topology of F[X]. Since the sets [F,U] are clopen, F[X] is zero-dimensional.

For any space X, let $F_n^{\#}[X] = \{F \subset X: |F| = n\}$. Then $\{F_n^{\#}[X]: n \in \omega\}$ is a left layering of F[X]. We can left-separate F[X] as outlined above. We can also put a partial order on F[X] which is simultaneously o-good and c-good: let $\{G: G \leq F\} = \{F: F < G\} = [F,X]$.

Letting $V(F) = [F,X]$ for all $F \in F[X]$ gives a clopen P-canonical cover for P = "point-finiteness," which shows that F[X] is hereditarily metacompact.

5.6. *Definition.* Let P be a property. A P-*canonical cover* of a space X is a function $V: X \to \mathcal{P}(x)$ such that (i) V(x) is a neighborhood of x for all $x \in X$, i.e. V is a neighbornet (ii) $\dot{V} = \{V(x): x \in V\}$ satisfies P and (iii) if $x \neq y$, $V(x) \neq V(y)$.

By abuse of language, I also refer to $\dot{V}$ as a P-canonical cover. The term "pf-canonical" will designate the special case P = point-finiteness.

5.7. *Theorem* [Ny_5]. If Y is a space with a pf-canonical cover, then Y is hereditarily metacompact, and its product with every [hereditarily] metacompact space is [hereditarily] metacompact. [For similar theorems for many other P's see [Ny_5].]

What makes this theorem tick is the way we can use a pf-canonical cover to construct other covers and bases. Each point of Y only belongs to V(y) for finitely many points y, so we can work inside each V(y), building local bases or refinements of covers, and each point will be interfered with by only finitely many others. For example, if Y is a lob-space, then Y will have a Noetherian base of point-finite rank, and if Y is globular, Y will have a Noetherian base of sub-infinite rank. Thus we have:

5.8. *Lemma.* Every F[X] is herediarily metacompact. If X is globular, so is F[X], and F[X] has a Noetherian base of subinfinite rank. If X has characteristic κ, so does F[X], and F[X] has a Noetherian base of point-finite rank.

That F[X] has a P-canonical cover can also be obtained by Theorem 4.9 of [Ny_5] from the fact that F[X] is σ-scattered, as is every space with a countable left layering.

If X is first countable, we have still more information on F[X]: it has a uniform base [GN]. This is one way of showing that F[X] is a Moore sapce whenever X is first countable [vD] [GN]. From the following theorem we also see that F[X] is σ-discrete, giving us yet another (and often very un-obvious) way of giving F[X] a left separation. [I adopt Junnila's term "[weakly] submetacompact" for "[weakly] θ-refinable."]

5.9. Theorem [Ny_5]. The following are equivalent for a σ-scattered space X.

1. X is weakly submetacompact and every closed subset is a G_δ.
2. X is σ-discrete.
3. X is submetacompact and every point is a G_δ.
4. X is submetacompact and has a G_δ-diagonal.

The problem of when F[X] has other properties is dealt with in [L], [BFL], and [P]. There are several interesting results on when it is normal and not paracompact [L], most of them consistency results. As to when it is paracompact (and hence, in the first countable case, metrizable), we have the results announced in [P] and [L], but perhaps the most sweeping result so far is:

5.10. Theorem [BFL]. If $\langle X, \leq \rangle$ is o-good, then $(F[X])^n$ is ultraparacompact for all finite n.

Now Pixley-Roy spaces themselves have o-good orders, so this theorem suggests iterating the operation: let $F_1[X] = F[X]$, and with $F_n[X]$ defined, let $F_{n+1}[X] = F[F_n[X]]$. If X is a first countable space with no, one, or infinitely many isolated points, this sequence settles down with $F_2[X]$ homeomorphic to $F_3[X]$. This is the corollary of the fact that if one adds "o-good" to the hypotheses on X, the sequence settles down one step earlier.

There is a way of characterizing all spaces of the form $F_2[X]$, X first countable: all are direct sums of countable σ-products of discrete spaces, but one does need to pay attention to the cardinalities. At any rate, they are σ-discrete and metrizable.

Much additional material on co-scattered spaces can be found in Arhangel'skii's papers [A_2], [A_3] and [A_4], where they are called "α-left." The articles also treat "α-expanded" spaces at length. These are the spaces which admit a c-good total order, so they include scattered, co-scattered, and generalized ordered (GO) spaces, and a surprising assortment of other spaces: Eberlein compact, metacompact Moore, paracompact with G_δ-diagonal, compact images of GO spaces, and open images of complete metric spaces. In [A_3] there is a neat proof, which works equally well for α-expanded, scattered, and co-scattered spaces, showing each category is preserved by perfect maps.

Section 6: Products of metacompact spaces

The concept of a point-additively Noetherian base generalizes that of a Noetherian base of sub-infinite rank, but still implies "hereditarily metacompact." [Ny_6]

6.1. Definition. A collection Γ of subsets of a set X is *point-additively Noetherian* if for each subcollection Γ' of Γ with nonempty intersection, there exists a finite subcollection F of Γ' such that $\cup F = \cup \Gamma'$.

In [Ny_6] it is shown how the following is a corollary of A.H. Stone's theorem [S] that the class of hereditarily compact spaces is closed under finite products:

6.2. Theorem. The finite product of point-additively Noetherian collections is again point-additively Noetherian.

6.3. Corollary. The class of spaces with point-additively Noetherian bases is finitely productive.

So this gives one condition under which the finite product of hereditarily metacompact spaces is again such a space; Theorem 5.7 gave another, the possession of a pf-canonical base. Here is a concept that generalizes both.

6.4. Definition. A pair $\langle \mathcal{B}, f\rangle$ is a *sheltering base* for a space X if $f: \mathcal{B} \to \mathcal{P}(X)$ is a function such that (i) $f(B) \subset B$ for all B, (ii) for each $x \in X$, $\{B: x \in f(B)\}$ is an open base at x, and (iii) for each $x \in X$, and each $\mathcal{B}' \subset \mathcal{B}(x)$, there exists a finite subcollection $\{B_1, \ldots, B_n\}$ of $\mathcal{B}'$ such that $f(B) \subset \cup_{i=1}^{n} B_i$.

Here again, $\mathcal{B}(x)$ denotes all members of $\mathcal{B}$ of which x is an element.

The example of point-additively Noetherian bases is the case where $f(B) = B$, while with a pf-canonical base V, one adds an open base for each point x inside V(x), omitting from each one the finitely many points y such that $x \in V(y)$, and then maps these local base elements to $\{x\}$ by f.

6.5. Theorem [Ny_7]. Every space with a sheltering base is (hereditarily) metacompact.

The product of two sheltering bases is defined in the obvious way: $f(B_1 \times B_2) = f(B_1) \times f(B_2)$. It too is a sheltering base, so:

6.6. Corollary. The finite product of spaces with sheltering bases is hereditarily metacompact.

Among the new spaces that come under the scope of this corollary are the various bowtie spaces with subsets of the plane as their underlying set and points on the x-axis having bowtie neighborhoods. For each bowtie neighborhood we let f(B) be the intersection of B with the x-axis.

6.7. Problem. Does every metacompact σ-space have a sheltering base?

In any event, it is possible to take any metacompact σ-space X and express it as the union of countably many closed subspaces X_n for which one can find a $\langle \mathcal{B}_n, f_n \rangle$ to serve as a sheltering base in X for the points of X_n. Then any *countable* product of metacompact σ-spaces has such a base, which implies it is hereditarily σ-metacompact; but also, it is a σ-space [BL], hence (hereditarily) countably metacompact, hence . . .

6.8. Theorem [Ny_7]. The countable product of metacompact σ-space is a hereditarily metacompact σ-space.

We can handle Σ-spaces [BL] if we are willing to give up "hereditarily," using compact sets instead of points. Given a cover $\mathcal{C}$ of a space X by compact sets, we can define a $\mathcal{C}$-*sheltering base* for X in the same way that we defined a sheltering base, but with $\mathcal{C}$ in place of X, C in place of x, $\cup f(B)$ in place of f(B) and $\mathcal{B}(C)$ denoting the set of all members of $\mathcal{B}$ containing C. The term "$\mathcal{C}$-sheltering" will also be used generically, in the absence of any specific $\mathcal{C}$.

Besides everything covered by sheltering bases, we can include all locally compact metacompact spaces in the class of spaces with $\mathcal{C}$-sheltering bases. More generally, every metacompact C-scattered space has a $\mathcal{C}$-sheltering base.

6.9. Definition. A space is *C-scattered* if every closed subspace Y has a point with a compact neighborhood in Y.

6.10. Theorem [Ny_7]. The finite product of spaces with $\mathcal{C}$-sheltering bases also have a $\mathcal{C}$-sheltering base, and is metacompact.

By the same kind of maneuvering that we used with σ-spaces, one obtains:

6.11. Thoerem [Ny_7]. The countable product of metacompact Σ-spaces is a metacompact Σ-space.

References

[A_1] A.V. Arhangel'skii, *Ranks of systems of sets and dimensionality of spaces*, Fund. Math. 52 (1963), 257-275 (in Russian).

[A_2] ________, *On left τ-separated spaces*, Vestnik Moskov Univ. Ser. I (1977), no. 5, 30-36 (in Russian).

[A_3] ________, *On topologies admitting a weak connection with orders*, Soviet Math. Dokl. 19 (1978), 77-81.

[A_4] ________, *On α-expanded spaces*, Soviet Math. Dokl. 19 (1978), 336-340.

[A_5] C.E. Aull, *Topological spaces with a σ-point-finite base*, AMS Proceedings 29 (1971), 411-416.

[A_6] ________, *A survey paper on some base axioms*, Topology Proceedings 3 (1978), 1-36.

[AF] A.V. Arhangel'skii and V.V. Filippov, *Spaces with bases of finite rank*, Math. USSR Sbornik 16 (1972), 147-158.

[B] H.R. Bennett, *A note on point-countability in linearly ordered spaces*, AMS Proceedings 28 (1971), 598-606.

[BFL] H.R. Bennett, W.G. Fleissner, and D.J. Lutzer, *Ultra-paracompactness in Pixley-Roy spaces*, to appear.

[BL] D.K. Burke and D.J. Lutzer, *Recent advances in the theory of generalized metric spaces*, in: Topology: Proceedings of the Memphis State Topology Conference, Marcel Dekker, 1976.

[D_1] R.P. Dilworth, *A decomposition theorem for partially ordered sets*, Ann. Math. 51 (1950), 161-166.

[D_2] S.W. Davis, *Spaces with linearly ordered local bases*, Topology Proceedings 3 (1978), 37-51.

[vD] E.K. van Douwen, *The Pixley Roy topology on spaces of subsets*, in: Set-Theoretic Topology, Academic Press, 1977.

[vDK] E.K. van Douwen and K. Kunen, *S and L subspaces of $P(\omega)$*, preprint.

[G_1] G. Gruenhage, *A note on quasi-metrizability*, Canad. J. Math. 29 (1977), 360-366.

[G_2] ________, *Some results on spaces having an ortho-base or a base of subinfinite rank*, Topology Proceedings 2 (1977), 151-160.

[GN] G. Gruenhage and P. Nyikos, *Spaces with bases of countable rank*, Gen. Top. Appl. 8 (1978), 233-257.

[GZ] G. Gruenhage and P. Zenor, *Metrization of spaces of countable large basis dimension*, Pac. J. Math. 59 (1975), 455-460.

[H_1] R.W. Heath, *Scrennability, pointwise paracompactness, and metrization of Moore spaces*, Canad. J. Math. 16 (1964), 763-770.

[H_2] R.E. Hodel, *Spaces defined by sequences of open covers which guarantee that certain sequences have cluster points*, Duke Math. J. 39 (1972), 253-263.

[H_3] ________, *Some results in metrization theory, 1950-1972*, in: Topology Conference, Lecture Notes in Math. no. 375, Springer-Verlag, 1974.

[IN] M. Ismail and P. Nyikos, *Countable small rank and cardinal invariants, II*, in preparation.

[JW] I. Juhász and W.A.R. Weiss, *On a problem of Sikorski*, Fund. Math. 100 (1978), 223-227.

[K] J. Kofner, *On Δ-metrizable spaces*, Math. Notes Acad. Sci. USSR 13 (1973), 168-174.

[L] D.J. Lutzer, *Pixley-Roy topology*, Topology Proceedings 3 (1978), 139-158.

[M_1] E. Michael, *A note on paracompact spaces*, AMS Proceedings 4 (1953), 831-838.

[M_2] A.S. Mishchenko, *Spaces with point-countable bases*, Soviet Math. Dokl. 3 (1962), 855-858.

[N] J. Nagata, *On dimension and metrization*, in: General Topology and Its Relation to Modern Analysis and Algebra, Academic Press, 1962.

[Ny_1] P.J. Nyikos, *Some surprising base properties in topology*, in: Studies in Topology, Academic Press, 1975.

[Ny_2] ________, *A survey of zero-dimensional spaces*, in: Topology: Proceedings of the Memphis State University Conference, Marcel Dekker, 1976.

[Ny_3] ________, *Some surprising base properties in topology, II*, in: Set-Theoretic Topology, Academic Press, 1977.

[Ny_4] ________, *Countable small rank and cardinal invariants*, in: General Topology and its Relations to Modern Analysis and Algebra IV, Part B, Prague, 1976.

[Ny_5] ________, *Covering properties on σ-scattered spaces*, Topology Proceedings 2 (1977), 509-542.

[Ny_6] ________, *On the product of metacompact spaces I: Connections with hereditary compactness*, Amer. J. Math. 100 (1978), 829-835.

[Ny_7] ________, *On the product of metacompact spaces II: Sheltering bases*, in preparation.

[NR_1] P.J. Nyikos and H.C. Reichel, *On the structure of zero-dimensional spaces*, Indag. Math. 37 (1975), 120-136.

[NR_2] ________, *On uniform spaces with linearly ordered bases II (ω_μ-metric spaces)*, Fund. Math. 93 (1976), 1-10.

[P] T. Przymusiński, *Normality and paracompactness of Pixley-Roy hyperspaces*, AMS Notices 26 (1979), A-284.

[R] H.C. Reichel, *Basis properties of topologies compatible with (not necessarily symmetric) distance-functions*, Gen. Top. Appl. 8 (1978), 283-289.

[S] A.H. Stone, *Hereditarily compact spaces*, Amer. J. Math. 82 (1960), 900-916.

[V] J.E. Vaughan, *Linearly stratifiable spaces*, Pac. J. Math. 43 (1972), 253-266.

PRODUCT SPACES

Teodor C. Przymusiński
Institute of Mathematics Polish
Academy of Sciences, Warsaw

In this article[1] we survey some of the results involving topological properties of product spaces obtained in the last decade. This area of research is very broad and develops rapidly, so the limited scope of this survey made it necessary to concentrate on a small number of selected topics. Accordingly, the article consists of three closely interrelated sections devoted, respectively, to covering properties of product spaces, extension of functions defined on product spaces and dimension of product spaces.

All spaces are assumed to be completely regular and all mappings are continuous. The classes of normal, paracompact and metrizable spaces are denoted by N, P and M, respectively.

1. *Covering properties of product spaces*

In this section we consider the preservation of covering properties in product spaces. We shall restrict ourselves to a discussion of paracompactness, normality and related covering properties.

This section is divided into two parts. In Part A, we discuss products of two factors and in Part B - countable products. The consideration of uncountable products is not interesting, since it is known that the product N^{ω_1} of ω_1 copies of natural numbers N is neither normal [S] nor

[1]The author wishes to express his gratitude to Professor K. Morita for his valuable comments.

ISBN: 0-12-584960-5

θ-refinable[1] [PP]. The last fact implies that if a product space $\mathbb{P}\{X_s: s \in S\}$ is θ-refinable, then all but countably many factors must be compact.

Part A: Products of two factors

Covering properties are not preserved by Cartesian products. Michael first gave an example of a separable metric space M and a Lindelöf space Y such that M × Y is not normal [M_1]. Alster and Engelking [AE] showed that there exists a paracompact space X such that X^2 is not subparacompact. The following example essentially shows that no covering property is preserved by Cartesian products of two factors even if one of the factors is metric separable and the other Lindelöf.

Example 1.1 [P_7]. A separable metric space M and a Lindelöf separable and first countable space Y such that M × Y is neither normal nor countably θ-refinable.[1]

Letting X be the free sum M ⊕ Y of spaces M and Y one gets a Lindelöf separable and first countable space Y such that X^2 is neither normal nor countably θ-refinable.

Remark. The space M in Example 1.1 is not complete which suggests a problem [P_7] whether there exists a complete separable metric space M and a Lindelöf space Y such that M × Y is not normal, or--equivalently--Lindelöf. Such spaces exist if CH is assumed [M_2]. □

In this situation the following general problem gains on importance.

---

[1]A space X is *(countably) θ-refinable* (see [WW]) if every (countable) open covering of X has an open refinement $\mathcal{U} = \bigcup_{n=1}^{\infty} \mathcal{U}_n$ consisting of countably many covers $\mathcal{U}_n$ such that for every $x \in X$ there exists an n such that x belongs to exactly one member of $\mathcal{U}_n$. Every metacompact and every subparacompact space is θ-refinable.

Problem 1. Suppose that K is a class of spaces. Characterize the class $N(K)$ (resp. $P(K)$) of those spaces X such that X × Y is normal (resp. paracompact) for every $Y \in K$.

This problem essentially belongs to Tamano, who first formulated it for the classes of paracompact and of metrizable spaces [T_2]. Table 1 below presents most important results obtained so far in answer to this problem.

K	$N(K)$	$P(K)$	Author(s)
metrizable	normal P-spaces	paracompact P-spaces	Morita [$Mo_{3,4}$]
compact	paracompact	paracompact	Dieudonné [D] Tamano [T_2]
paracompact P-spaces	paracompact P-spaces	paracompact P-spaces	Morita [Mo_5]
paracompact	?	?	
normal	discrete	X	Atsuji [At] M.E. Rudin [R_3]

Table 1

The two question marks in Table 1 require some explanation. Tamano proved the following important result.

Theorem 1.2 [T_2]. If X × Y is normal for every paracompact space Y, then X × Y is paracompact for every paracompact space Y.

In other words, this result states that the classes $N(P)$ and $P(P)$ coincide. Katuta [K_2] gave a very complicated external characterization of the class $N(P)$, but no internal characterization of this class is known. The diagram below presents several partial results,[2] due, respectively, to Dieudonné [D], Morita [Mo_4] and Telgársky [Te_1], [Te_2] and [Te_3].

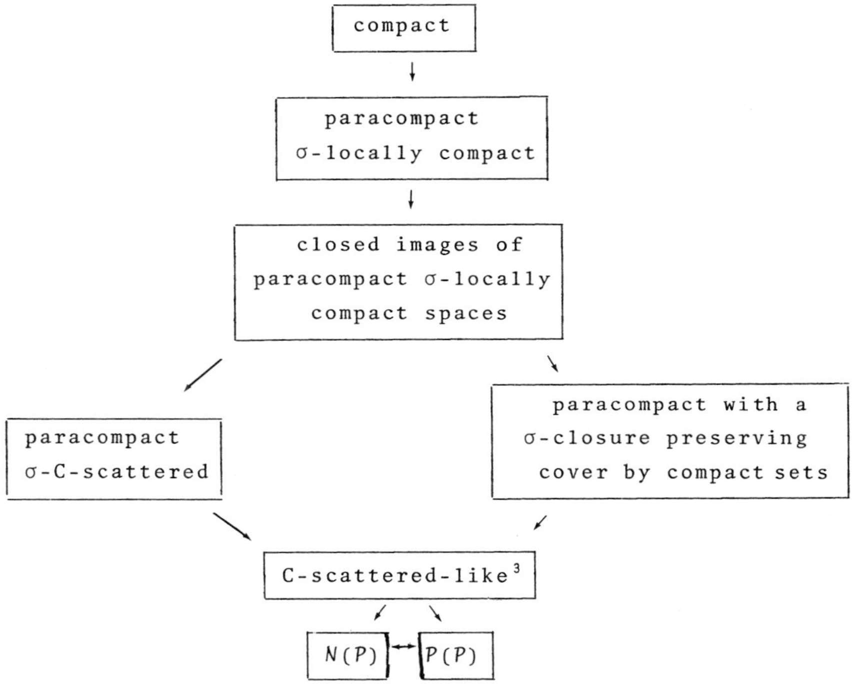

Some subclasses of the class $N(P)$, however, have nice characterizations. The characterization below of the class of metrizable members of $N(P)$ belongs to Morita.

Theorem 1.3 [Mo_4]. If X is metrizable or--more generally--a paracompact p-space, then $X \times Y$ is normal (paracompact) for every paracompact space Y if and only if X is σ-locally compact.

---

[2]The author is grateful to R. Telgársky who provided current information on this subject.

[3]A space X is *C-scattered-like* [Te_3] if there is a winning strategy for the first player in the following topological game: for each n=1,2,... the first player chooses a closed C-scattered subspace F_n of X and the second player chooses an open neighborhood U_n of F_n in X. The first player wins if $\bigcup_{n=1}^{\infty} U_n = X$.

Problem 2. Characterize other interesting subclasses of the class $N(P)$, e.g. the class of first countable and of separable members of $N(P)$.

It is interesting to investigate intrinsic topological properties of the class $N(P)$. One easily checks that the class $N(P)$ is perfect and closed under finite multiplication, i.e. $X \times Y \in N(P)$ if $X, Y \in N(P)$. On the other hand, since the space N^{ω} of irrationals does not belong to $N(P)$, the class $N(P)$ is not closed under countable products. The following problem, however, seems to be open.

Problem 3. Is the class $N(P)$ invariant under closed mappings?

Let us point out that all other classes $N(K)$ and $P(K)$ shown in Table 1 are invariant under closed mappings (cf. Theorem 1.9 and Problem 6).

The class $N(P)$ is also *reflexive*, i.e. $N(N(P)) = P$. In other words, a product space $X \times Y$ is normal for every space $Y \in N(P)$ if and only if X is paracompact. Similarly, the class $P(P)$ is reflexive, i.e. $P(P(P)) = P$. Naturally, also the class $N(N)$ is reflexive. With one possible exception (see problem below), no other class $N(K)$ or $P(K)$ represented in Table 1 is reflexive.

Problem 4 (cf. [Mo_9]). Is the class $N(M)$ reflexive? In other words: suppose that $X \times Y$ is normal for every normal P-space Y; is X metrizable?

This problem has a positive answer if X is separable.

Theorem 1.4 [CC]. For a separable space X the following conditions are equivalent:

(i) X is metrizable,

(ii) $X \times Y$ is normal for every normal P-space,

(iii) $X \times Y$ is normal for every perfectly normal space.

T. and K. Chiba also exhibit an example of a non-metrizable (and non-separable) space X such that $X \times Y$ is normal for every perfectly normal space.

A particularly interesting case of Problem 1 occurs when the class $\mathcal{K}$ consists of exactly one element. We shall write $\mathcal{N}(X)$ instead of $\mathcal{N}(\{X\})$ and $\mathcal{P}(X)$ instead of $\mathcal{P}(\{X\})$. Thus we have

$$\mathcal{N}(X) = \{Y: X \times Y \text{ is normal}\}$$

and

$$\mathcal{P}(X) = \{Y: X \times Y \text{ is paracompact}\}.$$

Clearly, if X is discrete, then the class $\mathcal{N}(X)$ (resp. $\mathcal{P}(X)$) coincides with the class of normal (resp. paracompact) spaces.

Tamano's Theorem 1.2 and results presented in Table 1 may lead to a conjecture (cf. [M_2]) that the classes $\mathcal{N}(X)$ and $\mathcal{P}(X)$ are not really different, i.e. that $\mathcal{N}(X)$ coincides with $\mathcal{P}(X)$ in the realm of paracompact spaces.

The following example refutes this conjecture.

Example 1.5 [P_7]. A separable and first countable paracompact space X such that X^2 is normal but not paracompact.

Thus $X \in \mathcal{N}(X) \cap \mathcal{P}$, but $X \notin \mathcal{P}(X)$. Earlier, consistent examples of such spaces, which were additionally perfectly normal, were obtained in [P_1] and [AZ].

Remark. Example 1.5 suggests the following problem:

Problem 5. Does there exist a Lindelöf space X such that X^2 is paracompact but not Lindelöf?

Such a space exists if CH is assumed [M_2]; naturally X cannot be separable. □

For paracompact p-spaces X it is true, however, that the class of paracompact members of $\mathcal{N}(X)$ coincides with $\mathcal{P}(X)$.

Theorem 1.6 [RS]. For a paracompact p-space X and a paracompact space Y the following conditions are equivalent:

(i) $X \times Y$ is normal,

(ii) $X \times Y$ is paracompact,

(iii) $X \times Y$ is countably paracompact.

A crucial role in the proof of Theorem 1.6 is played by the following important results (cf. Section 3).

Theorem 1.7 [RS]. If X is a non-discrete paracompact p-space and $X \times Y$ is normal, then $X \times Y$ is countably paracompact.

Theorem 1.8 [Mo_6]. If M is metrizable, Y is normal and $M \times Y$ is countably paracompact, then $M \times Y$ is normal.

Counterparts of Theorems 1.6, 1.7 and 1.8 are valid for open subsets of products with a metrizable factor (see [P_3]).

It is easy to see that classes $\mathcal{P}(X)$ are always perfect and that classes $\mathcal{N}(X)$ are always invariant under perfect mappings but not necessarily inversely invariant under perfect mappings. As to the invariance under closed mappings, the following result was proved by M.E. Rudin and Starbird.

Theorem 1.9 [R_2], [RS]. If X is a paracompact p-space, then the class $\mathcal{P}(X)$ is invariant under closed mappings. If X is either metrizable or compact, then the class $N(X)$ is invariant under closed mappings.

The following problems seem to be open.

Problem 6 [RS]. Is the class $N(X)$ invariant under closed mappings if X is a paracompact p-space? If X is a paracompact space?

Problem 7. Is the class $P(X)$ always invariant under closed mappings?

Problem 8 (Kunen). Let $f: X \to Y$ be a perfect mapping of a normal space X onto Y and let C be compact. Does $X \in N(C)$, if $Y \in N(C)$?

It should be mentioned that there exists a perfectly normal space X such that the class $N(X)$ is not invariant under closed mappings [RS]. Naturally, a positive answer to the second part of Problem 6 implies a positive answer to Problem 3.

Since, normality and paracompactness are not productible, classes $N(X)$ and $P(X)$ are not, in general, closed under finite multiplication. Moreover, under the assumption of CH, M. Wage [W_2] announced an example of a space X such that $X \in N(I)$, where I is the unit interval, and X^2 is normal, but $X^2 \notin N(I)$. On the other hand, M.E. Rudin and Starbird proved:

Theorem 1.10 [RS]. If C is compact, M is metrizable and spaces $X \times C$ and $X \times M$ are normal, then the space $X \times C \times M$ is normal. In other words, if $C \in N(X)$ and $M \in N(X)$, then $C \times M \in N(X)$.

Problem 9. Let Y and Z be paracompact p-spaces and suppose that $X \times Y$ and $X \times Z$ are normal. Is $X \times Y \times Z$ normal?

In other words: does $Y \times Z \in N(X)$, if $Y,Z \in N(X)$?

Not much more is known about classes $N(X)$ and $P(X)$ unless X is assumed to be compact. In this case, the class $P(X)$ coincides with the class of paracompact spaces, so it suffices to investigate classes $N(X)$. From now on, we assume that C is a *non-discrete compact space*.

A classical result of Dowker [Do_1] states that if $X \in N(C)$, then X is normal and countably paracompact. The famous example of M.E. Rudin shows that the first not necessarily implies the latter.

Example 1.11 [R_1]. There exists a normal space X which is not countably paracompact, i.e. such that $X \times I$ is not normal.

Another interesting result due to M.E. Rudin (see also Corollary 2.3 and [St_3]) strengthens Dowker's result.

Theorem 1.12 [R_2]. If $X \times C$ is normal, then X is countably paracompact and $w(C)$-collectionwise normal.

Remark. Let us mention without proof that a more general result is true (by $C(Y)$ we denote the space of continuous functions on Y with the compact-open topology):

Theorem 1.12a. If $X \times Y$ is normal and $C(Y)$ contains κ pairwise disjoint open sets, then X is κ-collectionwise normal. □

On the other hand, Morita proved

Theorem 1.13 [Mo_2]. If X is normal and $w(C)$-paracompact, then $X \times C$ is normal.

The above quoted results say, that the class $N(C)$ contains all normal $w(C)$-paracompact spaces and is contained in the class of countably paracompact $w(C)$-collectionwise normal spaces. For many compact spaces C the class $N(C)$ can be fully characterized in terms of κ-collectionwise normality and λ-paracompactness, for some infinite cardinals κ and λ. The table below gives examples of such characterizations, where D denotes the two-point discrete space, $A(\kappa)$ is a one-point compactification of the discrete space of cardinality κ and μ denotes the space of ordinals less than a given ordinal μ.

C	$N(C)$	Author
I^λ	normal and λ-paracompact	Morita [Mo_2]
D^λ	normal and λ-paracompact	Morita [Mo_2]
$\lambda+1$	normal and λ-paracompact	Kunen (quoted in [K_4])
$A(\kappa)$	κ-collectionwise normal and countably paracompact	Alas [A]
$A(\kappa)^\lambda$	κ-collectionwise normal and λ-paracompact	Katuta [K_3]

Table 2

It turns out that Morita's results quoted above can be significantly strengthened (cf. also [K_4]).

Theorem 1.14. If C is a dyadic space, then $X \times C$ is normal if and only if X is normal and $w(C)$-paracompact.

Proof. The sufficiency follows from Theorem 1.13. We prove the necessity. Let $\lambda = w(C)$. If $\lambda = \omega$, then the necessity follows from Dowker's result. If $cf(\lambda) > \omega$, then by [Ef] or [G] the space C contains a homeomorph of D^λ, hence also

$X \times D^{\lambda}$ is normal and thus X is λ-paracompact. If $\lambda > \omega$ and cf $\lambda = \omega$ then there exist λ_n such that $\lambda = \sup\{\lambda_n : n \in N\}$ and $\omega \leq \lambda_n < \lambda$. By ([Ef], Remark 3.12), for every $n \in N$ the space C contains a homeomorph of D^{λ_n}. Thus X is λ_n-paracompact for every $n \in N$. Since $\lambda_n \geq \omega$, X is also countably paracompact. The last two facts easily imply that X is λ-paracompact. □

In this context the following result of Katuta should be noted.

Theorem 1.15 [K_4]. If $X \times C$ is normal, then X is $<t(X)$-paracompact, where $t(X)$ denotes the tightness of X. Moreover, if $t(X)$ is not weakly inaccessible, then X is $t(X)$-paracompact.

The answer to the following problem is not known to the author.

Problem 10. Let C be compact. Does there exist a cardinal number λ such that $X \times C$ is normal if and only if X is $<\lambda$-paracompact and $w(C)$-collectionwise normal?

Let us finish with two results which are, in a sense, dual to Theorems 1.12-1.15.

Theorem 1.16 [P_2]. For every cardinal κ there exists a perfectly normal space X_κ such that $X_\kappa \times C$ is normal if and only if $w(C) \leq \kappa$ (i.e. spaces X_κ "measure" the weight of compact spaces).

Theorem 1.17 [No]. For every κ the space $\kappa^+ \times C$ is normal if and only if $t(C) \leq \kappa$ (i.e. spaces κ^+ "measure" the tightness of compact spaces).

Problem 11. What can be said about classes $N(M)$ and $P(M)$ for metrizable M? (Some results in this direction were obtained by Morita [Mo_5].)

Part B: Countable products

Nagami and Zenor proved the following interesting result (cf. Theorem 1.6).

Theorem 1.18 [N_4], [Z]. Suppose that X^n is normal (paracompact, Lindelöf) for every $n < \omega$. Then X^ω is normal (paracompact, Lindelof) if and only if X^ω is countably paracompact.

The assumption that X^ω is countably paracompact cannot be skipped as is shown by the following two examples.

Example 1.19 [P_7]. A separable and first countable space X such that X^n is Lindelöf for every $n < \omega$ but X^ω is not normal.

Example 1.20 [V]. A space X such that X^n is hereditarily paracompact for every $n < \omega$, but X^ω is not normal.

Example analogous to 1.19 has been first obtained by Michael under the assumption of CH [M_2].

Nagami [N_3] asked if the normality of X^ω implies paracompactness of X. N. Noble noticed that the countable product of the space of countable ordinals is normal and R. Pol observed that it suffices to consider the Σ-product of uncountably many real lines. Recently he obtained a better example.

Example 1.21 [Po] A first countable and locally metrizable non-paracompact space X such that X^ω is perfectly normal.

Such an example has been first obtained in [AP] under the assumption of MA + $\neg$CH.

It is known that X^ω is perfectly normal if and only if X^n is perfectly normal for every $n < \omega$. Heath [H] asked if there exists a space X such that X^2 is perfectly normal but X^ω is not. The following example answers this question in the positive under the assumption of CH.

Example 1.22 [P_8] (CH). For every $n < \omega$ there exists a separable and first countable space X such that X^n is perfectly normal and X^{n+1} is normal but X^{n+1} is not perfect.

Problem 12. Construct such a "real" example.

On the other hand, it is known that if C is compact and C^2 is perfectly normal, then C is metrizable [Š]. Katětov [Ka] proved that if C^3 is hereditarily normal, then C is metrizable and asked whether it suffices to assume that C^2 is hereditarily normal. The following example has been obtained under the assumption of MA + ¬CH.

Example 1.23 [Ny] (MA + ¬CH). A compact non-metrizable space C such that C^2 is hereditarily normal.

Problem 13. Construct such a "real" example.

Let us finish this section with an important result which is tightly related to the subject of our discussion.

It has been an open question for a long time whether Σ-products of metrizable spaces are normal (Corson). Recently, this question has been answered affirmatively and independently by Gul'ko [Gu] and M.E. Rudin [R_4]. Let us quote a theorem of Kombarov providing an interesting generalization of the results of Gul'ko and Rudin.

Theorem 1.24 [Ko]. For a Σ-product X of paracompact p-spaces the following conditions are equivalent:

(i) X is normal;

(ii) X is collectionwise normal;

(iii) all factors have countable tightness.

2. *Extension of functions defined on product spaces*

Before discussing extension properties of product spaces let us recall that a subset A of a space X is P^{κ}*-embedded* if every continuous function f: A → B into a Banach space of weight $\leq\kappa$ can be continuously extended over X (see [AS] or [P_5]). Instead of Banach spaces B of weight $\leq\kappa$ one can take the hedgehod $J(\kappa)$ with κ spikes [P_5]. We say that A is *P-embedded* if it is P^{κ}-embedded for every κ. Concepts of $P^{\aleph_0}$-embedding and of C-embedding coincide and a space X is κ-collectionwise normal if and only if all of its closed subsets are P^{κ}-embedded [Ga].

Extension properties of product spaces are closely related to covering properties of product spaces. If the product space X × Y is (collectionwise) normal, then clearly all closed subsets of X × Y are C-embedded (P-embedded). But it turns out that in order to extend functions defined on certain special subsets of X × Y it sometimes suffices to know that spaces X and Y have "sufficiently good" extension properties. The following theorem has been independently obtained by Morita and Hoshina [MH_2] and Przymusiński [P_5], but the equivalence of (i) and (iv) has been earlier proved by Alǒ and Sennott [AS].

Theorem 2.1 [MH_2], [P_5]. For a cardinal number κ and a subset A of X the following conditions are equivalent:[4]

(i) A is P^κ-embedded in X;

(ii) $A \times I^\kappa$ is C-embedded in $X \times I^\kappa$;

(iii) $A \times Z$ is C-embedded in $X \times Z$, for some compact space Z of weight κ;

(iv) $A \times Z$ is P^κ-embedded in $X \times Z$, for every compact space Z of weight κ.

The next two results follow immediately from Theorem 2.1.

Corollary 2.2 [MH_2], [P_5]. For a subset A of X the following conditions are equivalent:

(i) A is P-embedded in X;

(ii) $A \times \beta X$ is C-embedded in $X \times \beta X$;

(iii) $A \times Z$ is C-embedded in $X \times Z$, for some compact space Z of weight $\geq$ weight A;

(iv) $A \times Z$ is P-embedded in $X \times Z$ for every compact space Z.

Corollary 2.3 [St_3]. Let Z be a compact space of weight κ. A space X is collectionwise normal if and only if $F \times Z$ is C-embedded in $X \times Z$ for every closed subset F of X.

Rudin's Theorem 1.12 is an immediate consequence of 2.3. As an another application we also get a result of Tamano.

Corollary 2.4 [T_2]. A space X is collectionwise normal if and only if $F \times \beta X$ is C-embedded in $X \times \beta X$, for every closed subset F of X.

Corollaries 2.3 and 2.4 should be compared to the results of Morita (see Table 2) and Tamano [T_1] stating, respectively, that a space X is normal and κ-paracompact if and only if

---

[4] In all results presented in this section C-embedding can be always replaced by C*-embedding.

$X \times I^\kappa$ is normal and that a space is paracompact if and only if $X \times \beta X$ is normal.

The famous Borsuk homotopy extension theorem (see [B]) has been proved by Dowker [Do_2] to be valid if the product space $X \times I$ is normal. Recently, Morita [Mo_8] and Starbird [St_1] independently discovered that it suffices to assume that X is normal, which in view of Rudin's Example 1.11 is a substantial improvement. The essential part of their proof consists in showing that if X is normal, then the set $X \times \{0\} \cup F \times I$ is C-embedded in $X \times I$, for every closed subset F of X. The following result of Miednikov provides a nice generalization of this fact (by a rectangle we mean a set of the form $A \times B$; for further generalizations of the Borsuk homotopy extension theorem, see [MH_1], [P_4] and [Se]).

Theorem 2.5 [Mi]. If X is normal, then every subset of $X \times I$ being the union of finitely many closed rectangles is C-embedded in $X \times I$.

Theorem 2.1 and Corollary 2.2 suggest the following natural strengthening of the notion of P^κ-embedding. We shall say that a subset A of X is *strongly* P^κ*-embedded* in X if $A \times Y$ is P^κ*-embedded* in $X \times Y$, for every space Y of weight $\leq\kappa$. We say that A is *strongly P-embedded* in X if it is strongly P^κ-embedded in X for every κ, i.e. if $A \times Y$ is P-embedded in $X \times Y$, for every space Y.

Lemma 2.6. The following two statements are equivalent:

(i) A is strongly P^κ-embedded in X;

(ii) $A \times Y$ is C-embedded in $X \times Y$, for every space Y of weight $\leq \kappa$

Proof. It suffices to prove that (ii) implies (i). Let Y be a space of weight $\leq\kappa$ and let Z be a compact space of weight κ. By (ii) the set $A \times (Y \times Z)$ is C-embedded in $X \times (Y \times Z)$ and thus, by Theorem 2.1, $A \times Y$ is P^{κ}-embedded in $X \times Y$. □

Using our terminology and Lemma 2.6, Theorems 3 and 4 from [St_2] read as follows.

Theorem 2.7 [St_2]. Every closed subset of a compact space is strongly P-embedded.

Theorem 2.8 [St_2]. Every closed subset of a metric space is strongly P-embedded.

The proof of Theorem 2.7 is particularly easy if one applies the theorem from [LP] about the existence of continuous extenders.

Proof of Theorem 2.7. Let F be a closed subset of a compact space Z, let Y be an arbitrary space and $f: F \times Y \to B$ be a continuous mapping into a Banach space B. Since F is compact, the mapping $\Phi: Y \to C(F,B)$ defined by $\Phi(y)(z) = f(z,y)$ is continuous (see [E_1]). By [LP], there exists a continuous extender $e: C(F,B) \to C(Z,B)$, i.e. we have $e(\phi)|F = \phi$, for every $\phi \in C(F,B)$. The composition $\Psi = e\circ\Phi: Y \to C(Z,B)$ generates a continuous mapping $\tilde{f}: Z \times Y \to B$ defined by $\tilde{f}(z,y) = \Psi(y)(z)$ (see [E_1]) and clearly $\tilde{f}$ is an extension of f over $Z \times Y$. □

Obviously, Theorem 2.7 implies that every compact subset of a topological space is strongly P-embedded. Morita [Mo_{10}] strengthened this result by showing that every locally compact, paracompact and P-embedded subspace of a topological space is strongly P-embedded. His paper also contains the following example:

Example 2.9 [Mo_{10}]. Let X be the Michael line, i.e. the real line with all irrationals isolated. Then X is hereditarily paracompact and the subset Q of rationals is closed in X, countable and metrizable, but $Q \times Y$ is not C-embedded in $X \times Y$, where Y is the space of irrationals.

These results lead to the following problems:

Problem 14. Is every closed subset of a paracompact p-space strongly P-embedded?

Problem 15. Give internal characterization of strongly P-embedded subsets (such a characterization exists for P-embedded subsets).

Problem 16. Characterize those spaces all closed subsets of which are strongly P-embedded.

Problem 17. Is every Čech-complete, paracompact and P-embedded subset of a topological space strongly P-embedded?

3. *Dimension of product spaces*

In this section we discuss the dimension of product spaces. By a dimension we shall understand either the covering dimension dim or the large inductive dimension Ind or the small inductive dimension ind (for definitions, see [E_2]).

Let D denote one of the above dimensions. The most natural and important question arising in dimension theory of product spaces is the following: for what spaces X and Y the product theorem

$$D(X \times Y) \leq D(X) + D(Y)$$

is satisfied? Since the above formula is trivially false if

both spaces X and Y are empty, we shall always assume that at least one of the spaces in non-empty.

Thanks to various results obtained in the last decade we are now able to give a satisfactory answer to this question. Nevertheless, the emerging picture turns out to be much less harmonuous that we might wish.

Let us begin with the most important among dimension functions--the covering dimension dim.

Part A: Product theorems for the covering dimension

In [Pa_2] Pasynkov introduced the notion of a rectangular product and announced the following result which now appears to be the strongest product theorem for the covering dimension, implying, more or less directly, all other known product theorems.

Theorem 3.1 [Pa_2]. If $X \times Y$ is rectangular, then $\dim (X \times Y) \leq \dim X + \dim Y$.

A product space $X \times Y$ is *rectangular* if every cozero subset of $X \times Y$ has a σ-discrete (in $X \times Y$) covering consisting of cozero rectangles (i.e. products $U \times V$ of cozero subsets of X and Y). If $X \times Y$ is normal, then this notion coincides with the notion of an F-product introduced earlier by Nagata [Na].

The importance of Theorem 3.1 is a consequence of the next result also due to Pasynkov.

Theorem 3.2 [Pa_3]. The product space $X \times Y$ is rectangular in each of the following cases:

(1) X is a paracompact p-space and $X \times Y$ is normal;

(2) X is locally compact and paracompact;

(3) projection onto one of the factors is closed;

(4) X is a paracompact Σ-space[5] and Y is a paracompact P-space.

Theorem 3.2 in conjunction with Theorem 3.1 yields four interesting product theorems. Product theorem in case (3) has been first announced by Filippov [F_2] and in case (1) by Filippov [F_2] and Pasynkov [Pa_2] (see [F_3] for proofs). Both Filippov and Pasynkov assumed additionally that in (1) the product space $X \times Y$ is countably paracompact, however in view of Theorem 1.7 this assumption is superfluous. Product theorem in case (2) has been first proved by Morita [Mo_7] and in case (4) announced by Pasynkov in [Pa_2].

The following corollary is an immediate consequence of Theorems 3.1 and 3.2.

Corollary 3.3. The product theorem $\dim (X \times Y) \leq \dim X + \dim Y$ holds in each of the following cases:

(5) X is metrizable and $X \times Y$ is normal [Kd];

(6) X is compact [Mo_7];

(7) X is countably compact and Y is first countable;

(8) X and Y are paracompact p-spaces [F_2], [Pa_2].

Recently Morita obtained a result generalizing (2).

Theorem 3.4 [Mo_{10}]. The product formula $\dim (X \times Y) \leq \dim X + \dim Y$ holds if

(9) X is paracompact and σ-locally compact.

Moreover, Hoshina and Morita [HM] proved that Theorem 3.4 does not follow directly from Theorem 3.1. Namely, they showed that a space $X \times Y$ is rectangular for every space Y if and only if X is paracompact and locally compact.

---

[5] For a definition of Σ-spaces, see [N_2].

Nevertheless, Morita's proof of Theorem 3.4 consists in reducing (9) to the case of (6), so his result can also be considered as an indirect consequence of Theorem 3.1. To illustrate this point, we include a proof of Theorem 3.4 which simplifies Morita's original proof.

Proof of Theorem 3.4. Suppose that X is paracompact and σ-locally compact. Then, clearly, X is a σ-discrete union of compact subspaces, say $X = \bigcup_{n<\omega} \mathcal{F}_n$, where the family $\mathcal{F}_n$ is discrete in X and consists of compact sets. Let us put $F_n = \cup\mathcal{F}_n$. From (6) we infer that $\dim (F_n \times Y) \leq \dim F_n + \dim Y \leq \dim X + \dim Y$. Since $X \times Y = \bigcup_{n<\omega} (F_n \times Y)$, by Katětov's result (see [E_1]; Exercise 7.2.B) it suffices to show that for every n the subspace $F_n \times Y$ is C-embedded in $X \times Y$. But this is a simple consequence of collectionwise normality of X and Theorem 2.7. □

Let us mention one more product theorem that is also due to Morita.

Theorem 3.5 [Mo_1]. The product formula $\dim (X \times Y) \leq \dim X + \dim Y$ holds if

(10) $X \times Y$ is strongly paracompact (e.g. if $X \times Y$ is Lindelöf).

This result can also be reduced to (6). Indeed, by (6), $\dim (\beta X \times Y) \leq \dim \beta X + \dim Y = \dim X + \dim Y$. However, since $X \times Y$ is strongly paracompact, we have $\dim (X \times Y) \leq \dim (\beta X \times Y)$ [Mo_1] (cf. [E_2], Theorems 3.1.23 and 3.2.14).

In 1930 Pontryagin [Pt] constructed two compact metric spaces X and Y such that $\dim X = \dim Y = 2$ but $\dim (X \times Y) = 3$, which shows that in order to obtain a positive result ensuring the equality dim $(X \times Y) = \dim X + \dim Y$ strong assumptions on X and/or Y are necessary. An example of such a positive result is the following recent result of Morita.

Theorem 3.6 [Mo_{10}]. If X is a CW-complex,[6] then dim $(X \times Y) = \dim X + \dim Y$.

Let us now pass to examples illustrating the necessity of assumptions appearing in product theorems (1)-(10).

We begin with an example obtained under the assumption of CH by Wage [W_1] (announcement [W_2]).

Example 3.7. (CH) A separable and locally compact space X such that dim X = 0 and X^2 is perfectly normal,[7] but dim $X^2 > 0$.

The following two examples modifying Wage's example have been obtained by Przymusiński [P_6] without any set-theoretic assumptions beyond ZFC.[8]

Example 3.8 [P_6]. A first countable, separable and Lindelöf space X such that dim X = 0 and X^2 is normal, but dim $X^2 > 0$.[9]

---

[6] For a definition of CW-complexes see e.g. [B]. All CW-complexes are paracompact and σ-locally compact. Every (space homeomorphic to a) complex is a CW-complex.

[7] Perfect normality of X^2 can be achieved, using Kunen's technique, similarly as in [P_6]. The method suggested in [W_1] to make X^2 normal was incorrect.

[8] Lemma 1, due to M. Wage [W_1], which was quoted in [P_6], does not appear explicitely in [W_2], but the function sup: $C \to [-1,1]$ defined in [W_2], where Q is the set of rationals in $[-1,1]$, is lower semicontinuous and satisfies (1) and (2) of Lemma 1, with I replaced by C.

[9] As pointed out by M. Charalambous, the space X^2 can also serve as an example of an N-compact space of positive dimension. First example of such a space was constructed by Mrówka [Mr].

Example 3.9 [P_6]. A first countable, separable and locally compact space X such that dim X = 0 and X^2 is normal, but dim $X^2 > 0$.

In [W_2] Wage announced the existence of a separable metric space M and a first countable, separable and Lindelöf space Y such that dim M = dim Y = 0, but dim $M \times Y > 0$ and sketched their construction(such spaces can also be obtained as subspaces of the space X constructed in [P_6]; indeed, using the notation from [P_6], it suffices to put $M = (C_2, \tau_2)$ and $Y = (I, \tau_1)$). From (5) it follows, however, that in this case the product space $M \times Y$ cannot be normal.

Let us also mention that using methods from [W_1] and [P_6] Charalambous [C] constructed a normal space X of positive dimension being the inverse limit of a sequence of zero-dimensional Lindelof spaces. On the other hand, from results of Nagami [N_4] and Zenor [Z] it follows that if dim $X^n \le k$ for every $n < \omega$ and X^ω is normal, then dim $X^\omega \le k$.

The examples mentioned above answer most of the natural questions concerning the necessity of assumptions appearing in product theorems (1)-(10). Nevertheless, the following problems seem to be open and interesting.

Problem 18. Suppose that X or Y is countably compact. Is dim $(X \times Y) \le$ dim X + dim Y?

Problem 19. Suppose that $X \times Y$ is paracompact. Is dim $(X \times Y) \le$ dim X + dim Y?

Problem 20. (Fedorčuk). Does there exist a space X such that dim X = 0 and X^2 is normal, but X^2 is not countable-dimensional.[10]

Part B: Product theorems for the large inductive dimension

Since dimension Ind is defined only for normal spaces, when considering product theorems for the large inductive dimension we must always assume that the product space is normal. More importantly, we are also forced to assume that the finite sum theorem holds for both factors. Besides of these two (essential) differences, product theorems for Ind closely parallel product theorems for dim.

We say that a space X satisfies the *finite sum theorem* for dimension Ind if Ind $\bigcup_{i=1}^{n} F_i \leq \max\{\text{Ind } F_i\}_{i=1}^{n}$ for every finite sequence $\{F_i\}_{i=1}^{n}$ of closed subsets of X. The finite sum theorem holds in a relatively wide class of strongly hereditarily normal spaces introduced by Engelking [E_2]. Perfectly normal, hereditarily paracompact, totally normal and Dowker spaces are strongly hereditarily normal.

The following theorem, due to Pasynkov, corresponds to Theorem 3.1 and is the strongest product theorem known for the large inductive dimension. It is a strengthening of earlier results of Nagata [Na] and Pasynkov [Pa_1].

Theorem 3.10 [Pa_3]. If X × Y is rectangular and normal and if both factors satisfy the finite sum theorem for dimension Ind, then Ind (X × Y) $\leq$ Ind X + Ind Y.

The following corollary is an immediate consequence of Theorems 3.10 and 3.2.

---

[10] A space is *countable-dimensional* (in the sense of dim) if it is a union of countably many subspaces X_n such that dim $X_n < \infty$ (see [E_3]).

Corollary 3.11. Suppose that $X \times Y$ is normal and that both factors satisfy the finite sum theorem for dimension Ind.

Then the product formula $\mathrm{Ind}\,(X \times Y) \leq \mathrm{Ind}\,X + \mathrm{Ind}\,Y$ holds in each of the cases (1)-(8).

This result in cases (1) and (3) has been first announced by Filippov [F_2] (see [F_3] for proofs) and in case (8) it was first proved by Pasynkov [Pa_1]. Corollary 3.11 in case (2) is a strengthening of a result of Pasynkov [Pa_1] and in cases (4) and (5) it was first announced by Pasynkov [Pa_2] and generalizes a theorem obtained earlier by Nagami [N_1] and the following theorem due to Kirmura.

Corollary 3.12 [Ki]. If M is metrizable and Y is perfectly normal, then $\mathrm{Ind}\,(X \times Y) \leq \mathrm{Ind}\,M + \mathrm{Ind}\,Y$.

The author does not know whether Corollary 3.11 holds in cases (9) and (10), however, the following result is known.

Theorem 3.13 [K_1], [E_2]. If $X \times Y$ is strongly paracompact and strongly hereditarily normal, then $\mathrm{Ind}\,(X \times Y) \leq \mathrm{Ind}\,X + \mathrm{Ind}\,Y$.

The necessity of the assumption in 3.10 and 3.11 that the finite sum theorem is satisfied in both factors is clearly illustrated by the following example due to Filippov, which shows that without that assumption Corollary 3.11 is false in cases (1)-(6) and (8)-(10). Since for a normal space X $\dim X = 0$ if and only if $\mathrm{Ind}\,X = 0$, the necessity of other assumptions is illustrated by examples discussed in Part A.

Example 3.14 [F_1]. There exist compact spaces X and Y such that $\mathrm{Ind}\,X = \mathrm{ind}\,X = 1$ and $\mathrm{Ind}\,Y = \mathrm{ind}\,Y = 2$, but $\mathrm{Ind}\,(X \times Y) \geq \mathrm{ind}\,(X \times Y) \geq 4$.

Finally, let us mention the following problems.

Problem 21. Does there exist a metric space M and a (compact) space Y such that Ind (M × Y) > Ind M + Ind Y?

Part C: Product theorems for the small inductive dimension

The small inductive dimension ind behaves well only in the class of separable metric spaces, so it is much less important than the other dimensions. Basically, we have only two product theorems for dimension ind, but in view of Filippov's Example 3.14 we cannot expect much more.

Theorem 3.15 [Pa_1]. If both factors X and Y satisfy the finite sum theorem for dimension ind, then ind (X × Y) ≤ ind X + ind Y.

Theorem 3.16 [F_3]. If M is compact metric, then ind (X × Y) ≤ ind M + ind Y.

References

[A] O.T. Alas, *On a characterization of collectionwise normality*, Canad. Math. Bull. 14 (1971), 13-15.

[AS] R.A. Aló and L.I. Sennott, *Collectionwise normality and the extension of functions on product spaces*, Fund. Math. 76 (1972), 231-243.

[AE] K. Alster and R. Engelking, *Subparacompactness and product spaces*, Bull. Polon. Acad. Sci. 20 (1972), 763-767.

[AP] K. Alster and T.C. Przymusiński, *Normality and Martin's Axiom*, Fund. Math. 91 (1976), 124-130.

[AZ] K. Alster and P. Zenor, *An example concerning the preservation of the Lindelof property in product spaces*, in: Set-Theoretic Topology, Academic Press 1977, edited by G.M. Reed, 1-10.

[At] M. Atsuji, *On normality of the product of two spaces*, General Topology and Its Relations to Modern Analysis and Algebra, Part B, IV, Proc. of the Fourth Prague Top. Symposium 1976, Prague 1977, 25-27.

[B] K. Borsuk, *Theory of retracts*, Warszawa 1967.

[C] M.B. Charalambous, *An example concerning inverse limit sequences of normal spaces*, preprint.

[CC] T. Chiba and K. Chiba, *A note on normality of product spaces*, Sci. Rep. Tokyo Kyoiku Daigaku 12 (1974), 165-173.

[D] J. Dieudonné, *Une généralization des espaces compacts*, Journ. de Math. Pures et Appl. 23 (1944), 65-76.

[Do_1] C.H. Dowker, *On countably paracompact spaces*, Canad. J. Math. 3 (1951), 219-224.

[Do_2] ________, *Homotopy extension theorems*, Proc. London Math. Soc. 6 (1956), 100-116.

[Ef] B.A. Efimov, *Mappings and embeddings of dyadic spaces*, Mat. Sbornik 103 (1977), 52-68 (in Russian).

[E_1] R. Engelking, *General Topology*, Warszawa, 1977.

[E_2] ________, *Dimension Theory*, Warszawa-Amsterdam, 1978.

[E_3] ________, *Transfinite dimensions*, this volume.

[F_1] V.V. Filippov, *On the inductive dimension of the product of bicompacta*, Dokl. Akad. Nauk SSSR 202 (1972), 1016-1019 (in Russian).

[F_2] ________, *On the dimension of normal spaces*, Dokl. Akad. Nauk SSSR 209 (1973), 805-807 (in Russian).

[F_3] ________, *On the dimension of products of topological spaces*, Fund. Math. 105 (1979), to appear (in Russian).

[Ga] T.E. Gantner, *Extensions of uniform structures*, Fund. Math. 66 (1970), 263-282.

[G] J. Gerlits, *On subspaces of dyadic compacta*, Studia Sci. Math. Hungar. 11 (1976),

[Gu] S.P. Gul'ko, *On the properties of sets lying in Σ-products*, Dokl. Akad. Nauk SSSR 237 (1977), 505-508 (in Russian).

[H] R.W. Heath, *On p-spaces, q-spaces, r-spaces and "s"-spaces*, Proc. Auburn Top. Conf. March 1969, 123-134.

[HM] T. Hoshina and K. Morita, *On rectangular products of topological spaces*, Gen. Top. Appl., to appear.

[Ka] M. Katětov, *Complete normality of Cartesian products*, Fund. Math. 36 (1948), 271-274.

[K_1] Y. Katuta, *A note on the inductive dimension of product spaces*, Proc. Japan Acad. 42 (1966), 1011-1015.

[K_2] ________, *On the normality of the product of a normal space with a paracompact space*, Gen. Top. Appl. 1 (1971), 295-319.

[K_3] ________, *Paracompactness, collectionwise normality and product spaces*, Sci. Rep. Tokyo Kyoiku Daigaku 13 (1977), 165-172.

[K_4] ________, *Characterizations of paracompactness by increasing covers and normality of product spaces*, Tsukuba J. Math. 1 (1977), 27-43.

[Ki] N. Kimura, *On the inductive dimension of product spaces*, Proc. Japan Acad. 39 (1963), 641-646.

[Kd] Y. Kodama, *On subset theorems and the dimension of products*, Amer. J. of Math. 91 (1969), 486-497.

[Ko] A.P. Kombarov, *On the tightness and normality of Σ-products*, Dokl. Akad. Nauk SSSR 239 (1978), 775-778 (in Russian).

[LP] D.J. Lutzer and T.C. Przymusiński, *Continuous extenders in normal and collectionwise normal spaces*, Fund. Math. 102 (1978), 168-174.

[M_1] E. Michael, *The product of a normal space and a metric space need not be normal*, Bull. Amer. Math. Soc. 69 (1963), 375-376.

[M_2] ________, *Paracompactness and the Lindelöf property in finite and countable Cartesian products*, Comp. Math. 23 (1971), 119-214.

[Mi] L.E. Miednikov, *Embeddings of bicompacta into the Tychonoff cube and extension of mappings from subsets of products*, Dokl. Akad. Nauk SSSR 222 (1975), 1287-1290 (in Russian).

[Mo_1] K. Morita, *On the dimension of product spaces*, Amer. J. of Math. 75 (1953), 205-223.

[Mo_2] ________, *Paracompactness and product spaces*, Fund. Math. 50 (1961/62), 223-236.

[Mo_3] ________, *On the product of a normal space with a metric space*, Proc. Japan Acad. 39 (1963), 148-150.

[Mo_4] ________, *On the product of paracompact spaces*, Proc. Japan Acad. 39 (1963), 559-563.

[Mo_5] ________, *Products of normal spaces with metric spaces, I and II*, Math. Ann. 154 (1964), 365-382 and Sci. Rep. Tokyo Kyoiku Daigaku 8 (190) (1963), 87-92.

[Mo_6] ________, see the proof of the implication (3) → (4) in Theorem 1.3 in: T. Ishii, On product spaces and product mappings, J. Math. Soc. Japan 18 (1966), 166-181.

[Mo_7] K. Morita, *On the dimension of the product of Tychonoff spaces*, Gen. Top. Appl. 3 (1973), 125-133.

[Mo_8] ________, *On generalizations of Borsuk's homotopy extension theorem*, Fund. Math. 88 (1975), 1-6.

[Mo_9] ________, *Some problems on normality of product spaces*, General Topology and Its Relations to Modern Analysis and Algebra, IV, Proc. of the Fourth Prague Top. Symposium 1976, Part B, Prague 1977, 296-297.

[Mo_{10}] ________, *On the dimension of the product of topological spaces*, Tsukuba J. Math. 1 (1977), 1-6.

[MH_1] K. Morita and T. Hoshina, *C-embedding and the homotopy extension property*, Gen. Top. Appl. 5 (1975), 69-89.

[MH_2] ________, *P-embedding and product spaces*, Fund. Math. 93 (1976), 71-80.

[Mr] S. Mrówka, *Recent results on E-compact spaces*, Proc. Second Pittsburgh Intern. Conf., Lecture Notes 378, Springer-Verlag 1974, 298-301.

[N_1] K. Nagami, *A note on the large inductive dimension of totally normal spaces*, J. Math. Soc. Japan 21 (1969), 282-290.

[N_2] ________, *Σ-spaces*, Fund. Math. 65 (1969), 160-192.

[N_3] ________, *Normality of products*, Actes Congrés Intern Math. 2 (1970), 33-37.

[N_4] ________, *Countable paracompactness of inverse limits and products*, Fund. Math. 73 (1972), 261-270.

[Na] J. Nagata, *Product theorems in dimension theory I*, Bull. Polon. Acad. Sci. 15 (1967), 439-448.

[No] T. Nogura, *Tightness of compact Hausdorff spaces, and normality of product spaces*, J. Math. Soc. Japan 28 (1976), 360-362.

[Ny] P. Nyikos, *A compact non-metrizable space P such that P^2 is completely normal*, Topology Proceedings 2 (1977), 359-364.

[Pa_1] B.A. Pasynkov, *On inductive dimensions*, Dokl. Akad. Nauk SSSR 189 (1969), 254-257 (in Russian).

[Pa_2] ________, *On the dimension of products of normal spaces*, Dokl. Akad. Nauk SSSR 209 (1973), 792-794 (in Russian).

[Pa_3] B.A. Pasynkov, *On the dimension of rectangular products*, Dokl. Akad. Nauk SSSR 221 (1975), 291-294 (in Russian).

[Po] R. Pol, *A non-paracompact space whose countable product is perfectly normal*, Commentationes Math. 20 (1978), 435-437.

[PP] R. Pol and E. Puzio-Pol, *Remarks on Cartesian products*, Fund. Math. 93 (1976), 57-69.

[Pt] L.S. Pontryagin, *Sur une hypothèse fondamentale de la theorie de la dimension*, C.R. Acad. Paris 190 (1930), 1105-1107.

[P_1] T.C. Przymusiński, *A Lindelöf space X such that X^2 is normal but not paracompact*, Fund. Math. 78 (1973), 291-296.

[P_2] ________, *A note on collectionwise normality and product spaces*, Colloq. Math. 33 (1975), 65-70.

[P_3] ________, *Normality and paracompactness in subsets of product spaces*, Fund. Math. 91 (1976), 161-165.

[P_4] ________, *Collectionwise normality and absolute retracts*, Fund. Math. 98 (1978), 61-73.

[P_5] ________, *Collectionwise normality and extensions of continuous functions*, Fund. Math. 98 (1978), 75-81.

[P_6] ________, *On the dimension of product spaces and an example of M. Wage*, Proc. Amer. Math. Soc. (1979), to appear.

[P_7] ________, *Normality and paracompactness in finite and countable Cartesian products*, Fund. Math. 105 (1979), to appear.

[P_8] ________, *Products of perfectly normal spaces*, Fund. Math., to appear.

[R_1] M.E. Rudin, *A normal space X for which $X \times I$ is not normal*, Fund. Math. 73 (1971), 179-186.

[R_2] ________, *The normality of products with a compact factor*, Gen. Top. Appl. 5 (1975), 45-59.

[R_3] ________, *κ-Dowker spaces*, Czechoslovak Math. J. 28 (103) (1978), 324-326.

[R_4] ________, *Σ-products of metric spaces are normal*, preprint.

[RS] M.E. Rudin and M. Starbird, *Products with a metric factor*, Gen. Top. Appl. 5 (1975), 235-348.

[Se] L.I. Sennott, *Some remarks on M-embedding*, Proc. Univ. of Oklahoma Top. Conf., to appear.

[St_1] M. Starbird, *The Borsuk homotopy extension theorem without the binormality condition*, Fund. Math. 87 (1975), 207-211.

[St_2] ________, *Extending maps from products*, in: Studies in Topology, Academic Press 1975, edited by N.H. Stavrakas and K.R. Allen, 559-565.

[St_3] ________, *Products with a compact factor*, Gen. Top. Appl. 6 (1976), 297-303.

[S] A.H. Stone, *Paracompactness and product spaces*, Bull. Amer. Math. Soc. 54 (1948), 997-982.

[Š] V. Šneider, *Continuous images of Souslin and Borel sets; Metrization theorems*, Dokl. Akad. Nauk SSSR 50 (1945), 77-79 (in Russian).

[T_1] H. Tamano, *On paracompactness*, Pacific J. Math. 10 (1960), 1043-1047.

[T_2] ________, *On compactifications*, J. Math. Kyoto Univ. 1-2 (1962), 162-193.

[Te_1] R. Telgársky, *On the product of paracompact spaces*, Bull. Polon. Acad. Sci. 17 (1969), 533-536.

[Te_2] ________, *C-scattered and paracompact spaces*, Fund. Math. 73 (1971), 59-74.

[Te_3] ________, *Spaces defined by topological games*, Fund. Math. 88 (1975), 193-223.

[V] J.E. Vaughan, *Nonnormal products of ω_μ-metrizable spaces*, Proc. Amer. Math. Soc. 51 (1975), 203-208.

[W_1] M. Wage, *The dimension of product spaces*, preprint 1976, Department of Mathematics, Yale University.

[W_2] ________, *The dimension of product spaces*, Proc. Natl. Acad. Sci. USA 75 (1978), 4671-4672.

[WW] J. Worrell and H. Wicke, *Characterizations of developable topological spaces*, Canad. J. Math. 17 (1965), 820-830.

[Z] P. Zenor, *Countable paracompactness in product spaces*, Proc. Amer. Math. Soc. 30 (1971), 199-201.

S AND L SPACES

Mary Ellen Rudin
University of Wisconsin

An S space is a T_3 topological space which is hereditarily separable but not Lindelöf. An L space is a T_3 topological space which is hereditarily Lindelöf but not separable. The basic questions are:

(1) Does there exist an S space?

(2) Does there exist an L space?

The situation can be described in a nutshell:

If one replaces T_3 by T_2 one can answer the basic questions in the affirmative by giving very elementary examples; the same holds true for essentially all related questions.

On the other hand, despite ten years of concentrated effort (see the bibliography), the answers to the basic questions are unknown. It is known that it is consistent with the usual axioms for set theory that the answers be yes; and in every case where we know one answer to be yes, the other is also yes. But we do not know if the questions have the same answer in every model of ZFC and we do not know that the questions are independent of ZFC (Zermelo Frankel set theory together with the axiom of choice--the usual axioms for set theory). If one adds additional hypotheses, one can sometimes prove that the answers to these questions are independent of ZFC. Much of the activity you see evidence of in the bibliography is involved with these related problems,

ISBN: 0-12-584960-5

or with using the results of this activity to solve apparently unrelated problems. But the basic problems look harder (and more fundamental) with the years.

Despite being convinced personally that there are important problems, I feel I must include a paragraph of justification for so much effort being put into problems which sound so very esoteric. My good friend Istvan Juhász would say, these are pure set theoretic notions and the problems are very hard; these are reasons enough. These arguments carry weight with me; but the questions seem set theoretically esoteric as well as topologically esoteric. The telling mark with me has been the frequency with which I have worked on other, apparently unrelated problems in general topology where, to my surprise, the pathology of the problem turned out to be embedded in one or the other of these questions. Often general topology problems concern a trade off between limit point properties (like separability) and covering properties (like Lindelöfness). And the real difficulty is sometimes found in the existence of a very bad subspace. On these occasions S or L spaces have a way of showing up.

I. Formulating the problem

There is no obvious duality between separability and Lindelöfness. Spaces like $\beta N - N$ and the lexicographically ordered square are compact but not separable. Spaces like the Cantor tree or the tangent disk space are separable but not Lindelöf. However, these spaces all have uncountable discrete subspaces and are thus neither hereditarily Lindelof nor hereditarily separable. It was the study of uncountable discrete subspaces which first led de Groot [dG] and especially Hajnal and Juhász [HJ_1], [HJ_3] to S and L spaces.

A space X is said [Ju_1] to be *right separated* provided X is uncountable and, for some ordinal λ, $X = \{x_\alpha | \alpha < \lambda\}$ and every initial segment is open; X is *left separated* provided every initial segment is closed. A space is not Lindelöf provided it has a right separated subspace; it is not separable provided it has a left separated subspace; X is discrete provided X is both left and right separated. Thus problems (1) and (2) can be restated as:

(1*) Is there a T_3 right separated space with no left separated subspace?

(2*) Is there a T_3 left separated space with no right separated subspace?

This formulation of our problems makes the duality inherent in S and L spaces as well as the purely set theoretic nature of the problems abundantly clear. Also it makes the connection with discrete subspaces evident.

The study of hereditary separability and hereditarily Lindelöfness was not new when Hahnal and Juhász [HJ_2] first suggested this way of looking at our problems. For instance, I recall a homework problem from a topology class in the mid-1940s in which we proved (the simple and known) theorem that every hereditarily separable Moore space is Lindelöf and thus metric. But the special interest in these problems awaited the dual and set theoretic interpretation.

In [R_1] Roitman gives a purely set theoretic translation of the existence of S and L spaces; the translation is too awkward to be really useful; but the paper is worth reading for the rich collection of "well known" facts given there.

For example: if there are S or L spaces, there are 0-dimensional ones. I present here another translation due to Roitman and P. Nyikos:

(i) There is an S space if and only if for each $\alpha \in \omega_1$ we can choose a set $R_\alpha \subset [\alpha, \omega_1)$ so that $\alpha \in R_\alpha$ and, for every $S \subset \omega_1$, there is a countable $T \subset S$ such that, if $\alpha \in S$ and $A_\alpha \subset R_\alpha$ and $B_\alpha \subset (\omega_1 - R_\alpha)$ are finite, then there is a $t \in T$ with $A_\alpha \subset R_t$ and $B_\alpha \subset (\omega_1 - R_t)$.

(ii) There is an L space if and only if for each $\alpha \in \omega_1$ we can choose a set $L_\alpha \subset [0, \alpha]$ so that $\alpha \in L_\alpha$ and, for every $S \subset \omega_1$, if finite $A_\alpha \subset L_\alpha$ and $B_\alpha \subset (\omega_1 - L_\alpha)$ have been chosen for each $\alpha \in S$, then there is a countable $T \subset S$ such that, if $\alpha \in S$ there is a $t \in T$ with $A_t \subset R_\alpha$ and $B_t \subset (\omega_1 - R_\alpha)$.

It is trivial to check that these statements are equivalent to the existence of a 0-dimensional right separated S space and a 0-dimensional left separated L space, respectively. They amply illustrate my point that, while pure set theoretic translation is possible, it is neither pretty nor helpful in this case.

When our basic problems are asked in forms (1*) and (2*) one is almost forced to ask additional questions posed by Hajnal and Juhász:

(3) Are problems (1) and (2) the same? That is, in any model of set theory in which one can answer one of the questions, must the other of necessity have the same answer?

(4) If a T_3 space has no uncountable discrete subspace, must it be the union of a hereditarily separable

subspace and a hereditarily Lindelöf subspace?

(5) Are height and width ever different?

We do not know the answer to any of these questions.

II. Related cardinal function problems

The S and L problems are part of a vital movement in general topology to think of topological properties as cardinal functions [Ju_1], [Ru_3]. For a cardinal α one can define a space to be α-separable if α is the minimal cardinality of a dense set, α-Lindelöf if α is the minimal cardinal such that every open cover has a subscover of that cardinality. Generalizing right and left separated sets to have arbitrary cardinal length, the hereditary Lindelöf number of a space X (or height) is $\sup\{|A| \mid A \subset X$ and A is right separated$\}$. The hereditary separability number of X (or *width*) is $\sup\{|A| \mid A \subset X$ and A is left separated$\}$. The *spread* of X is $\sup\{|A| \mid A \subset X$ and A is discrete$\}$. An S space is characterized by being T_3 and having height ω_1 but countable width (or spread). An L space is characterized by being T_3 and having width ω_1 but countable height (or spread). Width, height, and spread can be compared with other cardinal functions [Ju_1].

There are clearly problems at limit cardinals as to whether the width, height, and spread are actually achieved. Also, if X is a T_3 space, $|X| \leq \exp$ (height X) $\leq$ exp (exp (density X)) and weight X $\leq$ exp (density X) $\leq$ exp (exp (spread X)). So there can be no L space of cardinality 2^c or S space of weight 2^c. Could there be an S space of cardinality 2^c or L space of weight 2^c?

Usually [HJ_5], [HJ_7] the results for cardinals like ω and ω_1 generalize so we concentrate on these cardinals, especially since we cannot answer the basic questions even there. But we have some hope that results involving S and L spaces will generalize for larger cardinals.

Hajnal and Juhász [HJ_1] show that spread, height, and width are usually achieved even for T_2 spaces. Roitman and Kunen [RK] show that if MA holds, spread and width are always achieved for T_2 spaces but that the existence of certain Lusin sets would cause them to fail for some cardinals of cofinality ω. In [R_2] Roitman shows that it is consistent with ZFC that there be a T_3 space of spread $\aleph_{\omega_1}$ with no discrete subspace of cardinality $\aleph_{\omega_1}$.

Roitman also shows in [R_2] that, if CH holds or there is a Souslin line, for example, then there is a T_3 space of spread ω which is not the union of a hereditarily Lindelöf and a hereditarily separable subspace. Thus the answer to problem (4) is consistently no. Roitman had previously shown [Ru_3] that there are simple T_2 spaces with this property.

Roitman's paper [R_3] discusses when a space which can be right and left separated (of cardinalities not necessarily ω_1) can be partitioned into a small number of discrete subspaces. This is another obvious generalization of the notions inherent in S and L spaces.

In their first paper on the subject [HJ_2], Hajnal and Juhász give examples of T_2 spaces with height α^+ but width α or width α^+ but height α for all cardinals α. In [HJ_4] and [HJ_5] they give forcing examples of T_3 (and thus S and L) spaces with these properties. In fact they show that the spaces can be of the special types we described earlier: an

S space of cardinality 2^c and an L space of weight 2^c exist in certain models of ZFC. In [T_5] Tall has a much simpler construction using Lusin sets to get L spaces of large weight. He actually shows that one can have a (0-dimensional) L space of weight κ for any $\aleph_2 \leq \kappa < 2^{\aleph}$ in a model which includes the continuum hypothesis but has $2^{\aleph}$ as large as you wish; the space has no subspaces of weight between $\aleph_1$ and κ which is also interesting.

III. Other related problems

T_3 Lindelöf spaces are always normal; so L spaces are always normal. Rather to everyones surprise, the first consistency examples of S spaces turned out to be normal. Normality is not usually an accident. So the question arose as to whether all S spaces had to be normal. I constructed sort of a machine [Ru_2] for turning normal S spaces into non-normal ones; and Jones [J] had a machine for turning non-normal S spaces into non-completely regular ones. Finally Tall [T_2] proved outright that if there is an S space, there is one which is not even completely regular.

We say that X is a *strong* S space if X^n is an S space for all $n \in \omega$; we define strong L spaces similarly. Zenor [Z] proves that there exist strong S spaces if and only if there exist strong L spaces; thus the problem analogous to (3) has a positive solution. Hajnal and Juhász [HJ_7] construct strong S and L spaces assuming CH but Kunen [K_2] shows that MA + $\neg$CH implies there are no strong S or L spaces. Thus the existence of strong S and L spaces is independent of ZFC. An interesting consequence of these theorems is one of Roitman's [R_5]: adding a Cohen or random real to a model of ZFC kills MA + $\neg$CH.

S and L topological *groups* were first constructed by Hajnal and Juhász [HJ_5], [HJ_7], [HJ_8]. These were strong S and L groups, the first constructed by forcing the later ones using only CH. The Kunen theorem, of course, says that their existence is independent of ZFC. Zenor [Z] shows that strong S and L spaces imply the existence of strong S and L groups. In [R_4] Roitman gives another proof of this fact and also shows, if CH holds, there are S and L groups whose squares are not S and L, respectively. It is unknown whether the existence of non-strong S and L groups is independent of ZFC.

IV. Consistency examples of S and L spaces

In this section we will work our way through a rather historical account of the known examples of S and L spaces listing the assumptions under which they were constructed and the resulting properties of the spaces. As one should expect, the assumptions get weaker, the properties weaker, and the constructions simpler with time.

A Souslin line is an L space, a very nice one being 1st countable, locally compact, locally connected, connected, linearly ordered, perfectly normal. I used a Souslin line [Ru_1] to construct the first example of an S space, not a particularly nice one.

Hajnal and Juhász [HJ_4], [HJ_5] then used forcing to construct a hereditarily α-Lindelöf non α-separable space of weight 2^{2^α} and a hereditarily α-separable none α-Lindelöf space of cardinality 2^{2^α} (as we previously mentioned) for each cardinal α. It is even true that every subspace of the α-Lindelöf space of cardinality less than 2^{2^α} is closed and discrete while the subspaces of cardinality 2^{2^α} all have weight 2^{2^α}. In [HJ_7] Hajnal and Juhász show that one can construct S and L spaces (without these properties) using only CH; the S space is a subspace of 2^c and the L space is constructed from the same subspace of 2^c.

Ostaszewski [O] wanted to construct a perfectly normal, countably compact space which was not compact. He was led to construct, using ◊, a perfectly normal, locally countable, locally compact, countably compact S space. The technique was simple and useful and became the basis for the construction of most later S spaces even those built using simpler hypotheses. Fedorchuk [F] used ◊ to construct a hereditarily normal, cardinality 2^c, S subspace of a compact space. Wage [W] used ◊ to construct an extremally disconnected (perfectly normal, Dowker, hereditarily normal) S space. Zenor and I [RZ] used ◊ to construct a countably compact, perfectly normal, manifold S space.

But the aim now became simple construction from minimal hypotheses, CH in particular. Zenor and I [RZ] constructed a perfectly normal manifold S space using CH. In [J], using Ostaszewski's technique and CH, Kunen constructed a simple S space refining the usual topology of the real line, one which was 1st countable, locally compact, perfectly normal and real compact. This S space was really just a clarification of the S space of Hajnal and Juhász in [HJ_7]. The same technique yielded a 1st countable, cardinality $\aleph_1$, Dowker S space from a Lusin set.

Tall observed [J] that (inherent in previous work of

H.E. White) a Sierpinski set or Lusin set in the density topology of the real line would lead to an L-space. Such sets exist assuming several different hypotheses, but especially the continuum hypothesis. This was used to prove there are Baire L spaces and L spaces embedded in $\beta N - N$ if one assumes CH. In [T_4] Tall shows that CH implies the existence of a homogeneous L space (again using the density topology). He shows also that MA + ¬CH implies that the width and height of a T_3 space need not be the same.

In [T_1] Tall sets up the problem of the existence of an L space with a point countable base. van Douwen, Tall, and Weiss then show [vDTW] that Lusin sets can be used to build such spaces. If CH holds there are Lusin sets in many spaces and such sets are most useful in the construction of various types of S and L spaces. However, Kunen [K_1] has shown that MA + ¬CH implies that there are no (useful) Lusin sets.

In (1) van Douwen and Kunen construct L and S subspaces of $P(\omega)$ with the Vietoris topology. The construction is especially nice because not only are the spaces dual but the construction is dual. The assumption (↓) is weaker than either the existence of a Souslin line or CH but again is killed by MA + ¬CH.

V. *Trying to kill S and L spaces*

We have already mentioned that MA + ¬CH kills (↓), the existence of a Souslin line, useful Lusin sets, Sierpinski sets in the density topology, and, of course CH and ◊: all of the hypotheses under which we have so far been able to construct S and L spaces. Also if MA + ¬CH, there are no strong S or L spaces. MA + ¬CH kills the existence of an

S space topology on a subset set of the real line which refines the usual topology as described in [J]. There can be no perfectly normal countably compact non-compact space as described by Ostaszewski in [O] if one assumes MA + $\neg$CH. Also MA + $\neg$CH kills the possibility of there being a perfectly normal nonmetrizable manifold [Ru_4]; this has been extended by Diana Lane to show that if MA + $\neg$CH, then every locally compact, locally connected, perfectly normal space is Lindelof. The most powerful theorem of this type is a most unexpected one of Szentmiklóssy [S] which shows that MA + CH implies there is no S subspace of a compact space or 1st countable L space. That MA + $\neg$CH implies that there could be no L subspace of a compact space was proved earlier by Juhász and Hajnal [Ju_1].

There are many related types of spaces we do <u>not</u> know how to kill; for instance perfectly normal S spaces, hereditarily normal S spaces, 1st countable S spaces, locally connected S or L spaces, extremally disconnected S spaces, L spaces of weight 2^c ... . But which are particularly important is not clear. That we need to know if there exist S and L spaces without any set theoretic assumption <u>is</u> clear.

It is hoped that Szentmiklóssy now has a proof that MA + $\neg$CH does <u>not</u> kill the existence of S spaces. So something different is called for if we wish to solve the basic problems.

The reader is encouraged to look up two other [Ju], [Ny] new surveys of the S and L space problem. Juhász is the real expert and his insight should be most interesting; Nyikos' summary will be brief and to the point. Nyikos is another person who has used S and L space ideas frequently; he also has a number of unpublished results, for instance giving necessary conditions for L spaces to have certain kinds of bases.

References

[vDK] E. van Douwen and K. Kunen, *L-spaces and S-spaces in P(ω)*, to appear.

[vDTW] E. van Douwen, F. Tall and W. Weiss, *Nonmetrizable hereditarily Lindelöf space with point-countable bases from CH*, Proc. Amer. Math. Soc., 64 1 (1977), 139-145.

[F] V.V. Fedorchuk, *Fully closed mappings and the consistency of some theorems of general topology with the axioms of set theory*, Math. USSR, 28 (1976), 1-26 or Mat. Sbornik 99 (1976), 3-33.

[dG] J. de Groot, *Discrete subspaces of Hausdorff spaces*, Bull. Acad. Pol. Sci. 13 (1965), 537-544.

[HJ_1] A. Hajnal and I. Juhász, *On discrete subspaces of topological spaces*, Indag. Math. 29 (1967), 343-356.

[HJ_2] ________, *On hereditarily α-Lindelöf and hereditarily α-separable spaces*, Annales Univ. Sci. Budapest, XI (1968), 115-124.

[HJ_3] ________, *Discrete subspaces of topological spaces II*, Indag. Math. 31 (1969), 18-30.

[HJ_4] ________, *A consistency result concerning hereditarily α-separable spaces*, Indag. Math. 35 (1973), 301-307.

[HJ_5] ________, *A consistency result concerning α-Lindelöf spaces*, Acta Math. Acad. Sci. Hungary 24 (1972), 307-312.

[HJ_6] ________, *On first countable non-Lindelof S spaces*, Coll. Math. Soc. János Bolyai 10 Kesthely Hungary (1973), 837-852.

[HJ_7] ________, *On hereditarily α-Lindelöf and α-separable spaces II*, Fund. Math. 81 (1974), 147-158.

[HJ_8] ________, *A separable normal topological group need not be Lindelöf*, Gen. Top. Appl. 6 (1976), 199-205.

[J] ________, *Hereditarily separable, non-completely regular spaces*, Topology Conference VPISU (1973), Lecture Notes in Math. 375 Springer-Verlag, 149-153.

[Ju] I. Juhász, *Cardinal functions in topology*, Math. Centre Tracts 34 (1975).

[JKR] I. Juhász, K. Kunen and M.E. Rudin, *Two more hereditarily separable non-Lindelöf spaces*, Can. J. Math. 28, 5 (1976), 998-1005.

[K_1] K. Kunen, *Luzin spaces*, Topology Proceedings 1 (1976), 191-199.

[K_2] ________, *Strong S and L spaces under MA*, Set-Theoretic Topology (1977) Academic Press, 265-268.

[KR] K. Kunen and J. Roitman, *Attaining the spread at cardinals of cofinality* ω, Pac. J. Math. 70 1 (1977), 199-205.

[KT] K. Kunen and F. Tall, *Between Martin's axiom and Souslin's hypothesis*, Fund. Math. (to appear).

[Ny] P. Nyikos, *Topology Proceedings*, Classic Problem, 1978 (to appear).

[O] A. Ostaszewski, *On countably compact, perfectly normal spaces*, London Math. Soc. (2) 14 (1976), 505-516.

[R_1] J. Roitman, *A reformulation of S and L*, Proc. Amer. Math. Soc., 69 2 (1978), 344-348.

[R_2] ________, *The spread of regular spaces*, Gen. Top. Appl. 8 (1978), 85-91.

[R_3] ________, *Partitioning spaces which are both right and left separated*, to appear.

[R_4] ________, *Easy S and L groups*, to appear.

[R_5] ________, *Adding a random or Cohen real: topological consequences and the effect on Martin's axiom*, to appear.

[Ru_1] M.E. Rudin, *A normal hereditarily separable non-Lindelöf space*, Illinois J. Math. 16 4 (1972), 621-626.

[Ru_2] ________, *A nonnormal hereditarily separable space*, Illinois J. Math. 18 (1974), 481-483.

[Ru_3] ________, *Lectures on set theoretic topology*, CBMS regional conference series 23 (1975), 25-31.

[Ru_4] ________, *The undecidability of the existence of a perfectly normal nonmetrizable manifold*, Houston J. Math. 5 (1979).

[RZ] M.E. Rudin and P. Zenor, *A perfectly normal nonmetrizable manifold*, Houston J. Math. 2, 1 (1976), 129-134.

[S] Z. Szentmiklóssy, *S-spaces and L-spaces under Martin's axiom*, to appear.

[T_1] F. Tall, *On the existence of nonmetrizable hereditarily Lindelöf spaces with point countable bases*, Duke Math. J. 41 (1974), 299-304.

[T_2] ________, *A reduction of the hereditarily separable non-Lindelöf problem*, Set-Theoretic Topology, Academic Press (1977), 349-351.

[T_3] ________, *The density topology*, Pacific J. Math. 75, 1 (1976), 275-284.

[T_4] ________, *Normal subspaces of the density topology*, Pacific J. Math. 75, 2 (1978), 579-588.

[T_5] ________, *Some applications of a generalized Martin's Axiom*, Proceedings 1978 Budapest Topology Conference, to appear.

[W] M. Wage, *Extremally disconnected spaces*, Topology Proceedings, Auburn 1 (1976), 181-187.

[Z] P. Zenor, *Hereditary m-separability and the hereditary m-Lindelöf property in product and function spaces*, to appear.

LARGE CARDINALS FOR TOPOLOGISTS

Franklin D. Tall[1]
University of Toronto

0. *Introduction*

When I was asked to write a survey paper for this volume, I decided it was time to put into print a point I'd been urging upon (set-theoretic) topologists for the past few years, namely that it was time for them to learn more set theory. In the past decade topologists have made extensive use of Martin's Axiom and $\diamond$ (probably more than set theorists have), not to mention partition calculus, but when it comes down to actually doing some sophisticated set theory rather than quoting a by now well-known axiom, their ignorance has often stymied them. As a result, set theorists who know barely any topology have been cleaning up the so-called hard problems. Until recently, topologists had the excuse that there were no good set theory texts available, but this is no longer true. See the Appendix. It seemed appropriate therefore to write a survey that would nudge topologists in the direction of forcing and large cardinals by giving them the flavour of the two subjects, pointing out suitable references, and exhibiting topological applications. Unfortunately, due to constraints of time and space, only one of these topics could even be attempted, and thus the forcing

[1]The preparation of this paper was assisted by Grant A-7354 of the Natural Sciences and Engineering Research Council of Canada.

ISBN: 0-12-584960-5

survey has had to be postponed. I claim no proprietary rights. I also will gladly leave to others the tasks of eventually surveying topological applications of the structure theory of ideals [BTW] and of infinitary games. I'd like to thank Jim Baumgartner for many discussions concerning the subject matter herein.

"Large cardinals" are hypothetical cardinal numbers with interesting combinatorial and model-theoretical properties. There are several reasons why topologists should become acquainted with them. Some of the combinatorics arises naturally from topological problems; some of the large cardinal axioms are so strong--verging on inconsistency--that it sometimes seems one can prove anything with them; and perhaps most interesting, various combinatorial propositions concerning small cardinals--e.g. $\aleph_1$--are equiconsistent with the existence of large cardinals. For example there is a model of set theory in which there are no Kurepa trees if and only if there is a model in which there is an inaccessible cardinal.

Since we will be attempting to use as little formal logic as we can get away with, we shall be as combinatorial as possible, although this does result in a neglect of much of the subject. Another point we shall largely neglect is the set theorist's preoccupation with the relative strength of large cardinal assumptions. Instead, we shall concentrate on what the cardinals can do for us. However, in keeping with tradition, we shall survey the cardinals more or less in order of increasing strength. NOTE: set theorists usually think of a large cardinal property P as being at least as strong as P' if the consistency (with ZFC) of P implies the consistency of P'. This does not entail that P implies P'.

Before proceeding further, we should point out to the topologist that the assumption of the existence of a large cardinal is qualitatively different from such assumptions as Martin's Axiom or $\Diamond$. Call a cardinal number *weakly inaccessible* if it is uncountable, regular, and not a successor cardinal. A cardinal number κ is *inaccessible* if it is uncountable, regular and for all cardinals $\lambda < \kappa$, $2^\lambda < \kappa$. Thus, inaccessible cardinals are weakly inaccessible, and under GCH the two notions coincide.

In ZFC one can show that if κ is an inaccessible cardinal, then V_κ is a model of ZFC, where V_κ is defined by induction as follows. $V_0 = 0$, $V_{\alpha+1} = \{x: x \subseteq V_\alpha\}$, and if λ is a limit ordinal, $V_\lambda = \bigcup\{V_\alpha: \alpha < \lambda\}$. The point is that V_κ is closed under the operations required by the axioms. If κ is the least inaccessible cardinal, V_κ is a model of "there are no inaccessible cardinals." Hence it can't be proved in ZFC that inaccessible cardinals exist. Indeed it can't even be shown in ZFC (unless ZFC is inconsistent) that the existence of inaccessible cardinals is consistent with ZFC! Let "I" stand for "there is an inaccessible cardinal". We shall show that the assumption that ZFC proves "if ZFC is consistent so is ZFC + I" leads to a contradiction. If ZFC proves this, so does ZFC + I. But ZFC + I proves ZFC is consistent. Hence ZFC + I proves ZFC + I is consistent. But the Second Gödel Incompleteness Theorem rules out the possibility of any sufficiently strong theory--such as ZFC + I--proving its own consistency.

A slight extension of the above argument proves that not even the existence of *weakly* inaccessible cardinals can be shown consistent with ZFC. The point is that Gödel observed that any model of ZFC includes a submodel with the same ordinals in which GCH holds.

The theory of large cardinals (which are all at least weakly inaccessible) is marked by occasional sleepless nights as inconsistencies dance in one's head. Think of spending years deriving consequences from a proposition which is then proved false! Shades of Piltdown man. However there does seem to be a consensus among set theorists that many large cardinals are quite harmless--after all, by the same argument we gave, $\aleph_0$ cannot be proved to consistently exist in ZFC minus the Axiom of Infinity. On the other hand, "Reinhardt cardinals" were seriously considered before Kunen [Ku_1] refuted their existence. In any event we shall henceforth ignore these foundational questions and refer the reader to the texts [KM], [J], [Dr].

1. *Inaccessible cardinals*

The weakly inaccessible and inaccessible cardinals are the smallest large cardinals--why should they concern the topologist? First of all, they often constitute a case to worry about if one is trying to prove by induction a result involving all cardinals. Second, they arise in what Juhász [Ju] calls "sup = max" problems. For example, suppose for each $\lambda < \kappa$ that a space X admits a collection of λ disjoint open sets. Does X have such a collection of size κ? The answer is yes if κ is singular but no if κ is weakly inaccessible. See Juhász for a proof. We shall encounter another example of this sort later: if X is λ-collectionwise Hausdorff for all $\lambda < \kappa$, is X κ-collectionwise Hausdorff?

The most interesting uses of merely (weakly) inaccessible cardinals are in equiconsistency results of the form alluded to earlier. Recall a Kurepa tree is a tree of height $\aleph_1$ with countable levels and at least $\aleph_2$ branches. Allowing the levels to have size $\leq\aleph_1$ gives us a *Canadian tree*. I admit responsibility for this apparently irrevocably adopted term, but I refrain from committing to print the pun that inspired it, since it was not one of my best. Observe that the binary tree on ω_1 is a Canadian tree if CH holds. Results of Silver [Si] (or see [J]) established that the consistency (we henceforth drop "with ZFC") of an inaccessible cardinal is equivalent to the consistency of there being no Kurepa trees. One can have CH or not as one pleases in the latter model. Mitchell [M] improved Silver's result to exclude all Canadian trees; of course in his model CH fails. The latest improvement, due to Baumgartner [Ba_3], again from an inaccessible, yields a model for MA plus not CH plus no Canadian trees. This is interesting because my student Peter Davies has shown that these assumptions imply among other things that there is no linearly ordered Baire space of cardinality $\aleph_1$ without isolated points [D].

To discuss the Silver, Mitchell, or Baumgartner results adequately, one of course has to assume some rudimentary knowledge of forcing. We refer the virgin reader to the Appendix for directions, while reassuring him that he will not lose continuity by skipping over all mention of that arcane technique.

The crucial step in Silver's destruction of Kurepa trees is the observation that countably closed forcing adds no new cofinal branches to trees of height ω_1 with levels of power less than continuum. The point is that if there were a new branch, by looking at the tree of possibilities for points to be on the branch, one could inductively arrive at $2^{\aleph_0}$ points, all on the same level. With this observation, one can then construct the desired model by starting with a model of GCH and *Lévy-collapsing* the inaccessible down to $\aleph_2$. (The Lévy collapse is a forcing construction used very frequently in efforts to get small cardinals to possess some of the combinatorial properties of large ones. It consists of iteratively collapsing all cardinals λ less than the large cardinal κ down to $\aleph_1$ via countable partial functions from $\aleph_1$ to λ. It follows by a chain condition argument from the inaccessibility of κ that it remains a cardinal in the new model, so it must become $\aleph_2$.) Suppose T were a tree with countable levels and height $\aleph_1$ in the final model. Since the cardinality of T is $\aleph_1$, standard arguments tell us that T appears at some initial stage in the iteration. At that stage T had δ many branches, where $\delta < \kappa$ by inaccessibility. From then on T gets no new branches, while δ gets collapsed to $\aleph_1$. Thus T has no more than $\aleph_1$ branches.

Such iterations (actually in this case a simple product) are common when one wishes to force a universal statement. One argues that a counter-example, if it exists, must appear at some stage. But at that stage it was destroyed.

Let $\chi(\omega_1,FC)$ be the sup of the character (as if a point) of homeomorphs of ω_1 in first countable spaces. By mimicking Silver's argument, Fleissner [F_2] establishes the consistency of $\chi(\omega_1,FC) = \aleph_1$, assuming the consistency of an inaccessible. He proves that countably closed forcing does not increase the character of a particular copy of ω_1 in a first countable space, and then collapses as above.

We omit Mitchell's argument since Baumgartner's is more interesting. He interlaces the usual countable chain condition iteration used to obtain Martin's Axiom with forcing designed to kill Canadian trees. He kills such a tree by simply collapsing the cardinality of its number of branches down to $\aleph_1$. He prevents its resurrection by then countable chain condition forcing it to be "special," a condition that is preserved by later extensions and which ensures it has $\leq\aleph_1$ branches. As before the inaccessible is collapsed to $\aleph_2$ by this iteration.

Another topological application of the collapse of an inaccessible is due to Kunen [Ku4]. It is well-known that every closed subset of $\{0,1\}^{\aleph_0}$ (the Cantor set) is either countable or has cardinality continuum. Is an analogous result true for $\{0,1\}^{\aleph_1}$? Kunen shows that CH + $2^{\aleph_1} > \aleph_2$ does not decide the question. More precisely, he proves first that there is a model in which CH + $2^{\aleph_1} > \aleph_2$ holds and yet there is a closed subset of $\{0,1\}^{\aleph_1}$ of power $\aleph_2$. To do this, observe that if one starts with a model of GCH plus there is a Kurepa tree and then adjoins $>\aleph_2$ subsets of ω_1 via countable partial functions, then by Silver's result quoted earlier, the Kurepa tree grows no new branches. The set of its branches can then be considered as a closed subset of $\{0,1\}^{\aleph_1}$. On the other hand, by starting with a model of GCH, Lévy-collapsing an inaccessible, and then adjoining $>\aleph_2$ Cohen subsets of ω_1 via countable partial functions as before, Kunen obtains a model in which CH plus $2^{\aleph_1} > \aleph_2$ holds, but every closed set of power $\geq\aleph_2$ has power $2^{\aleph_1}$. The proof proceeds in two steps: Kunen first observes that it suffices to show that in this model every Canadian tree has $2^{\aleph_1}$ branches, and then he verifies that that in fact holds. For the first step, use CH, compactness, and the fact that $\{0,1\}^{\aleph_1}$ has a basis of cardinality $\aleph_1$ to build a tree of height and cardinality $\aleph_1$ such that each point in the given closed set corresponds to a branch of the tree. This can be done by inductively splitting the closed set into disjoint closed sets. Since different points yield different branches, it suffices to show that if the tree is Canadian, it has $2^{\aleph_1}$ branches.

To do this then, first note that since the cardinality of the tree T is $\aleph_1$ it only depends on an initial segment of the collapse and an initial segment of the Cohen subsets. Without loss of generality then, we may assume it appears in the ground model. The argument then splits into two cases depending on whether or not the tree has new branches in the final model. If not, then it has a total of $\leq\aleph_1$ branches, since the cardinality of its set of branches was collapsed. If it does get new branches, we claim that then the Cohen forcing gives it new branches and indeed each Cohen subset gives it a different branch and thus it must have $2^{\aleph_1}$ branches. The point is that as in Silver's argument, if b is a new branch, one can inductively construct a tree S of conditions deciding the members of b. Any cofinal branch through S determines an interpretation of the name for b, i.e. determines a cofinal branch of T. Different cofinal branches of S determine different branches of T. Since the binary tree on ω_1 is embedded in S, every Cohen subset--indeed every subset--of ω_1 determines a branch of S and hence of T. Thus T must have $2^{\aleph_1}$ branches.

2. *Weakly compact cardinals, indescribable cardinals*

Weakly compact cardinals should be familiar to topologists if only because they should all have read Kunen's article on combinatorics [Ku_2]. Any of the following equivalent statements (and many more) can serve as a definition of a cardinal κ being weakly compact. Convention: when we define a large cardinal we shall always assume it is regular and uncountable.

a) $\kappa \rightarrow (\kappa)^n_\lambda$ for every $n < \omega$ and every $\lambda < \kappa$,

b) κ is inaccessible and has the *tree property*, i.e. every tree of height κ with levels of cardinality $<\kappa$ has a branch of length κ,

c) κ is inaccessible and whenever $\mathcal{A}$ is a collection of $\leq\kappa$ subsets of κ, there is a κ-complete filter $\mathcal{F}$ on κ which is *non-trivial* (i.e. $0 \notin \mathcal{F}$ and all complements of singletons are in $\mathcal{F}$) and for every $A \in \mathcal{A}$, either A or $\kappa - A \in \mathcal{F}$,

d) the topology on $\{0,1\}^\kappa$ obtained from boxes fixing $<\kappa$ coordinates is κ-*compact*, i.e. each open cover has a subcover of power $<\kappa$,

e) every κ-compact Hausdorff space such that each point has character $<\kappa$ has cardinality $<\kappa$.

There are many other equivalents. See, e.g. [Ku_1], [Dr], [J], [KM], [KT], [CN], [P].

For a typical application of weak compactness, let us prove via b) a result of Fleissner. The proof he gives in [F_3] uses another combinatorial variant of weak compactness.

Theorem. Let κ be weakly compact. If each point of X is of character $<\kappa$ and if for each $\lambda < \kappa$, X is λ-collectionwise Hausdorff, then X is κ-collectionwise Hausdorff.

Proof. We inductively build a tree such that the points at the α'th level are certain basic open sets about the α'th point in the discrete collection. We assume by induction that the open sets along any branch are disjoint. Because X is λ-collectionwise Hausdorff for all $\lambda < \kappa$, at each level α some of the branches can be continued. We continue those branches in all possible ways using basic open sets about the α'th point. The resulting tree has height κ. Each level has cardinality $<\kappa$ since κ is inaccessible and the character of the space is $<\kappa$ at each point. The cofinal branch then yields the desired separation.

A tree argument was also used by Kunen and Parsons [KP] in their characterization of which ordinal spaces have normal absolutes. Among other results, they proved that the absolute of α is normal and not paracompact if and only if the cofinality of α is a weakly compact κ, and there is a β such that $\alpha = \beta \dotplus \kappa$.

One variant of weak compactness that is very important does require some logic. Let $\Pi^m_n(\kappa)$ ($\Sigma^m_n(\kappa)$, respectively) be the class of formulas in the usual language for set theory which are of the following form:

i) all quantifiers are bounded by some $V_{\kappa+j}$, $j \leq m$ (i.e. rather than "$(\forall x)$", we have "$(\forall x \in V_{\kappa+j})$"),

ii) all quantifiers are at the front of the formula,

iii) there are at most n alternating blocks of quantifiers of the $V_{\kappa+m}$ form, starting with a universal (existential, respectively) quantifier.

κ is Π^m_n-*indescribable* if whenever $\Phi \in \Pi^m_n(\kappa)$ has one free variable and $R \subseteq V_\kappa$, then if the structure $\langle V_{\kappa+m}, \in, R\rangle$ is a model of $\Phi(R)$, then the structure $\langle V_{\alpha+m}, \in, R \cap V_\alpha\rangle$ is a model of $\Phi(R \cap V_\alpha)$ for some $\alpha < \kappa$. Define Σ_n^m-indescribable analogously. Saying that a cardinal κ is Π_n^m- or Σ_n^m-indescribable then means what it says: κ can't be characterized by a sentence of that complexity; anything that can be said about κ by such a sentence holds for a smaller α. The relation of this notion to weak compactness is that κ is weakly compact if and only if

f) κ is Π_1^1-indescribable.

There is a hierarchy of indescribable cardinals--see the texts. To give some idea of the uses of indescribability, we prove the following

Theorem. Suppose κ is Π_2^1-indescribable and for every $\lambda < \kappa$, every normal space of character $\leq\lambda$ is λ-collectionwise Hausdorff. Then every normal space of character $\leq\kappa$ is κ-collectionwise Hausdorff.

Proof. Suppose on the contrary that there exists a normal space of character $\leq\kappa$ which is not κ-collectionwise Hausdorff. By some straightforward topology (see [T_1, 2.2.1]) one can construct a space with the same properties having cardinality κ such that each point not in the discrete collection is isolated. Thus our new space Y can be completely described by a function $\mathcal{U}$: $(\kappa \times 2) \times \kappa \to \mathcal{P}(\kappa \times 2)$ ($\mathcal{P}(X)$ is the collection of subsets of X), where $\mathcal{U}(\alpha,0,\beta)$ is the β'th open set about the α'th point in the discrete collection, and $\mathcal{U}(\alpha,1,\beta) = \{<\alpha,1>\}$, i.e. $\kappa \times \{1\}$ is the set of isolated points. Let $\mathcal{U}' = \{<\alpha,i,\beta,\mu>: \mu \in \mathcal{U}(\alpha,i,\beta)\}$. Then $\mathcal{U}(\alpha,i,\beta) = \{\mu: (\exists\alpha,i,\beta)(<\alpha,i,\beta,\mu> \in \mathcal{U}')\}$. Thus $\mathcal{U}$ has been coded as a subset $\mathcal{U}'$ of $\kappa \times \kappa \times \kappa \times \kappa$. By a canonical bijection, $\mathcal{U}'$ may be coded as a subset " of κ and hence of V_κ. Let $\Phi_0(\mathcal{U}")$ say that $\mathcal{U}$ is a basis for a topology on $\kappa \times 2$ in which $\kappa \times \{0\}$ is closed discrete and $\kappa \times \{1\}$ is isolated. We write down the formula for $\mathcal{U}$ and leave to the reader the task of translating to $\mathcal{U}''$.

$$(\forall\alpha \in \kappa)(\forall i \in 2)(\forall\beta \in \kappa)(\langle\alpha,i\rangle \in \mathcal{U}(\alpha,i,\beta))$$
$$\& \; (\forall\alpha \in \kappa)(\forall i \in 2)(\forall\beta,\beta' \in \kappa)(\exists\beta'' \in \kappa)(\mathcal{U}(\alpha,i,\beta'') \subseteq \mathcal{U}(\alpha,i,\beta) \cap \mathcal{U}(\alpha,i,\beta'))$$
$$\& \; (\forall\alpha \in \kappa)(\exists\beta \in \kappa)(\forall\alpha' \in \kappa)(\alpha' \neq \alpha \rightarrow \langle\alpha',0\rangle \notin \mathcal{U}(\alpha,0,\beta))$$
$$\& \; (\forall\alpha \in \kappa)(\forall\beta \in \kappa)(\forall\gamma \in \kappa)(\forall i \in 2)(\langle\gamma,i\rangle \in \mathcal{U}(\alpha,1,\beta) \rightarrow \gamma = \alpha).$$

Let $\Phi_1(\mathcal{U}'')$ say that Y is not κ-collectionwise Hausdorff, i.e.

$$(\forall f \in {}^{\kappa}\kappa)(\exists\alpha_0,\alpha_1 \in \kappa)(\alpha_0 \neq \alpha_1 \; \& \; \mathcal{U}(\alpha_0,0,f(\alpha_0)) \cap \mathcal{U}(\alpha_1,0,f(\alpha_1)) \neq 0\}.$$

Then $\Phi_1(\cdot)$ is Π^1_1. Let $\Phi_2(\mathcal{U}'')$ say that Y is normal, i.e.

$$(\forall Z \in {}^{\kappa}2)(\exists f \in {}^{\kappa}\kappa)(\forall\alpha_0,\alpha_1 \in \kappa)(Z(\alpha_0) \neq Z(\alpha_1) \rightarrow \mathcal{U}(\alpha_0,0,f(\alpha_0)) \cap \mathcal{U}(\alpha_1,0,f(\alpha_1)) = 0).$$

Then $\Phi_2(\cdot)$ is $\Pi^1_2(\kappa)$ and so the conjunction Φ of Φ_0, Φ_1 and Φ_2 is Π^1_2. In the definition of indescribability one may in fact assume that κ is reflected down to a cardinal λ; doing so in this case, one may then define from $\mathcal{U}'' \cap V_\lambda$ a normal space of character $\leq\lambda$ which is not λ-collectionwise Hausdorff, contradicting the hypothesis of the Theorem.

The foregoing proof embodies the beauty (or perhaps the cynicism?!) of large cardinal theory--if you wish to prove a result of the form "$<\kappa$ implies $\leq\kappa$", merely define κ so that it works! It is considered bad form however to define large cardinals which can easily be proved to be inconsistent.

Π^1_1-indescribability is worth knowing about because of its crucial role in many forcing arguments in which a weakly compact cardinal is collapsed down to $\aleph_2$. Results obtained this way may less easily be ignored, since they yield phenomena at $\aleph_1$ or $\aleph_2$. For example consider the problem of Arhangel'skiĭ concerning the existence of a Lindelöf T_2 space with points G_δ of cardinality greater than continuum. Shelah [S_2], [Pr], [HJ] solved half the problem by forcing to obtain a model of GCH in which there is such a space of cardinality $\aleph_2$. He partially solved the other half [S_2] by collapsing a weakly compact and obtaining a model of CH in which $2^{\aleph_1} < \aleph_2$ and there is no such space of cardinality $\aleph_2$. The two main lemmas in his second proof are

Lemma 1. If $\langle X,\mathcal{T}\rangle$ is Lindelöf in a model of CH, and if countably closed forcing can destroy the Lindelöfness of $\langle X,\mathcal{T}\rangle$ (i.e. the topology $\mathcal{T}$ generates on X in the extension is not Lindelöf), then the simplest countably closed forcing, viz. countable partial functions from ω_1 to ω_1, destroys the Lindelöfness of X.

Lemma 2. Assume CH. Let $X = \{x_\alpha: \alpha < \omega_2>\}$ be a topological space in which points are G_δ's. Then there is a closed unbounded $C \subseteq \omega_2$ such that if $\beta \in C$ has uncountable cofinality, then the subspace $\{x_\alpha: \alpha < \beta\}$ is not Lindelöf.

Shelah's model is obtained by Lévy-collapsing the weakly compact κ down to $\aleph_2$ and then adjoining $\aleph_3$ ($= \kappa^+$) subsets of $\aleph_1$ with countable conditions. The basic idea of the proof is to use the Π^1_1-indescribability of κ to get a $\beta \in C$ such that $\{x_\alpha: \alpha < \beta\}$ is Lindelöf if X is, contradiction. Of course it isn't so simple, since in the final model κ is no longer weakly compact. In fact one has to reflect down the sentence that says X is forced to be Lindelöf (plus assorted other necessities). To make sure this sentence is sufficiently simple, i.e. $\Pi^1_1(\kappa)$, one needs to get by with talking about adjoining subsets of ω_1 via countable conditions, rather than about arbitrary countably closed forcing. This is where the first Lemma comes in.

Shelah was led to collapse a weakly compact because of

Theorem. The cardinality of Lindelöf T_2 space with points G_δ is not weakly compact.

We shall prove this result in the section on measurable cardinals, for reasons that will become clear.

There are other useful ways of collapsing a weakly compact (or other large cardinal) besides the straightforward Lévy collapse. Silver [Mi] and Baumgartner and Laver [BL] each add weakly compact many reals to obtain a model in which $\aleph_2$ has the tree property. The crucial property of both forcing extensions is that if f is a function with domain an ordinal α of uncountable cofinality, and if a condition p for each $\beta < \alpha$ forces $f|\beta$ to be in the ground model, then p forces f to be there. Here's a sketch of how this kills $\aleph_2$-Aronszajn trees. Suppose there were an $\aleph_2$-Aronszajn tree T in the final extension. The statement that this is forced is Π^1_1 as before and so by the weak compactness of the cardinal which becomes $\aleph_2$, this statement gets reflected down to some ordinal δ. So the restriction of the tree to δ is forced (via the δ'th partial order) to have no branch of length δ. Now T really does have a branch of length δ since its height is $\aleph_2$. To obtain a contradiction, it suffices to show that that branch is forced to appear via the δ'th partial order. But all the restrictions of that branch appear in the tree restricted to

δ, so the whole branch does.

The restriction property of these models should have applications in topology but I have so far not found any.

A final note about weak compactness that should be mentioned is Jensen's deep result [Je] that V = L implies a cardinal κ is weakly compact if and only if there are no κ-Souslin trees.

3. *Ramsey cardinals, ineffable cardinals*

Our next large cardinal is defined by a partition relation which is a natural strengthening of weak compactness. (Recall from e.g. [Ku_2] that $\kappa \to (\kappa)^n_\lambda$ all $n < \omega$, $\lambda < \kappa$ if and only if $\kappa \to (\kappa)^2_2$.)

Definition. κ is *Ramsey* if $\kappa \to (\kappa)^{<\omega}_2$, i.e. whenever $f: [\kappa]^{<\omega} \to 2$, there is an $X \subseteq \kappa$ of size κ such that for every n, $[X]^n$ is homogeneous for f.

In other words, colour the finite subsets of κ with two colours. Then there is a large subset X of κ such that for each n, all its n-element subsets get coloured the same way. This does *not* mean that e.g. the one element subsets are coloured the same as the pairs.

Since topologists often deal with collections of finite sets (think e.g. of paracompactness), one would expect Ramsey cardinals to have many topological applications. I have only seen one paper make use of them however--an article in *Soviet Mathematics (Doklady)* which has exasperatingly disappeared from my files.

We have mentioned previously that the existence of an inaccessible is equiconsistent with the negation of Kurepa's

Hypothesis, viz. there are no Kurepa trees. For a regular uncountable κ, let KH_κ be the assertion that there is an $F \subseteq P(\kappa)$ with $|F| \geq \kappa^+$ such that $|\{X \cap \alpha: X \in F\}| \leq |\alpha|$ for every infinite $\alpha < \kappa$. (The reader is invited to find an equivalent tree version.) Jensen proved that KH_κ holds under $V = L$ except for certain large cardinals he called *ineffable*. More precisely, if $V = L$ and κ is not ineffable, KH_κ holds; if κ is ineffable, KH_κ fails without additional hypotheses. See Devlin [De_2] for proofs. We give two equivalent combinatorial definitions for ineffability and leave it as an exercise for the reader to apply them in topology.

Definition. A regular uncountable cardinal κ is *ineffable* if either of the following (equivalent) conditions holds:

a) for any sequence $\{S_\alpha: \alpha < \kappa\}$ with $S_\alpha \subseteq \alpha$ for every $\alpha < \kappa$, there is an $S \subseteq \kappa$ such that $\{\alpha < \kappa: S_\alpha = S \cap \alpha\}$ is stationary in κ,

b) whenever $f: [\kappa]^2 \to 2$, there is a stationary $X \subseteq \kappa$ which is homogeneous for f.

By b) ineffable cardinals are weakly compact. It can be shown that measurable cardinals (see next section) are ineffable (and Ramsey as well). Ineffable cardinals are Π^1_2-indescribable ([JK] or see [KM]) so our collectionwise Hausdorff result of the previous section applies. Although $\diamondsuit^+_\kappa$ does not hold for ineffable κ, $\diamondsuit_\kappa$ does [JK]. One might therefore in view of [F_1] conjecture that for ineffable κ normal spaces of character $\leq\kappa$ which are $<\kappa$-collectionwise Hausdorff are κ-collectionwise Hausdorff, which would make our indescribability result less interesting. However this is not known because it is open whether $\diamondsuit_\kappa$ for stationary systems or indeed even $\diamondsuit_\kappa$ for stationary sets of ordinals of cofinality ω holds.

The reader unfamiliar with collectionwise Hausdorff results might wonder why we didn't drop the normality to obtain a Π_1^1 result. The reason is of course that there are spaces of character $\leq\lambda$ which are not λ-collectionwise Hausdorff, so that the hypothesis would have been vacuous.

4. *Measurable cardinals*

The large cardinals most familiar to topologists are the *measurable* cardinals.

Definition. A cardinal κ is *measurable* if there is a two-valued κ-additive measure defined on all its subsets.

Measurable cardinals have been somewhat of an annoyance to topologists working in rings of continuous functions. They would like to assert that every discrete space is realcompact, but one of measurable cardinality trivially isn't since it bears a countably complete non-principal ultrafilter of functionally closed sets. One can of course assume V = L from which it follows that there are no measurable cardinals, or work in the model of set theory V_κ, where κ is the least measurable cardinal. That model has no measurable cardinals. However perhaps a change in attitude is called for--think of that discrete space as interesting pathology rather than as an exception to a theorem.

The study of types of ultrafilters on ω has been extensive, see e.g. [CN]. Similar study can be made of types of ultrafilters on a measurable cardinal, see e.g. [K]. The topology involved is minimal. Countably complete non-principal ultrafilters are a good source for counterexamples in topology; it would be difficult to catalog all their uses since they are scattered thinly in the literature. I do wish to mention one however because it is probably not known in the West. Malyhin [M1] proves that if there is a measurable cardinal, then there is a Baire T_1 space X which is *irresolvable*, i.e. there do not exist disjoint dense subsets of X. Presumably because he is involved in demonstrating various related results, his proof is quite lengthy. We shall give a trivial proof of which he surely must have been aware. Let $\mathcal{U}$ be a countably complete non-principal ultrafilter on κ. Then $\mathcal{U} \cup \{0\}$ is a T_1 topology on κ. Suppose D is dense in this topology. Then it meets each member of $\mathcal{U}$ and hence is in $\mathcal{U}$. It follows that the intersection of countably many dense sets is dense, so the space is Baire and irresolvable. There are other relationships between Baire irresolvable spaces and large cardinals--see [T_2].

Another easy use of countably complete ultrafilters concerns the Lindelöf space problem mentioned earlier. The following result appears in Price's writeup [Pr] of Shelah's work.

Theorem. Let κ be the least measurable cardinal. Every Lindelöf space with points G_δ has cardinality $<\kappa$.

Price requires the space be T_2 but that is not necessary. For each point x in the Lindelöf space X, let $\{U(x,n): n < \omega\}$ be such that $\{x\} = \bigcap\{U(x,n): n < \omega\}$. Suppose there is a $Y \subseteq X$ of cardinality κ. Let $\mathcal{U}$ be a countably complete non-principal ultrafilter on Y. Then for each $x \in X$, there is an $n(x)$ such that $U(x,n(x)) \cap Y \notin \mathcal{U}$. The open cover $\{U(x,n(x)): x \in X\}$ has a countable subcover $\{U(x_k,n(x_k)): k < \omega\}$.

Then $Y = \bigcup\{U(x_k, n(x_k)) \cap Y: k < \omega\}$. By countable completeness, some $U(x_k, n(x_k)) \cap Y \in \mathcal{U}$, contradiction. The same proof establishes that Lindelöf spaces with points G_δ cannot have weakly compact cardinality, using variant c) of weak compactness. Price claims in fact that such (T_2) spaces cannot have cardinality $\geq$ the first weakly compact and says this can be proved using variant c) in place of measurability, but I do not see how to do this, and suspect it is not true in view of Shelah's inability to rule out cardinals $>\aleph_2$ in his collapse discussed earlier. It is however reasonable to suppose that by collapsing a suitably large cardinal, Shelah's result can be improved to force all Lindelöf spaces with points G_δ to have cardinality $\leq\aleph_1$.

The translated survey [AC] investigates a number of topological problems in which large cardinals naturally arise. Its extensive bibliography is particularly useful for its many Soviet references. Among other noteworthy features is a proof via infinitary languages of Arhangel'skiĭ's theorem [A] on the cardinality of first countable compact Hausdorff spaces. However the paper contains a number of errors. In particular, the reader should be warned that it is still very much open (through no fault of the authors) whether "uncountable Martin's Axiom" is consistent, even assuming large cardinals. One natural problem (which dates back to [Mz]) that the authors consider is the question of when sequentially continuous maps from product spaces are continuous. A typical result is the following

Theorem. κ is measurable if and only if there is a discontinuous sequentially continuous map $f: \{0,1\}^\kappa \to \{0,1\}$.

Another question--which is perhaps more interesting for the large cardinal theory used to answer it rather than intrinsically--is whether the κ-compactness of N^κ is equivalent to κ being closed embedded in N^κ, where N is the countable discrete space. It turns out that the answer depends on such questions as whether or not the first measurable cardinal is the first *strongly compact* (see definition in next section) one. That question has different answers in different models of set theory.

Mŕowka has studied similar embedding questions extensively, distinguishing various classes of cardinals and formulating equivalent conditions in terms of matrices of sets, existence of measures, etc. Among his many papers, perhaps the relatively discursive [Mr] is a good place to begin.

There is a combinatorial assertion about ω_1 which is equiconsistent with the existence of a measurable cardinal. Let I be an ideal on ω_1 which is *non-trivial*, by which I mean $\omega_1 \notin I$, each singleton is in I, and I is closed under unions of size $<\aleph_1$. Let $S \subseteq \omega_1$ not be in I. $\mathcal{W}$ is an *I-partition* of S if $\mathcal{W} \subseteq P(S) - I$ is maximal with respect to: $Y \cap Z \in I$ for distinct $Y, Z \in \mathcal{W}$. I is *precipitous* if whenever $\{\mathcal{W}_n\}_{n<\omega}$ is a sequence of I-partitions of some $S \notin I$ such that $\mathcal{W}_{n+1}$ refines $\mathcal{W}_n$ for all n, then there exist $X_n \in \mathcal{W}_n$ for all n such that $X_n \supseteq X_{n+1}$ and $\bigcap_{n<\omega} X_n \neq \emptyset$.

Theorem [JMMP]. The existence of a measurable cardinal is equiconsistent with the proposition that the non-stationary ideal on ω_1 is precipitous.

Precipitous ideals do not as yet have applications to topology, but I expect them to, which is why I mention them.

There is quite a considerable interplay between large cardinal theory and the study of combinatorial properties of ideals such as saturation and precipitousness. [J] is a good source as is [BTW], wherein the ideals are the primary object of study. Ideal theory may be expected to have an increasing number of applications in topology--see e.g. [Ta] and [BGKT].

There are other large cardinals whose definitions involve measures. One that has been around for a long time is the real-valued measurable cardinal. A cardinal κ is *real-valued measurable* if there is a κ-additive *real-valued* measure defined on all the subsets of κ such that μ has no atoms, i.e. any subset of κ of positive measure can be split into two disjoint sets of positive measure. In contrast to the other large cardinals mentioned so far, real-valued measurable cardinals are $\leq 2^{\aleph_0}$ and hence *not* inaccessible, although they are weakly inaccessible. Solovay [So] proved however that real-valued measurable cardinals are equiconsistent with measurable cardinals. His ideas lead to Nyikos' Product Measure Extension Axiom, of which more later.

Van Douwen [vD] calls a cardinal κ *weakly Ulam-measurable* if there is a countable family $\mathcal{U}$ of non-principal ultrafilters on κ such that any countable subcollection of $\cap\mathcal{U}$ has non-empty intersection. Baumgartner has proven that in contrast to real-valued measurable cardinals, the first non-measurable wUm must be less than $2^{\aleph_0}$. They are however weakly inaccessible. WUmen arose in connection with van Douwen's investigation of a property intermediate between realcompactness and Lindelöfness. Call a space X *strongly realcompact* if for each separable compact subset S of $\beta X-X$ there is a G_δ-set $B \subseteq \beta X$ such that $S \subseteq B \subseteq \beta X-X$. Then the discrete space κ is strongly realcompact if and only if κ is not a wUm. Van Douwen also proves (and this is the point that interests him) that if κ is strongly realcompact, then for $Y \subseteq \beta\kappa - \kappa$, if Y is countable and $\overline{Y}$ is compact, then $\overline{Y} = \beta Y$.

A point I wish to emphasize concerning van Douwen's work is that "large" cardinals can be "small" enough so that they can't be safely ignored by topologists.

5. *Strongly compact cardinals, supercompact cardinals*

Strongly compact cardinals burst upon the topological scene in 1977 when the results of Kunen and Nyikos [Ku_5], [N] established that if the existence of such cardinals is consistent, so is the normal Moore space conjecture. As is usual in large cardinal theory (and general topology!) there are a number of equivalent definitions. For example, a cardinal κ is *strongly compact* if either of

a) every κ-complete filter can be extended to a κ-complete ultrafilter,

b) every product of κ-compact spaces is κ-compact,

We see immediately from a) that strongly compact cardinals are measurable, and from b) that strongly compact cardinals are weakly compact. (In fact, measurable cardinals are weakly compact--use variant c) of weak compactness.) The general theme with strongly compact cardinals is that they give the same sorts of results as weakly compact or measurable ones, but with stronger conclusions. Again, they can be used directly or collapsed. A typical example of an extension of a weakly compact result is Fleissner's [F_3].

Theorem. Let X be κ-collectionwise Hausdorff for all $\lambda < \kappa$, κ strongly compact, and suppose each point of X has character $<\kappa$. Then X is collectionwise Hausdorff.

Fleissner's proof proceeds as follows. Let Y be a discrete collection of points in X. For each $y \in Y$, let $\{B(y,\nu): \nu < \chi(y)\}$ be a neighborhood base at y. For any set Z and any cardinal μ, $P_\mu(Z)$ is defined to be the collection of subsets of Z of cardinality $<\mu$. For each $A \in P_\kappa(Y)$, let S_A be a separation of A by basic open sets. For each $y \in Y$, let $P_y = \{A \in P_\kappa(Y): y \in A\}$. $\{P_y: y \in Y\} \cup \{\{A: B(y,\nu) \in S_A\}: y \in Y, \nu < \chi(y)\}$ generates a κ-*complete field* $\mathcal{A}$ on $P_\kappa(Y)$ (i.e. $\mathcal{A}$ is a subfamily of $P_\kappa(Y)$ closed under complementation and unions of cardinality $<\kappa$). $\{P_y: y \in Y\}$ generates a κ-complete filter $\mathcal{F}$ on $\mathcal{A}$. Let $\mathcal{U}$ be a κ-complete ultrafilter extending $\mathcal{F}$. For each y there is $\nu(y)$ such that $\{A: B(y,\nu(y)) \in S_A\} \in \mathcal{U}$ because $P_y \in \mathcal{U}$ and $\mathcal{U}$ is κ-complete. $\{B(y,\nu(y)): y \in Y\}$ is the required separation.

An example of how strong compactness enables the strengthening of a result obtained from measurability is Kunen's proof of the consistency of Nyikos' Product Measure Extension Axiom from the assumption of a strongly compact cardinal. PMEA asserts that for each cardinal λ there is a $2^{\aleph_0}$-additive real-valued measure on $P(2^\lambda)$ extending the usual product measure. Nyikos proved that PMEA implies every normal Moore space is metrizable. It is easy to see that PMEA implies $2^{\aleph_0}$ is real-valued measurable. Solovay [So] (see the texts) had observed that by adjoining κ many random reals (i.e. forcing with the measure algebra of 2^κ), one could make $2^{\aleph_0}$ real-valued measurable, assuming κ was measurable. In parallel, Kunen [Ku_5] demonstrates that if strongly compact many random reals are adjoined, PMEA is obtained. Thus measurability gives a result for $\aleph_0$ and strong compactness extends it to all larger cardinals.

I do not know of any topological applications of the usual Lévy collapse of a strongly compact cardinal. Fleissner [F_3] conjectures that there are "$\leq\aleph_1$-collectionwise Hausdorff implies collectionwise Hausdorff" results obtainable in this fashion. He and Shelah [S_2] have gotten some such results by Lévy-collapsing an even larger cardinal--a *supercompact* one.

Definition. An ultrafilter $\mathcal{U}$ *over* $P_\kappa(\lambda)$ (i.e. the elements of $\mathcal{U}$ are subsets of $P_\kappa(\lambda)$) is *normal* if it satisfies

(i) $\mathcal{U}$ is a κ-complete ultrafilter,

(ii) for every $\alpha \in \lambda$, $P_\alpha = \{X: \alpha \in X\} \in \mathcal{U}$,

(iii) if f is a function defined on a set in $\mathcal{U}$ such that $\{X: f(X) \in X\} \in \mathcal{U}$, then there is an $\alpha \in \lambda$ such that $\{X: f(X) = \alpha\} \in \mathcal{U}$.

κ is *supercompact* if for every $\lambda \geq \kappa$, there is a normal ultrafilter over $P_\kappa(\lambda)$.

Supercompactness, like strong compactness, is used to prove results of the general form that if a proposition holds up to κ, then it holds for all cardinals $\geq\kappa$. Again, it is often useful to collapse a supercompact. For example Shelah [S_1] proves that in the model obtained by collapsing a supercompact to $\aleph_2$, if X is a locally countable space which is $<\aleph_2$-collectionwise Hausdorff, then it is collectionwise Hausdorff. Fleissner [F_4] obtains the consistency of a number of topological results by Lévy-collapsing a supercompact, in particular that "point finite analytic additive families are σ discretely decomposible" (sic). It would take us too far afield to give the definitions, let alone the proof; suffice it to say that Fleissner obtains a model of "SC(ω_2)" which "roughly says that ω_2 retains a few properties of a supercompact cardinal".

When one is actually trying to prove something by using supercompactness, one unfortunately is unlikely to get by with the combinatorial definition we have given. The "real definition" follows.

Definition. κ is *supercompact* if for each $\lambda \geq \kappa$ there is an elementary embedding $j_\lambda: V \to M$ such that

a) j_λ has critical point κ,

b) ${}^\lambda M \subseteq M$.

There are many concepts involved in this definition which are unfamiliar to topologists. By convention M is taken to be a subclass of the class V of all sets which is *transitive* (members of members of M are members of M) and contains all the ordinals. j_λ is a (by convention non-identical) function which preserves all relations definable in the language of set theory, i.e. whatever can be said that is true about sets $x_1,\ldots,x_n$ is true in M about $j_\lambda(x_1),\ldots,j_\lambda(x_n)$. To say that j_λ has *critical point* κ means that κ is the least ordinal such that $j_\lambda(\alpha) > \alpha$. ${}^\lambda M$ is just the collection of all λ-sequences of elements of M. The power of supercompactness comes from the interplay of κ and $j_\lambda(\kappa)$. It would take up too much space to sketch a supercompact collapse argument here. The interested reader is advised to consult [SRK] or the texts first, and then [S_1] and [F_4].

Supercompact cardinals satisfy a strong reflection principle which makes understandable the kinds of results we have mentioned. Define Σ_n (respectively, Π_n) as we defined Σ_n^m before, but without any bound on the quantified variables. Say that Σ_n (respectively, Π_n) *relativizes down to* V_κ if whenever $a \in V_\kappa$ and $\Phi(a)$ holds, where Φ is Σ_n, then $\Phi(a)$ holds in V_κ. Then it can be shown that if κ is supercompact, Σ_2 relativizes down to V_κ. One may of course consider cardinals κ for which $\Sigma_n(\Pi_n)$ relativizes down to V_κ.

Some exciting developments have recently taken place in the theory of iterated forcing, mainly due to Baumgartner and Shelah. Both have come up with iteration axioms that combine Martin's Axiom ideas with large cardinal collapse techniques, such as in the MA plus not CH plus no Canadian trees result mentioned earlier. The strongest such axioms thus far have recently been conceived by Baumgartner using supercompact collapse. Those topologists who hope they can get away with axioms and thereby avoid forcing will be disappointed to learn that the statements of many of these axioms do involve forcing.

6. *Towards inconsistency*

Our selection of large cardinals has included only those that have had or should have topological applications. The largest cardinals which set theorists have so far considered which seem likely to have such applications are the *huge* cardinals. We omit the (model-theoretic) definition--see the texts--for the main interest is in the consistency results these cardinals enable one to demonstrate for small cardinals. These have been of two sorts. First, by assuming the consistency of a huge cardinal with a supercompact below it, Magidor [M] established the consistency of GCH holding up to $\aleph_\omega$ but failing at $\aleph_\omega$. His methods can probably be used to establish the consistency of topological statements of a similar sort--see for example, the problem stated in [CEG] concerning the density of λ-box products. Incidentally, the failure of 2^{κ} for singular κ to be as small as it can be, i.e. the least cardinal $\geq \sup\{2^{\lambda}: \lambda < \kappa\}$ and of cofinality $>\kappa$, implies the consistency of measurable cardinals [DJ]. See [KM] for a discussion of related results, especially *Jensen's Covering Theorem* which may be expected to have topological applications. We omit discussion of it since it assumes certain large cardinals do *not* exist, and hence belongs in a different survey.

In the other (prior) application of hugeness Kunen [Ku_2] proved that if it is consistent that there is a huge cardinal, then it is consistent that there be an $\aleph_2$-saturated ideal over $\aleph_1$. (An ideal over κ is λ-*saturated* if given any λ subsets of κ which are not in the ideal, there are two of them whose intersection is not in the ideal, i.e. there do not exist λ "I-almost disjoint" subsets of κ). Kunen's result does not as yet have topological applications, but I expect it to. Of great interest is the possibility of the non-stationary ideal over $\aleph_1$ being $\aleph_2$-saturated. A recent result of van Wesep [vW] establishes the consistency of this proposition but uses rather outlandish hypotheses to which we now turn.

Infinitary games have been used in topology since the 1930's, most recently by Telgársky, Galvin, and Gruenhage. It is not our intention to survey this area, but it appears promising. Set-theorists have intensively studied such games. For a typical example, let player I pick a natural number a_0, player II pick an a_1, I pick an a_2, etc. Suppose a set $B \subseteq {}^{\omega}\omega$ is given in advance. Does some player have a *winning strategy* for the B-*game*, i.e. can he ensure by his choices (made in full knowledge of all previous choices) that the sequence $\{a_n\}_{n<\omega}$ is in B, regardless of what the other player does? If so, the game is said to be *determined*. The answer of course depends on the complexity of B. A deep result of Martin [Mar] is that all Borel games on ${}^{\omega}\omega$ (or the real line) are determined. A typical Axiom of Choice diagonalization enables the construction of non-determined games on ${}^{\omega}\omega$ [GS]. The radical solution of Mycielski-Steinhaus [MS] is to assume as an axiom, the *Axiom of Determinateness* that all games on ${}^{\omega}\omega$ are determined. Set theorists have investigated AD for a number of years, deriving all sorts of bizarre consequences. Those like the author who have no intention of even considering the abandonment of the Axiom of Choice have ignored these efforts. However a number of set theorists--most notably van Wesep--have assumed the consistency of ZF with AD (and some other technical hypotheses) and then gone on to prove the consistency *with ZFC* of assorted propositions. It turns out that the assumption of the consistency of AD in this context acts like a very large cardinal assumption, e.g. in the improvement of Kunen's result referred to above. Indeed another improvement actually has topological consequences: H. Woodin has recently extended van Wesep's work to obtain an ideal I on ω_1 such that the Boolean algebra $P(\omega_1)/I$ has a dense set of power $\aleph_1$. Such an ideal can be used to construct a Baire irresolvable space of power $\aleph_1$ [T_2].

We have now come sufficiently close to inconsistency that the author and a fortiori the reader can doubtless withstand no more. We close with the usual hope of propagandistic survey writers: may this survey rapidly become obsolete.

Appendix: Guide to the Literature

The first place to look when attempting to learn about large cardinals is [KM]. They provide a comprehensive up-to-date overview of the entire subject. The proofs are rather sketchy, so after giving the paper a once-over, for intensive study one should turn to the appropriate sections in [J]. If you get stuck on some of the proofs or exercises, retreat to (the somewhat dated) [Dr] for all the details. Other useful sources are [Bo], [CK], [De_i], $i \leq 3$. [SKR] is the basic source for supercompactness and elementary embeddings. [CN] will satisfy any ultrafilter cravings, and also has a chapter on some of the older large cardinals and their topological equivalents.

Of course, before starting large cardinals one should make sure one's set theory is up to snuff. For an introduction to combinatorics, I recommend [Ku_2]. [W] carries the subject further. The all-purpose set theory text and reference is [J]. For forcing I prefer Kunen's lucidly written and beautifully organized text [Ku_6] which should be available next year. This book will become *the* graduate set theory text. Highly recommended for novices is Baumgartner's noble attempt [Ba_2] to reduce forcing to combinatorics by an axiomatic approach. Finally, if one knows absolutely no logic, Barwise's introductory article [B] in the *Handbook of Mathematical Logic* is a good place to start.

REFERENCES

[A] A.V. Arhangel'skiĭ, *The power of bicompacta with first axiom of countability*, Soviet Math. Dokl. 10 (1969), 951-955.

[AC] M. Ya. Antonovskii and D.V. Chudnovskii, *Some questions of general topology and Tikhonov semifields, II*, Russian Math. Surveys 31 (1976), 69-128.

[B] J. Barwise, *An introduction to first-order logic*, 5-46 in *Handbook of Mathematical Logic*, ed. J. Barwise, North-Holland, Amsterdam, 1977.

[Ba_1] J.E. Baumgartner, *A new class of order types*, Ann. Math. Logic 9 (1976), 187-222.

[Ba_2] ________, *Independence proofs and combinatorics*, Proc. Symp. Pure Math. 34 (1979), 35-46.

[Ba_3] ________, *Iterated forcing*, lecture notes from Cambridge Summer School, 1978.

[BGKT] S. Broverman, J. Ginsburg, K. Kunen, and F.D. Tall, *Topologies determined by σ-ideals on* ω_1, Can. J. Math. 30 (1978), 1306-1312.

[BL] J.E. Baumgartner and R. Laver, *Iterated perfect-set forcing*, preprint.

[Bo] W. Boos, *Lectures on large cardinal axioms*, 25-88 in *Logic Conference, Kiel 1974*, Lect. Notes Math. 499, Springer-Verlag, Berlin, 1975.

[BTW] J.E. Baumgartner, A.D. Taylor and S. Wagon, *Structural properties of ideals*, Dissert. Math., to appear.

[CEG] F.S. Cater, P. Erdös and F. Galvin, *On the density of λ-box products*, Gen. Top. Appl. 9 (1978), 307-312.

[CK] C.C. Chang and H.J. Keisler, *Model Theory*, North-Holland, Amsterdam, 1973.

[CN] W.W. Comfort and S. Negrepontis, *The Theory of Ultrafilters*, Springer-Verlag, Berlin, 1974.

[D] P. Davies, *Small Baire spaces and σ-dense posets*, Thesis, University of Toronto, in preparation, 1979.

[De_1] K.J. Devlin, *Some weak versions of large cardinal axioms*, Ann. Math. Logic 5 (1973), 291-325.

[De_2] ________, *Aspects of Constructibility*, Lect. Notes Math. 354, Springer-Verlag, Berlin, 1973.

[De_3] ________, *Indescribability properties and small large cardinals*, 89-114 in *Logic Conference, Kiel, 1974*, Lect. Notes Math. 499, Springer-Verlag, Berlin, 1975.

[DJ] T. Dodd and R.B. Jensen, *The core model*, circulated notes.

[Dr] F.R. Drake, *Set Theory*, North-Holland, Amsterdam, 1974.

[vD] E.K. van Douwen, *Strongly realcompact spaces and weakly Ulam-measurable cardinals*, in preparation.

[F_1] W.G. Fleissner, *Normal Moore spaces in the constructible universe*, Proc. Amer. Math. Soc. 46 (1974), 294-298.

[F_2] ________, *The character of* ω_1 *in first countable spaces*, Proc. Amer. Math. Soc. 62 (1977), 149-155.

[F_3] ________, *On* λ *collection Hausdorff spaces*, Top. Proc. 2 (1977), 445-456.

[F_4] ________, *An axiom for nonseparable Borel theory*, Trans. Amer. Math. Soc., to appear.

[GS] D. Gale and F.M. Stewart, *Infinite games with perfect information*, 245-266 in *Contributions to the Theory of Games*, ed. H.W. Kuhn and A.W. Tucker, Ann. Math. Stud. 28, v. 3, Princeton, 1953.

[HJ] A. Hajnal and I. Juhász, *Lindelöf spaces á la Shelah*, preprint.

[J] T.J. Jech, *Set Theory*, Academic Press, New York, 1978.

[Je] R.B. Jensen, *The fine structure of the constructible hierarchy*, Ann. Math. Logic 4 (1972), 229-308.

[JK] R.B. Jensen and K. Kunen, *Some combinatorial properties of L and V*, mimeographed manuscript.

[JMMP] T.J. Jech, M. Magidor, W. Mitchell, and K. Prikry, *On precipitous ideals*, to appear.

[Ju] I. Juhász, *Cardinal Functions in Topology*, Math. Centre, Amsterdam, 1971.

[K] A. Kanamori, *Ultrafilters over a measurable cardinal*, Annals Math. Logic 11 (1977), 315-356.

[KM] A. Kanamori and M. Magidor, *The evolution of large cardinal axioms in set theory*, 99-276 in *Higher Set Theory*, Lect. Notes Math. 669, Springer-Verlag, Berlin, 1978.

[KP] K. Kunen and L. Parsons, *Projective covers of ordinal sub-spaces*, Topological Proceedings 3 (1978), 407-428.

[KT] H.J. Keisler and A. Tarski, *From accessible to inaccessible cardinals*, Fund. Math. 13 (1964), 225-308.

[Ku_1] K. Kunen, *Elementary embeddings and infinitary combinatorics*, J. Symb. Logic 36 (1971), 407-413.

[Ku_2] ________, *Combinatorics*, 371-402 in *Handbook of Mathematical Logic*, ed. J. Barwise, North-Holland, Amsterdam, 1977.

[Ku_3] K. Kunen, *Saturated ideals*, J. Symb. Logic 43 (1978), 65-76.

[Ku_4] ________, *Cardinalities of closed subspaces of* 2^{ω_1}, handwritten manuscript.

[Ku_5] ________, *Measures on* 2^λ, handwritten manuscript.

[Ku_6] ________, *Set Theory*, North-Holland, Amsterdam, to appear.

[M] M. Magidor, *On the singular cardinals problem II*, Ann. Math., to appear.

[Ma] V.I. Malyhin, *Products of ultrafilters and irresolvable spaces*, Math. USSR Sb. 19 (1973), 105-116.

[Mar] D.A. Martin, *Borel determinacy*, Ann. Math. 102 (1975), 363-371.

[Mi] W. Mitchell, *Aronszajn trees and the independence of the transfer property*, Ann. Math. Logic 5 (1972), 21-46.

[Mr] S. Mŕowka, *Recent results on E-compact spaces and structures of continuous functions*, 168-221 in *Proc. Univ. of Oklahoma Top. Conf.*, Norman, Oklahoma, 1972.

[MS] J. Mycielski and H. Steinhaus, *On a mathematical axiom contradicting the axiom of choice*, Bull. Acad. Polon. Sci. Sér. Sci. Math. Ast. Phys. 10 (1962), 67-71.

[Mz] S. Mazur, *On continuous mappings of Cartesian products*, Fund. Math. 39 (1952), 229-238.

[N] P. Nyikos, *A provisional solution to the normal Moore space problem*, Proc. Amer. Math. Soc., to appear.

[P] I.I. Parovičenko, *The branching hypothesis and the correlation between local weight and power to topological spaces*, Soviet Math. Dokl. 8 (1967), 589-591.

[Pr] R. Price, *Concerning Arhangel'skii's problem*, mimeographed handwritten manuscript.

[S_1] S. Shelah, *Remarks on λ-collectionwise Hausdorff spaces*, Top. Proc. 2 (1977), 583-592.

[S_2] ________, *On some problems in general topology*, preprint.

[Si] J.H. Silver, *The independence of Kurepa's conjecture and two-cardinal conjectures in model theory*, Proc. Symp. Pure Math. 13(1) (1971), 383-390.

[So] R.M. Solovay, *Real-valued measurable cardinals*, Proc. Symp. Pure Math. 13(1) (1971), 397-428.

[SRK] R.M. Solovay, W.N. Reinhardt, and A. Kanamori, *Strong axioms of infinity and elementary embeddings*, Ann. Math. Logic 13 (1978), 73-116.

[T_1] F.D. Tall, *Set-theoretic consistency results and topological theorems concerning the normal Moore space conjecture and related problems*, Thesis, University of Wisconsin, Madison, 1969; Dissert. Math. 148 (1977), 1-53.

[T_2] ________, *Baire irresolvable spaces and the theory of ideals*, in preparation.

[Ta] A.D. Taylor, *Diamond principles, ideals and the normal Moore space problem*, Can. J. Math., to appear.

[vW] R. van Wesep, *The non-stationary ideal on ω_1 can be ω_2-saturated*, handwritten manuscript.

[W] N.H. Williams, *Combinatorial Set Theory*, North-Holland, Amsterdam, 1977.

WEAKLY COMPACT SUBSETS OF BANACH SPACES

Michael L. Wage
Institute for Medicine and Mathematics, Ohio University and Harvard Medical School

The fields of analysis, general topology, and set theory have another happy reunion in the study of weakly compact subsets of Banach spaces. This survey covers some of the basic results and recent advances relating to weakly compact sets from a set theoretic topologist's point of view.

Lindenstrauss [L] gave the name Eberlein Compact (abreviated EC) to those topological spaces that are homeomorphic to a weakly compact subset of a Banach space. The basic results on EC's are given in section 1. Section 2 covers the Banach space and measure theoretic aspects of EC's, while Section 3 is concerned with topological properties. Definitions not given below can be found in the standard references [DS] and [E].

Definitions: All spaces are assumed Hausdorff. A family of sets, G, called *point finite* if each point in $\cup G$ is contained in only finitely many elements of G. If G is the union of countably many points finite families, then G is called σ-point finite. If A is a subset of a *density* of a space X is the least cardinal, κ, such that X contains a dense subset of cardinality κ. The *cellularity* of X is the supremum of all cardinals κ such that X contains κ disjoint nonempty open sets. A space with countable cellularity is said to be ccc. The *weight* of X is the least

ISBN: 0-12-584960-5

cardinality of a basis. The minimal cardinal κ such that

$$\forall A \subset X\ \forall x \in \overline{A}\ \exists B \subset A\ (x \in \overline{B} \wedge |B| \leq \kappa)$$

is called the *tightness* of X. The *character* of X is the least cardinal, κ, such that each point in X has a neighborhood base of cardinality at most κ. Note that in a compact space each G_δ point has a countable neighborhood base. The notation $|\cdot|$ stands for either the cardinality or absolute value and will be made clear by context. Many other definitions are given within the text below.

1. *What is an EC?*

The EC spaces are topological spaces that are homeomorphic to weakly compact subsets of Banach spaces. This definition doesn't give me much of an intuitive feel for EC's. The important theorems and examples given in this section will make the basic structure and properties of these spaces clear.

Our first theorem gives three useful characterizations of EC's. The space of all real valued functions vanishing at infinity on a set Γ is denoted $c_0(\Gamma)$, i.e. $c_0(\Gamma) = \{f: \Gamma \to \mathbb{R} \mid \forall \varepsilon > 0\ |f(\gamma)| > \varepsilon$ for only finitely many $\gamma \in \Gamma\}$. The weak topology and the topology of pointwise convergence are denoted by T_w and T_p respectively.

Theorem 1.1. The following are equivalent for a compact space K.

(1) K is an EC.

(2) K is homeomorphic to a subset of $(c_0(\Gamma), T_w)$.

(3) K is homeomorphic to a subset of $(c_0(\Gamma), T_p)$.

(4) K has a σ-point finite separating family of open F_σ subsets.

The characterization (2) is a very deep result of Amir and Lindenstrauss [AL, p. 35] and shows that every EC can be found as a weakly compact subset from the restricted class of Banach spaces of the form $c_0(\Gamma)$. This restriction is of tremendous value since the spaces of the form $c_0(\Gamma)$ are much easier to work with than arbitrary Banach spaces (see section 2). S.P. Gulko recently obtained a topological proof of this result. Characterization (3) is a reformulation of (2) that avoids the weak topology (see e.g. [MR]). Rosenthal's characterization (4) does not mention a linear structure at all and is tailor made for the general topologist, see [Ro]. Recall that a family, G, of subsets of K is called *separating* if for each $k_1, k_2 \in K$ there exists a $G \in G$ such that G contains exactly one of k_1 and k_2. Thus the term separation is used in the T_0 sense. Separation in the T_1 sense is sometimes called *strong separation*, i.e. G strongly separates the points of K if for each $k_1, k_2 \in K$ there exists a $G \in G$ such that $k_1 \in G$ but $k_2 \notin G$. Since a compact space is metrizable if and only if it has a strongly separating σ-point finite family of open F_σ subsets (see e.g. [Ro]), the term "strongly" severs the non-metrizable from the metrizable EC's.

Theorem 1.2. EC spaces are preserved under

a) closed subsets,

b) countable products, and

c) continuous images.

Both (a) and (b) are easy consequences of the characterizations given in Theorem 1.1 above. The proof of (c) is difficult and is the main result in [BRW]. Two slightly different discussions of (c) can be found in [MR] and [G].

The following theorem gives basic topological properties of EC's to shed more light on their structure.

Theorem 1.3. If K is an EC, then

i) if $L \subset K$ and $k \in \overline{L}$, then there is a sequence in L that converges to k,

ii) if L is a closed subset of K then L is metric if and only if L is separable, if and only if L is ccc, and

iii) K has a dense G_δ metrizable subset of G_δ points.

Part (i) of the above theorem is a famous result of Eberlein and Smulian and is the origin of the name "Eberlein compact." With the aid of the Amir-Lindenstrauss result (Theorem 1.1(2)) it is easy to prove (i): Let K be an EC and suppose $L \subset K$ and $k \in \overline{L}$. Without loss of generality we can assume $K \subset c_0(\Gamma)$ for some Γ. We want to find a sequence, k_n, in L that converges to k. Choose $k_1 \in L$ arbitrarily and proceed recursively. Having chosen $k_1, \ldots, k_{n-1}$, we choose k_n as any element of L such that $|k_n(\gamma) - k(\gamma)| < 2^{-n}$ for each γ in the set $\{\gamma: |k_j(\gamma) - k(\gamma)| > 2^{-n}$ for some $j < n\}$. Note that this is possible since all points of K vanish at infinity, so that the above set of coordinates is finite. Then the sequence $\{k_n\}$ is as desired. The proof of (ii) is also easy and is left to the reader (or see [Ro] and [AL]). Variations of (iii) were given in [AL, Theorem 5], [BRW,p.318], [N], and a more general result was proved by P. Kenderov of Bulgaria.

The three theorems above help us to determine whether a given compact space is an EC-Compact metric spaces and one point compactifications of discrete spaces are easily seen to be EC. (For each of these spaces one can easily construct a family satisfying Rosenthal's condition 1.1(4). In fact, such a family can easily be found for more general spaces such as the one point compactification of a locally compact metric space and the one point compactification of a disjoint union of EC's. See [L] for a different proof that these spaces are EC.) Although the countable product of EC's is an EC, Theorem 1.3 (iii) shows us that no uncountable product of nontrivial spaces is an EC (since in such a product there are no G_δ points). Uncountable ordinal spaces (e.g. $\omega_1 + 1$) are not EC since they do not satisfy (i) of Theorem 1.3. A Souslin line, βN, and the lexicographically ordered square all fail to satisfy (ii) of Theorem 1.3, and hence are not EC.

2. *Banach Spaces*

The concept of an EC space originated in the study of Banach spaces and continues to play an important role in the field. In this section we will review a few of the basic Banach space and measure space results concerning EC's and indicate how these results relate to set theoretic topology.

Definition. A Banach space is said to be *weakly compactly generated* (abbreviated WCG) if it has a weakly compact subset whose closed linear span is the whole space.

Separable Banach spaces, reflexive Banach spaces, and spaces of the form $c_0(\Gamma)$ are all WCG. $L_1(\mu)$ is WCG if and only if μ is σ-finite. WCG spaces are preserved under quotients, but not under the taking of subspaces, [R]. Both ℓ_∞ and $\ell_1(\Gamma)$ (for uncountable Γ) fail to be WCG since each weakly compact subset of these non-separable spaces is separable (see [D, p. 33]). The fact that WCG Banach spaces have many non-trivial projections on separable subspaces allows a number of results concerning separable Banach spaces to be extended to WCG spaces, see [L].

The characterization of EC's given in Theorem 1.1 (2) is actually a corollary of the following deep result of Amir and Lindenstrauss, [AL].

Theorem 2.1. Let B be WCG. Then there exists a set Γ and a bounded one-to-one linear operator from B into $c_0(\Gamma)$.

This theorem was well known for certain classes of spaces. For separable spaces the proof is trivial, and if $B = L_p(\mu)$, where μ is the Haar measure on a compact abelian group G and $1 \leq p \leq \infty$, then the Fourier transform is the desired operator into $c_0(\Gamma)$ (with Γ the dual group of G). The operator guaranteed by the theorem is generally not invertable, but it can be constructed so that its restriction to a weakly compact subset is a homeomorphism (with respect to the weak topologies), and hence the characterization in theorem 1.1 (2) follows. Since weakly compact subsets of $c_0(\Gamma)$ are much easier to work with than general weakly compact sets, the Amir-Lindenstrauss theorem has many corollaries, some of which we state below. Recall that C(K) denotes the Banach space of all continuous real valued functions on K with the sup-norm.

Corollary 2.2. Let K be a compact Hausdorff space. The following are equivalent.

a) K is EC.

b) C(K) is WCG.

c) The unit ball of [C(K)]* is EC.

The first two thirds of Corollary 2.2 immediately imply that if K is EC and H is compact Hausdorff, and if C(K) and C(H) are isomorphic Banach spaces, then H is also an EC. The next corollary, also proved in [AL], is not as immediate.

Corollary 2.3. Let X be a continuous image of a separable metric space and let K be EC. Then the space of all continuous functions from X to K is Lindelöf in the topology of pointwise convergence.

Other corollaries of the Amir-Lindenstrauss relate to convexity. For example, if B is WCG, then both B and B* are isomorphic to a strictly convex space. Also, every weakly compact convex subset of a Banach space is the closed convex hull of its exposed points.

Independence results are becoming more common in the field of Banach spaces as the questions posed become more set theoretic. In [Ro,p.97], Rosenthal used Martin's axiom plus the negation of the continuum hypothesis to prove the statement in Theorem 2.4 below, and used the continuum hypothesis to construct a counterexample.

Theorem 2.4. The following statement is consistent with and independent of the standard axioms for set theory: Each closed nonseparable linear subspace of $L^1(\mu)$ (for some probability measure μ) contains a nonseparable weakly compact subset.

Measure theory has two ties with theory of EC spaces. One tie is through the Banach spaces generated by measures, as shown by the theorem above and the next result of Dunford and

Pettis, [DS,p.508]. The other tie is through the structure of measures on EC's.

Theorem 2.5. Let μ be a measure and let T be a weakly compact operator from $L_1(\mu)$ to a Banach space X. Then T maps every weakly compact set into a norm compact set.

The support of a measure is the intersection of all closed sets whose complement have measure zero. Grothendieck used Theorem 2.5 to obtain:

Theorem 2.6. Every finite positive regular measure on an EC has separable support.

The last two results of this section were proved by W. Schachermayer in [S] and fit in well with Theorem 2.6. A Borel measure, μ, on a topological space is called a *Radon measure* if $\mu(B) = \sup\{\mu(K): K \subset B$ and K is compact$\}$ for each Borel set B. A Radon space is a topological space on which every Borel measure is a Radon measure. See [Sc] for more on Radon spaces.

Theorem 2.7. Let K be an EC whose density is less than the first measurable cardinal. Then K is a Radon space.

Corollary 2.8. Let X be an Eberlein compactifiable topological space whose density is less than the first measurable cardinal, then each Borel measure on X is τ-additive, i.e. each Borel measure μ on X satisfies $\mu(\cup_{\alpha\in A}U_\alpha) = \sup\{\mu(U_\alpha): \alpha \in A\}$ for each increasing collection of open sets $\{U_\alpha: \alpha \in A\}$.

3. *Topological Properties*

The cardinal functions of EC's are quite well behaved. The partial ordering of the cardinal functions of infinite compact Hausdorff spaces is given by Juhász on the insert of

[Ju]. This insert, together with the following theorem, allow one to easily deduce almost all of the common cardinal functions of EC's. For example, if K and λ are as in the theorem below, the minimal cardinality of a dense set in K must be λ, since this minimal cardinality is bounded above by the weight of K and below by the cellularity of K.

Theorem 3.1. Let K be an infinite EC, and let λ be the smallest infinite cardinal such that K embedds in $c_0(\Gamma)$ for some set Γ of cardinality λ. Then

a) the cellularity of $X = \lambda$.

b) the weight of $X = \lambda$,

c) $\lambda \leq |X| \leq \lambda^\omega$,

d) the tightness of $X = \omega$,

e) $\omega \leq$ the character of $X \leq \lambda$.

Proof. For each of the proofs below we assume that $K \subset c_0(\Gamma)$ where $|\Gamma| = \lambda$.

a) We follow the proof given in [BRW, p. 318]. Let $\mu = \lambda$ if λ is regular and $\mu < \lambda$ if λ is singular. Then since λ is minimal, there is an $\varepsilon > 0$ such that $\mathcal{A} = \{A \subset \Gamma\colon \exists k \in K \ \ \forall \gamma \in A \ \ |k(\gamma)| > \varepsilon\}$ has cardinality at least μ. $\mathcal{A}$ can contain no infinite increasing chain A_n, since any limit point of the corresponding x_n's would be at least ε on the infinite set $\cup A_n$. Thus every $A \in \mathcal{A}$ is contained in a maximal element of $\mathcal{A}$, so if $\mathcal{B}$ denotes the set of maximal elements of $\mathcal{A}$, $|\mathcal{B}| \geq \mu$ also. For each $B \in \mathcal{B}$, the set $U_B = \{k \in K\colon \forall \gamma \in B \ \ |k(\gamma)| > \varepsilon\}$ is open and nonempty, and the maximality of $B, B' \in \mathcal{B}$ guarantees $U_B \cap U_{B'} = \emptyset$ if $B \neq B'$. Hence $\{U_B\colon B \in \mathcal{B}\}$ is a collection of disjoint open sets in K of cardinality μ, and the result follows.

b) The standard basis for $c_0(\Gamma)$ (consisting of all sets of the form $\{k \in K: \forall\gamma \in G \ \ k(\gamma) \in I_\gamma\}$ where G is a finite subset of Γ and each I_γ is an open interval in $\mathbb{R}$ with rational endpoints) has cardinality λ, hence the weight of K is at most λ. Since the weight is always at least as large as the cellularity, we conclude the weight of K is exactly λ.

c) Each point of K is uniquely determined by a countable set of the form $\{(\gamma,r): \gamma \in \Gamma, r \in \mathbb{R}\}$. Since there are only $\lambda^\omega \cdot 2^\omega = \lambda^\omega$ sets of this form, $|K| \leq \lambda^\omega$. The fact that the cardinality of a space must always be as large as the cellularity gives us the inequality $\lambda \leq |K|$.

d) This follows directly from the stronger Eberlein-Smulian theorem, Theorem 1.3 (i).

e) Every singleton in $c_0(\Gamma)$ can be written as the intersection of $|\Gamma|$ open sets.

Remark. The bounds given in (c) and (e) above are best possible. The only non-trivial case is the first half of (e). The Alexandroff duplicate of the unit internal provides an example of a first countable EC for which $\lambda > \omega$: Let K = $I \times \{0,1\}$ be the union of two copies of the unit interval. Let each point in $I \times \{1\}$ be isolated and declare each set of the form $(U \times \{0,1\})\backslash\{(x,1)\}$, where U is a usual open set and $x \in I$, to be open. K is an EC since the family $\{\{(x,1)\}: x \in I\} \cup \{U \times \{0,1\}: U$ is a rational interval in $I\}$ is a σ-point finite separating family of open F_σ sets. It is clearly first countable, but its weight, and hence λ, is 2^ω. Other examples of such spaces are given in [L, p. 253]. In [BRW, p. 321] an example of a non-metrizable EC is given in which every metrizable closed subset is a G_δ.

It is reasonable to ask whether the properties (i), (ii) and (iii) of Theorem 1.3 characterize the EC's. In [Ro], Rosenthal asked whether each space having a separating, σ-point finite family of open sets is an Eberlein compact. Both of these questions are answered in the negative with the following example from [W]. Rosenthal's question has been independently answered in [Si], where it is also shown that every compact space with a point finite separating collection of open subsets is an EC.

Example. There exists a compact topology on $\omega_1 + 1$ which satisfies (i), (ii) and (iii) of Theorem 1.3, has a separating σ-point finite family of open sets, and yet is not an Eberlein compact.

Construction. Let Λ be the set of all limit ordinals in ω_1 and $D = \omega_1 - \Lambda$. Declare each point of D to be open. For each $\lambda \in \Lambda$, choose $d_{n,\lambda} \nearrow \lambda$ with $d_{n,\lambda} \in D$, and let $\{\lambda\} \cup \{d_{n,\lambda} | n > m\}$ be the mth basic open set around λ.

With this topology, $D \cup \Lambda$ is a locally compact space. $\{\omega_1\} \cup (D \cup \Lambda$ - a compact subset of $D \cup \Lambda)$.

The resulting topology on $\omega_1 + 1$ is compact and T_2. Properties (i), (ii) and (iii) are easy to verify. It is clear that this space has a separating, σ-finite family of open sets once we note that $|\Lambda| \leq 2^{\omega}$ and the real line has a countable basis. Let $f: \Lambda \to \mathbb{R}$ be one-to-one and $\mathcal{B}$ be a countable basis for $\mathbb{R}$. Then $\{\{d\} | d \in D\} \cup \{f^{-1}(B) \cup D | B \cup \mathcal{B}\}$ is the desired cover. The space is not an Eberlein compact since the pressing down lemma shows it has no separating, σ-point finite family of open F_σ's. □

Notice that the above example is the union of three metric subspaces (in fact three discrete subspaces: D, Λ, and $\{\omega_1\}$). In answer to a question raised in correspondence with A.V. Arhangelskii, M.E. Rudin, [Ru], has shown that a compact Hausdorff space which is the union of <u>two</u> metric subspaces is an EC.

The next theorem gives two more topological characterizations of EC's that were proved in [MR]. A family G of subsets of a space K is said to be F-*separating* if for each $x,y \in K$ there is a $G \in G$ such that G contains one of x and y and $K\setminus\overline{G}$ contains the other.

Theorem 3.2. The following are equivalent for a compact Hausdorff space K.

1) K is EC.

2) K has a σ-point finite F-separating family of open subsets.

3) K has a σ-point finite collection G of open subsets such that if A and B are disjoint closed subsets of X, then there exists a finite set $F \subset G$ such that F separates any a in A from any b in B.

There are two subclasses of EC's that, as well as being important for the study of EC's, are independently interesting.

Definition. An EC, K, is called *strong* if it embeds in $c_0(\Gamma)$ in such a way that $k(\gamma) = 0$ or $k(\gamma) = 1$ for all $k \in K$ and $\gamma \in \Gamma$.

Definition. An EC, K, is called *uniform* if it embeds in $c_0(\Gamma)$ in such a way that there is a function $N(\varepsilon)$ such that for all $k \in K$ and $\varepsilon > 0$, $|\{\gamma: |k(\gamma)| \geq \varepsilon\}| \leq N(\varepsilon)$.

The strong EC's are just those EC's that embed in $c_0(\Gamma)$ as a set of characteristic functions. It is easy to tell which sets of characteristic functions determine an EC: Let Γ be a set and let K be a collection of finite subsets of Γ. Then the space of characteristic functions of the members of K is an EC if K contains no infinite increasing chain and whenever $A \in K$, then all subsets of A are members of K, [BS, p. 139]. This condition is sufficient, but not quite necessary, since K need not be closed under all subsets of its members to determine a strong EC. More on strong EC's can be found in the references. The following result is proved in both [BRW, p. 310] and [Si].

Theorem 3.3. Every EC is a continuous image of a closed subset of a countable product of strong EC's.

The uniform EC's were introduced in [BS]. Our next theorem gives characterizations of uniform EC's that were obtained in [BS] and [BRW].

Theorem 3.4. The following conditions on a compact Hausdorff space K are equivalent.

a) K is a uniform EC.

b) K embeds as a weakly compact subset in a Hilbert space.

c) There is a dense range operator from a Hilbert space into C(k).

d) K embeds as a weakly compact subset in a super-reflexive space.

e) There is a dense range operator from some super-reflexive space into C(k).

f) B(C(X)*) is a uniform EC.

An example of an EC that is not a uniform EC is also given in [BS]. The example is difficult, and is based on the next proposition. Perhaps one of the standard topological spaces could be modified to produce an easier counterexample.

Proposition 3.5. Let K be a weakly compact subset of a Hilbert space, and let $D \subset K$ be a discrete set with a unique limit point k. Then for every $d \in D$ there exists a relatively open subset V_d of K such that:

1) For all $d \in D$, $d \in V_d$.

2) There is a countable partition $D = \cup D_n$ such that if $d_1, \ldots, d_{n+1}$ are distinct elements in D_n, then $\cap_{j=1}^{n+1} V_d = \emptyset$.

It was shown in [BRW], p. 316] that the continuous image of a uniform EC is also a uniform EC. Moreover, every uniform EC is the continuous image of a closed subset of the countable product of one point compactifications of discrete spaces. This yields:

Theorem 3.6. The class of uniform EC's is the smallest class of spaces that (a) contains the one point compactification of each discrete space and (b) is closed under continuous images, countable products, and closed subspaces.

The above theorem shows that the *uniform* EC's are *topologically* generated by the one point compactifications of discrete spaces. This contrasts with the next theorem, from [L,p.249], which states that, in some sense, the one point compactifications of discrete spaces are enough to *analytically* do the job of *all* EC's. Perhaps the conditions in Theorem 3.6 can be varied slightly to topologically generate all EC's from the one point compactifications of discrete spaces.

Theorem 3.7. Every WCG Banach space is generated by a subset which is homeomorphic (in the weak topology) to the one point compactification of a discrete set.

References

[AL] D. Amir and J. Lindenstrauss, *The structure of weakly compact subsets in Banach spaces*, Ann. of Math. 88 (1968), 35-46.

[BRW] Y. Benyamini, M.E. Rudin, and M.L. Wage, *Continuous images of weakly compact subsets of Banach spaces*, Pacific J. of Math., vol. 70, no. 2, 1977, 309-324.

[BS] Y. Benyamini and T. Starbird, *Embedding weakly compact sets into Hilbert space*, Israel J. of Math., vol. 23, no. 2, 1976, 137-141.

[D] M.M. Day, *Normed Linear Spaces*, Berlin: Springer-Verlag, 1958.

[DS] N. Dunford and J.T. Schwartz, *Linear Operators*, Part I, Interscience Pub., New York, 1958.

[E] R. Engelking, *Topology*, Polish Scientific Publishers, Warsaw, 1976.

[G] S.P. Gul'ko, *On properties of Σ-products*, Soviet Math. Pokl., Vol. 18 (1977), no. 6, 1438-1442.

[J] R.C. James, *Weakly compact sets*, Trans. Amer. Math. Soc., 113 (1964), 129-140.

[Ju] I. Juhász, *Cardinal functions in topology*, Math. Centre tract 34, Amsterdam, 1971.

[L] J. Lindenstrauss, *Weakly compact sets--their topological properties and the Banach spaces they generate*, Annals of Math. Studies 69, Princeton Univ. Press (1972), 235-273.

[MR] E. Michael and M.E. Rudin, *A note on Eberlein compacts*, Pacific J. of Math., to appear.

[N] I. Namiok, *Separate continuity and joint continuity*, Pacific J. of Math. 51 (1974), 515-531.

[Ro] H.P. Rosenthal, *The heredity problem for weakly compactly generated Banach spaces*, Compositio Math., vol. 28, Fasc. 1, 1974, 83-111.

[Ru] M.E. Rudin, *Eberlein compacts and metric subspaces*, preprint.

[S] W. Schachermayer, *Eberlein-Compacts et espaces de Radon*, preprint.

[Sc] L. Schwartz, *Radon measures on arbitrary topological spaces*, Oxford University Press, 1973.

[Si] P. Simon, *On continuous images of Eberlein compacts*, CMUC, to appear.

THE DEVELOPMENT OF GENERALIZED BASE OF COUNTABLE ORDER THEORY[1]

H.H. Wicke and J.M. Worrell, Jr.
Department of Mathematics and Institute for Medicine and Mathematics, Ohio University, Athens

1. *Introduction*

In 1966 at the International Congress of Mathematicians meeting in Moscow, USSR, we made a presentation in which we advanced the thesis, after several years of detailed exploration, that the base of countable order property [Arh_2] reflected principles of topological uniformization with extensive application and expressing a substantial content of what is topologically fundamental in metrizability [WiW_1]. A frutiful exchange took place with Arhangel'skiĭ in which the topic of p-spaces [Arh_3], was reviewed. In 1967, after extensive investigations, we announced at a meeting of the American Mathematical Society a theory of non-first-countable topological structure, generalizing the base of countable order uniformization and including the p-space uniformization (for appropriate separation) as a special case [WoW_3]. Entering as one component of a basis for making this announcement, in effect a large thesis, was the realization that virtually the main body of classical metrization theory extended to

[1]The University of Texas, the National Science Foundation, and the U.S. Atomic Energy Commission at Sandia Laboratories provided the principal financial support for the work of the authors in their researches reported here through about 1970. Major support was provided after 1970 by Ohio University.

ISBN: 0-12-584960-5

characterizations of the Hausdorff perfect preimages of metrizable spaces [Wo_{14}]. Wicke was invited to give a Kempner Colloquium Lecture on this subject at the University of Colorado in Boulder in 1967 [Wi_4]. There, in support of the thesis, certain of the main mapping invariances and characterizations were reviewed. In 1968, at an invited hour talk at the Annual Topology Conference, held in that year at the University of Houston, Worrell stressed the fundamentality of the base of countable property in comparing certain of the spatial conceptions of R.L. Moore and A.V. Arhangel'skiǐ [Wo_{17}] At the time of the Kempner Lecture, techniques sufficing for the development of a substantial part of Generalized Base of Countable Order Theory, as reviewed here, had been identified and published [Wi_2], [Wo_7], [Wo_{10}], [Wo_9], [WoW_1], [WiW_2] submitted for publication [Wo_{16}], [Wo_{15}], [WoW_1],, made accessible to public review [Wo_1] or reported to the American Mathematical Society typically after presentation in the Department of Mathematical Research at Sandia Laboratories [Wo_4], [Wo_{13}], [Wo_{11}] [Wo_5], [Wo_8], [Wo_3], [WoW_2], [Wo_{12}]. Of particular bearing is a presentation we made to the American Mathematical Society in 1965, giving a characterization of the essentially T_1 spaces having bases of countable order as the essentially T_1 spaces having primitive bases in which closed sets are sets of interior condensation [WoW_2]. We showed that for regular T_o spaces, the property of having a primitively complete base is equivalent with that of satisfying Aronszajn's sequence condition of [Aro]. The announcement gave finite intersection property formulations for conditions also formulatable with respect to underlying primitive sequence properties of monotonicity of the base of countable order topological uniformization. Strongly convinced that primitive sequences, and certain explicit techniques we had applied using them, underlay a general theory of topological uniformization, we conducted a long range program of research, at Sandia Laboratories, of systematic exploration of these modes of topological uniformization.

A comparatively straightforward test case was provided in the application of these techniques to an analysis of the M-spaces of Morita [Mor_1]. The general structural formulations were readily generalized, and this was reported at the First International Topology Conference at Pittsburgh, in 1970 [Wi_7]. At this meeting Worrell announced the existence of a map of a completely regular p-space into a completely regular non-p-space [Wo_{21}], [Wo_{22}]. This result had important implications for the theory, for it showed that certain approaches to defining topological uniformization in reference to suggestive developable-like appearing properties of finite intersection of the p-space concept are still not satisfactory if one uses as one criterion that the property be preserved by perfect mappings (between appropriately separated spaces). In the course of his presentation of this meeting, Worrell discussed certain paracompactnesss-like properties defined with respect to compactifications, including a characterization of Čech complete developable spaces [Wo_{13}] Wicke reviewed certain of these topics in an invited talk at the Annual Topology Conference in 1971, held at Houston [Wi_6]

Diagonal theorems of Wicke provided important links of the non-first-countable- to the first-countable uniformizations [Wi_8], [Wi_9], [Wi_{11}]. The general theory was put on a basis of a relaxation of regularity implying Axiom T_2 for T_0 spaces, in a formulation utilizing uniformization principles of primitive sequences [WiW_6] and [WiW_7]. Reports on topological structures defined by primitive structures were given at the Charlotte, N.C. Topology Conference and the International Congress at Vancouver, B.C. in 1974 [WiW_{12}], [WiW_8].

As of 1974, one of the single most important questions was whether the primitive base property, even for the regular cases, is preserved by perfect mappings of spaces having the property. (For the case of continuous mappings which are both open and closed, the answer is in the affirmative.) Burke's report on the invariance of the property at the annual Topology Conference, held at Ohio University in 1979 is a landmark for generalized base of countable order theory [Bu_2]. This result, which can be extended to perfect mappings of regular primitive β_b- and β_c-spaces, respectively, utilizing techniques of primitive sequences, provides one of the most important components supporting our thesis, stated at Moscow in 1966 and extended in the Kempner Colloquium Lecture in 1967.

2. *Some of the main concepts and definitions introduced in the development of base of countable order theory*

We list here some of the main concepts and definitions introduced in the development of base of countable order theory, or which can be interpreted as explicitly influencing its development (as in the case of a sequence condition introduced by N. Aronszajn [Aro].)

Complete Aronszajn space. In *Fundamenta* in 1931, Aronszajn published a paper on the construction of arcs. He shows that a connected and locally connected space S is arcwise connected if it satisfies the following sequence condition: There exists a sequence G_1, G_2,... of bases for S such that if P is a point of an open set D, Y is a point distinct from P, and g_1, g_2,... is a decreasingly monotonic sequence of sets such that each g_n belongs to G_n and contains P, then some $\overline{g}_n$ is a subset of D which does not contain Y. He also showed that the so-called "long line" is an example of such a space S. In our search of the literature we have not found any papers of Aronszajn in sequal to this 1931 paper developing any further theorems concerning this base concept. Nevertheless, R.L. Moore was aware of the paper and brought the sequence condition to the attention of various of his students over the years. In this way, Aronszajn's paper was brought to the attention of various, one might suppose, of several generations of Moore's students, some of whom attempted to develop theorems using the sequence condition. In effect, Worrell interpreted Moore as non-rhetorically asking the question whether it is possible to develop a serious topological theory based on Aronszajn's sequence condition. One wonders if for Moore the main heuristically guiding consideration focused in the arc theorem.

Out of whatever mathematical considerations may have occasioned his doing so, Aronszajn placed a gem in the rough in the literature that was alluring to R.L. Moore and occupied his attention intermittantly over a quarter of a century. In honor of Aronszajn we have called spaces satisfying his sequence condition complete Aronszajn spaces.

Complete Aronszajn sequences. We call certain sequences satisfying certain modifications of Aronszajn's sequence condition of [Aro] complete Aronszajn sequences [WiW_2]. For the regular T_0 case, these sequence conditions are equivalent for a space.

Point regular base. A base B for a space S is said to be *point-regular* if and only if it is true that if P is a point common to each member of an infinite subcollection B' of B, then B' is a base for S at P [Al_2]. Under another terminology, such a base is called a *uniform* base.

Base of countable order. Arhangel'skiĭ introduced the base of countable order concept into the literature in a distinguished paper on metrization [Arh_2]. He shows that a Hausdorff space is metrizable if and only if it is paracompact and has a base of countable order.

σ point-finite base. In the above paper [Arh_2], Arhangel'skiĭ shows that a regular space S has a point regular (σ-uniform) base if and only if 1) S has a σ point-finite base and 2) a closed sets of S are inner limiting sets. The definition of a σ point-finite base is introduced in [SZ].

θ-refinable space. Worrell and Wicke introduced the definition of θ-refinability in [WoW_1], having previously presented the concept under a different terminology to the American Mathematical Society in 1964. It has been suggested that the term *submetacompact space* be used for θ-refinable spaces [J].

θ-base. Worrell and Wicke introduced the definition of a θ-base in [WoW_2], having previously presented the concept under a different terminology to the American Mathematical Society in 1964.

Absolute θ-refinability, absolute subparacompactness, absolute metacompactness [Wo_{13}]. Worrell introduced embedding formulations of these properties as generalizations of a property of fully normal [Tu] spaces. At the First Pittsburgh International Congress on topology, 1970, he emphasized the importance of these formulations, while giving embedding formulations of the Nagata-Smirnov Theorem. For definitions, see [WiW_{14}].

Sets of interior condensation. We introduced the definition of *set of interior condensation* in [WiW_5], [WiW_6]. This generalizes the definition of an inner limiting set. The terminology W_δ has been recently suggested for these sets [CČN].

Sets of absolute interior condensation. The concept of a set of absolute interior condensation generalizes that of an absolute G_δ [WiW_5].

δθ-base. A definition of δθ-base is given in [Au_2].

Monotonically complete base, countably monotonically complete base. We give definitions in [WiW_2], [WiW_6]. One can find precursors of the concept of monotonic completeness in the literature [C], [W].

Monotonically complete base of countable order. We give a definition in [WiW_2]. If a regular countably monotonically complete space has a base of countable order, it has a monotonically complete base of countable order [WiW_6].

Uniformly monotonically complete mapping, uniformly countably monotonically complete mapping. We give definitions in [WiW_2], [WiW_6]. The concept was discussed in our presentation at the International Congress of Mathematicians in Moscow, USSR in 1966 [WiW_1].

Primitive bases. We introduced a formulation of primitive bases in a characterization of essentially T_1 spaces having bases of countable order [WoW_2]. An equivalent formulation which constitutes a definition now frequently used is given in [WiW_{11}]. The primitive base concept generalizes those of a σ-locally finite base, a σ point-finite base, and a θ-base. See Section 12.

Complete primitive base. A concept of a complete primitive base is used in [WoW_2]. We give a formal definition in Section 12. See Section 12 for a formulation of *primitive basis compactness.*

Primitive sequence, primitive part, primitive representative of a primitive sequence, primitive core of a primitive representative. We give definitions in [WiW_{12}].

p-space. Arhangel'skiĭ gives definitions in [Arh_3]. He gives formulations both with respect to Wallman compactifications and Stone-Čech compactifications.

Space of point-countable type. A definition is given in [Arh_3].

M-space. Morita gives a definition in [Mor_1].

q-space. Michael gives a definition in [Mi_3].

$\omega\Delta$-space. A definition is given in [Bo].

β_b-space, λ_b-space. We give definitions in [WoW_8]. For the regular cases, these spaces are equivalent with the μ-spaces and complete μ-spaces of earlier reports in the Notices. See also [Wi_6].

β_c-space, λ_c-space. We give definitions in [WoW_8]

Primitive β_b*-space, primitive* β_c*-space.* Analogously as with primitive spaces, there are equivalent monotonic and finite intersection formulations. The later type are given here. A primitive β_b-sequence (respectively, β_c-sequence) for a space S is a sequence H_1, H_2,... of well-ordered collections of open sets covering S such that if P is a point and for each n, h_n is the first element of H_n that contains P, then (1) the common part M of the closures $\bar{h}_n$ of the sets h_n is compact (respectively, countably compact) and (2) every open set that includes M includes some h_n. A *primitive* β_b*-space* (respectively, β_c*-space*) is a space having a primitive β_b-sequence (respectively, β_c-sequence).

Sets of primitive interior condensation. A point set M of a space S is said to be a set of primitive interior condensation of S if and only if there exists a sequence H_1, H_2,... of well-ordered collections of open sets covering M such that if P is a point of M and for each n, h_n is the first element of H_n containing P, then the common part of the sets h_n is a subset of M.

Representative of a sequence G_1, G_2,... *of collections of sets.* A representative of a sequence G_1, G_2,... of collections of sets is a sequence g_1, g_2,... such that for each n, g_n belongs to G_n [WiW_7].

Monotonically contracting sequence of collections of open sets. We give a definition in [WiW_7].

Monotonically contracting collection of open sets. We give a definition in [WiW_7].

σ-distributively point-finite, and point-countable collection of open sets. We give definitions in [WoW_4].

Weak θ-refinability. A definition of weak θ-refinability is given in [BeL].

δθ-refinability, weak δθ-refinability. Definitions are given in [AU_1], [WiW_{15}], and [Smi].

Quasi-developable space. A definition is given in [Be].

$[\alpha,\beta]^r$-regular refinability, weak $[\alpha,\beta]^r$-regular refinability. Hodel and Vaughan defined refinement concepts, which generalize δθ-refinability and weak δθ-refinability, which they call (with respect to appropriate notation) $[\alpha,\beta]^r$-refinability and weak $[\alpha,\beta]^r$-refinability [HoV]. These concepts have an important bearing on base of countable order theory. See [WoW_{10}] and [WiW_{19}].

Local-implies-global property. A space is said to have a property Q locally if and only if each point P of a open set D belongs to some open subset of D containing P which has property Q. A property Q is said to be a local-implies-global type property if and only if it is true that if a space S has property Q locally then S has property Q [WiW_{13}]. The base of countable order property is a local-implies-global type property [WoW_1]. Monotonic completeness, the β_b-, β_c-, λ_b-, λ_c-space properties, the primitive β_b- and β_c-space properties, primitive base properties and the set of interior condensation properties are also of this type [WiW_{13}].

Regularly refinable space. A definition is given in [Wo_{15}] The concept is essentially implicit in the statement of the Alexandroff-Urysohn metrization theorem [AlU_1]. Also see [Al_3] for the original definition.

Essentially T_1*-space*. A definition of essentially T_1 space is given in [WoW_1]. The concept had been introduced under a different terminology in [D].

Pararegular space. We give a definition in [WiW_7], [WiW_6].

A representative of a sequence $G_1, G_2, \ldots$ *of collections of sets controlling a representative of a sequence* $V_1, V_2, \ldots$ *of collections of sets*. If $G_1, G_2, \ldots$ and $V_1, V_2, \ldots$ are sequences of collections of sets, a representative $g_1, g_2, \ldots$ of $G_1, G_2, \ldots$ is said to control a representative $v_1, v_2, \ldots$ of $V_1, V_2, \ldots$ if and only if each term of $g_1, g_2, \ldots$ includes all except finitely many terms of $v_1, v_2, \ldots$.

Property A. A space S is said to have *Property A* if and only if for every monotonically contracting sequence $G_1, G_2, \ldots$ of collections of open coverings of S there exists a sequence $W_1, W_2, \ldots$ of collections of open coverings of S such that (1) for each n, W_n refines G_n and (2) every representative of $W_1, W_2, \ldots$ which has a point common to all its terms is controlled by a decreasingly monotonic representative of $G_1, G_2, \ldots$ [WoW_9],[WiW_{14}]. In this definition $W_1, W_2, \ldots$ is a *sequential refinement* of $G_1, G_2, \ldots$ [Wo_{25}]. Every θ-refinable topological space has property A [WiW_{14}].

Property A'. A space S is said to have *Property A'* if and only if for every monotonically contracting sequence $G_1, G_2, \ldots$ of open coverings of S there exists a sequence $W_1, W_2, \ldots$ of monotonically contracting open coverings of S such that every representative of W_1, W_2,... with the finite intersection property is controlled by a decreasingly monotonic representative of G_1, G_2,... . Every absolutely θ-refinable space has Property A' [WiW_{14}].

3. Some of the underlying point of view of base of countable order theory

Topological uniformizations are being sought which are not dependent on specifications of transfinite weight of a space and which generalize certain of the main topological content of metrizability. The obtaining of satisfying general uniformization for first countability is hypothesized to be an appropriate step in making the transition to spaces not necessarily first countable. In the making the latter transition, spaces of point-countable type and q-spaces are viewed as appropriate bridging concepts in their abstraction of first-countability.

The concept of the "long line" as pathological was rejected [Wo_{17}]. The view is by contrast that not only the "long line", but all arcs and their connected subsets must fall within the concept of natural topological structure for a satisfactory theory of topological uniformization. To the extent that various paracompactness-like properties preclude these spaces for uniformizations in which they appear, such properties are aspectival to an overall theory. The technical difficulty in analyzing such continua constitute a challenge and topological frontier.

It is taken as a guiding hypothesis that identification of correct topological uniformizations will be linked with invariances of the uniformizations under the actions of perfect mappings as a necessary, but not sufficient, feature. An additional requirement is the preservation of the uniformization by appropriately complete open continuous mappings (for appropriate separation).

Identification of correct completeness properties is viewed as topologically central. The view of what constitutes the character of a correct concept of topological completeness is analytic in esprit. Topological uniformization is closely linked with topological completeness, so that certain of the uniformizations, the base of countable order property to illustrate, are regarded as instances of topological completeness. In this particular instance, the hereditary character of the base of countable order property is regarded as reflecting a Baire category type content which can be explicitly elicited in certain of the uniformizations applying to non-first-countable structure [WoW_1]. Heuristic links with compactification theory are provided in observing [WiW_{17}] that the concept of a monotonically complete base abstracts that of compactness (≡ perfect compactness) as formulated in Chapter I of Moore's *Foundations* and that of a countably monotonically complete base abstracts that of countable compactness. The hypothesis is made that these heuristic links are underscored by a technical content, one reflection of which is a modification of a classical orientation as to what constitutes natural separation.

4. *Some of the main techniques of base of countable order theory*

The dominating techniques are methods of primitive

sequences. Our presentation at Moscow in 1966 [WiW_1] to the American Mathematical Society in 1967 [WoW_3] and of Wicke's Kempner Colloquium Lecture at the University of Colorado in 1967 [Wi_4] have at their basis an hypothesis of the scale of application of these methods.

Techniques of [Wo_9], [Wo_{10}] and [Wo_{15}] established methods analyzing certain paracompactness-like properties, establishing certain invariances for closed mappings, and providing transitions from hypothesis of local cardinal restriction (such as separability, Lindelöf's property) to global hypothesis in establishing properties of σ-distributed neighborhood or point finitude from properties of σ-distributive point-countability. These have applications to the separation of closed mapping invariances of certain types of topological uniformization having certain paracompactness-like properties (for illustration, metacompactness) into invariances of a base of countable order type uniformization and of a paracompactness-like property.

Techniques of [WoW_1] provide general methods for eliciting certain kinds of uniformizations from conditions of σ-distributive point finitude with the use of primitive sequences.

Techniques of [Wo_1] provide general methods for analysis of paracompactness-like properties under hypothesis of local cardinality restrictions and global refinement properties formulated with respect to transfinite inequalities. In [Wo_{15}] a global hypotheses (regular refinability) replaces certain of the hypotheses of local cardinality restrictions of [Wo_1]

A technique of Wicke in [Wi_2], [WiW_2] provides a general method of construction of open continuous images of certain paracompact spaces. These constructions provide a link of certain of the uniformizations of generalized base of countable order theory to paracompact spaces by means of open continuous mappings.

A technique of [Wo_5] has wide application in the analysis of closed mappings of spaces satisfying certain paracompactness-like conditions with particular application to generalized base of countable order theory.

5. *Characterizations of developability*

At the threshold of a successful transition from a theory of the countable structure of metrizable space and certain abstractions of features of the spaces (perfect normality [U_2], [$\check{C}_1$]) to that of non-first-countable structure, and indeed even topologically uniformizable first countable structure of sufficient scope, is a general problem of obtaining freedom from naive reliance on the hypothesis, on implication, for a space that closed sets are innerlimiting sets. In the research of [WoW_1] a prime technical problem was the identification of a sufficiently general paracompactness-like property such that (1) it would generalize metacompactness and (2) it would be technically tractable in the absence of the hypothesis that closed sets are inner limiting sets. A solution was provided by a class of spaces called θ-*refinable* [WoW_1] and which it has been suggested more recently be called *submetacompact spaces* [Ju].

Using techniques of primitive sequences, we obtained a characterization of essentially T_1 spaces having point regular bases as the essentially T_1 metacompact spaces having bases of countable order [WoW_1] and of developable spaces as essentially

T_1 spaces having bases of countable order as analogues of Arhangel'skiĭ's characterization of metrizable spaces as the Hausdorff paracompact space having bases of countable order [Arh_2]. Moreover, Arhangel'skiĭ's theorem itself was extended in that in all three of these characterizations the theorem that spaces having bases of countable order locally have this global property can be applied [WoW_1]. These characterizations can then be put in a form generalizing Smirnov's characterization of metrizable spaces as the Hausdorff paracompact locally metrizable spaces [Sm].

The character of these theorems as analogues was reinforced by theorems of Worrell giving Axiom 0 space [Mo_4] characterizations of metacompactness and θ-refinability as variations of the Tukey property of full normalcy [Tu]. Furthermore, subparacompactness [Bu_1] was shown not to imply absolute θ-refinability even under the hypothesis of metacompactness [Wo_9].

It has repeatedly been shown that techniques of [WoW_1] used to establish these characterizations can be extended to certain analogous problems in non-first countable structure. For illustration, we have shown that the above Arhangel'skiĭ metrization theorem can be generalized to an analogous characterization of the Hausdorff paracompact p-spaces, or, equivalently, to a characterization of the Hausdorff perfect preimage of metrizable spaces [Wo_{14}]. Analogous characterizations of θ-refinable regular p-spaces can be obtained to those above of developability.

The σ-point finite base property and the θ-base property of [WoW_1] are regarded as analogues of the Nagata-Smirnov σ-locally finite base property which in themselves have a special significance and take on interest as instances of primitive bases [WoW_2], [WiW_9]. In this regard, Lutzer and Bennett's work [BeL] in characterizing spaces having θ-base with respect to a sequence condition generalizing developability and designating spaces satisfying this sequence condition (equivalently, of spaces having a θ-base) as <u>quasi</u> developable [Be] is particularly sound. The theorem of [WoW_1] characterizing developable spaces as spaces with θ-bases where closed sets are inner limiting sets may properly be viewed as a theorem of separation of properties. The so-called "Michael line" is an example of a paracompact Hausdorff space with θ-base, indeed a σ point finite base. Moreover, there exist Lindelöf, non-separable regular T_0 spaces which are quasi-developable (equivalently, have θbases) on which are not developable [Be].

Embedding formulations of certain paracompactness-like properties can be applied to extensions of the work of [WoW_1 to obtain characterizations of Čech complete developable space [Wo_4] complete Moore spaces [WiW_{14}], and subspaces of these spaces utilizing a characterization of the subspaces of complete Aronszajn spaces [Wo_{24}]. See also [Ph_5]. The general point was emphasized by Worrell in the presentation at the Pittsburgh International Conference on Topology, 1970. The Čech complete developable spaces are the completely regular T_0 spaces S having monotonically complete bases of countable order which can be θ-refinably embedded in $\beta(S)$ [WiW_{14}]. Complete Moore spaces can be similarly characterized with respect to θ-refinable embeddings of regular T_0 spaces having bases of countable order in their Wallman compactifications. M.E. Estill Rudin, has given an example of a developable complete Aronszajn space which is not a complete Moore space [Ru_2].

There are characterizations of locally second countable developable spaces in which the hypothesis of θ-refinability is relaxed to δθ-refinability. To illustrate, a regular T_0 locally second countable space is a Moore space if and only if it is δθ-refinable [Wo_{26}]. Techniques of [Wo_1] can be applied to obtain a relaxation of δθ-refinability to some general refinement conditions formulated with respect to transfinite inequalities of an $[\alpha,\beta]^r$-refinability type. These characterizations, with appropriate modifications taking into account non-separability of certain Lindelöf spaces, generalize for characterizations of locally Lindelöf regular T_0 θ-refinable p-spaces [Wo_{26}].

With these types of characterizations of regular T_0 θ-refinable p-spaces one can obtain characterizations of Moore spaces with application of diagonal theorems of Wicke in which the hypothesis that the diagonals of the space S are locally sets of interior condensation enters to imply the base of countable order property [Wi_8] or in which the hypothesis that the diagonals of the spaces S are locally sets of primitive interior condensation enter to imply the primitive base property [Wi_9]. We illustrate these applications in the following characterizations of Moore spaces and complete Moore spaces.

Theorem. A regular T_0 space S is a Moore space if and only if (1) locally, S is a β_c-space, (2) S is θ-refinable, and (3) locally, the diagonal of S is a set of interior condensation.

Theorem. A regular T_0 space S is a complete Moore space if and only if (1) locally, S is a β_c-space, (2) S can be θ-refinably embedded in its Wallman compactification, and

(3) locally, the diagonal of S is a set of primitive interior condensation.

In [WoW_1] equivalences are established for essentially T_1 spaces for the base of countable order property for a space S and the existence of a sequence G_1, G_2,... of bases for S such that if P is a point common to all elements of a monotonically decreasing representative g_1, g_2,... of G_1, G_2,..., then $\{g_1, g_2, \ldots\}$ is a base for S at P [WoW_1]. In [WiW_2], an equivalence is given for a regular T_0 space S of being a complete Aronszajn space and having monotonically complete bases of countable order locally. These theorems have played an important role in providing a means for making the transition from the topologically uniform first countability property of spaces having bases of countable order to the β_b- and β_c-space uniformizations. While it is readily evident that locally θ-refinable regular T_0 β_c-spaces have a base of countable order-like property formulatable with respect to the above definition of a monotonically contracting collection of open sets, the question as to whether the property of being a regular T_0 β_c-space can, in general, be given this type of formulation simply in terms of monotonically contracting collections of open sets and convergence is not known to us.

Properties A and A' respectively, abstract θ-refinability and absolute θ-refinability. They require of completely regular T_0 β_b and λ_b-spaces respectively satisfying these conditions the p-space property and Čech completeness [WiW_{14}]. They do not require δθ-refinability of such spaces. But, for essentially T_1 spaces having bases of countable order and regular T_0 spaces having monotonically complete bases of countable order, having Properties A and A' respectively, is equivalent with developability and being a complete Moore space [WiW_{14}]. With applications of certain of the diagonal theorems of Wicke, one can obtain for the characterizations of Moore spaces and complete Moore spaces in which no explicit condition of θ-refinability, or even weak δθ-refinability enters formally in the hypotheses:

Theorem. A regular T_0 space S is a Moore space if and only if (1) locally, S is a β_c-space, (2) S has property A, and (3) locally, the diagonal of S is a set of interior condensation.

Theorem. A regular T_0 space S is a complete Moore space if and only if (1) locally, S is a λ_c-space, (2) S has property A', and (3) locally, the diagonal of S is a set of primitive interior condensation.

6. Open continuous mappings of spaces having bases of countable order

The concept of interiority of a continuous mapping in an earlier form was introduced and emphasized by H. Weyl [Wey]. Hausdorff showed that the *metrizable* open continuous images of complete metric spaces are metrically topologically complete [Ha]. Michael showed that if ϕ is an open continuous mapping of a metrizable space S onto a paracompact Hausdorff space R such that for some topology preserving metric δ for S it is true that if P is a point of R the restriction of δ to $\phi^{-1}(P)$ is a complete metric for $\phi^{-1}(P)$ in the relative topology, then R is metrizable [Mi_1]. A major improvement on Hausdorff's result follows as a corollary to Michael's theorem: If R is a Hausdorff open continuous image of a metrically topologically complete space then (1) R is metrizable and (2) R is metrically topologically complete. There, to a large extent as viewed in the literature to about 1962, the theory seemed to stop in terms of the eliciting of certain kinds of invariances. Even compactness of an open continuous mapping of a metrizable space onto a regular T_0 space R does not imply metrizability of R.

From the standpoint of the uniformizations of base of countable order theory, it was evident in restrospective review of the literature that neither in Hausdorff's theorem nor in Michael's did metrizability of the domain as such enter. In both instances, the base of countable order property did enter under an hypothesis explicitly or implicitly requiring a uniform monotonic completeness condition for the mapping. In Hausdorff's theorem the property of having a monotonically complete base of countable order was preserved. In both instances one could obtain metrizability by appropriate applications of an Arhangel'skiǐ metrization theorem Similarly, by applying certain characterization theorems of [WoW_1] one could obtain conditions of sufficiency that the range have a point regular base or that it be developable.

Wicke showed that every regular T_0 space having a monotonically complete base of countable order is an open continuous image of a metrically topologically complete space [Wi_2] This result can be put in the form that every complete Aronszajn space R is an open continuous image of a metrically topologically complete space. Wicke was able to show, moreover, that if R is a T_1 space having a base of countable order, there exists an open continuous uniformly monotonically complete mapping ψ of a metrizable space S where range is R.

The result can be put in the equivalent form that there exists such a ψ such that for some topology preserving metric δ for S it is true that for each point P of R, δ is a topology preserving complete metric for $\psi^{-1}(P)$ [WiW_2].

Ponomarev had characterized the first countable T_1 spaces as the open continuous images of metrizable spaces [Po]. Thus it was clear that the condition of uniform monotonic completeness on an open continuous mapping of a regular T_0 space S having a base of countable order onto a T_1 space as a condition of sufficiency for preservation of the base of countable order property was not superfluous [WiW_2]. We showed that for spaces S which are subspaces of complete Aronszahn spaces, an open continuous mapping ψ of S onto a T_1 space preserves the base of countable order property if and only if ψ has a range preserving uniformly monotonically complete open continuous extension f whose domain E is a regular T_0 space having a base of countable order such that f is uniformly monotonically complete [WiW_2].

The mapping constructions are such as to readily permit demonstration that every T_1 space having a point regular base is an open continuous compact image of a metrizable space [WiW_2]. Of course, collectionwise normalcy of the range would then imply metrizability. But in general, compactness of the mapping, a very strong condition, would imply neither preservation of metrizability nor, where the domain is metrically topologically complete, of Čech completeness.

Worrell obtained a construction of a collectionwise normal T_1 countably compact space R having a base of countable order such that: (1) R is not Čech complete, (2) R has topologically equivalent local and global structure, (3) R is homogeneous [Wo_{11}]. By a theorem of Wicke cited, R is an open continuous image of a metrically topologically complete space [WiW_2] Indeed, the space R is regularly refinable. It may be seen that, in a very strong sense, Čech completeness is not in general preserved by the actions of open continuous mappings whose domains are normal Čech complete spaces and where ranges are normal T_1 spaces. Similar observations apply to metric topological completeness, metrizability, to the property of being a Complete Moore space and that of being a Moore space, and to the property of being a p-space, and the property of being a semimetric space. (Since countably compact completely regular T_0 p-spaces are Čech complete, R is nowhere locally a p-space.)

In a context of an emphasis on uniform spaces in the sense of A. Weil, it seems natural to ask whether Hausdorff's theorem generalizes for open continuous mappings of complete uniform spaces: Does it follow that if ψ is an open continuous mapping of a complete uniform space S onto a uniform space (in the sense of Weil), then $\psi(S)$ has a topology preserving complete uniformity? The answer is in the negative. The countably compact space R above has no topology preserving complete uniformity anywhere locally though it is uniform in the sense of Weil and is an open continuous image of a space which has a topology preserving complete uniformity [WoW_5] Also see [Ko] for a relevant example.

The characterization of the regular T_0 open continuous images of complete metric spaces as the regular T_0 spaces having monotonically complete bases of countable order is extended to a characterization of the Hausdorff open continuous images of complete metric spaces as the basically complete spaces [WiW_{17}].

This characterization of the Hausdorff open continuous images of complete metric spaces is generalized to characterizations of the open continuous images of Čech complete paracompact spaces as the Hausdorff λ_b-spaces [Wi_6], [WiW_7], [WiW_5]. The pararegular T_0 open continuous images of pararegular T_0 M-spaces which are inverse quasi perfect images of complete metric spaces are characterized as the pararegular T_0 λ_b-spaces [WiW_7]. The pararegular T_0 uniformly β_b-complete open continuous images of Hausdorff paracompact p-spaces are characterized as the pararegular T_0 β_b-spaces [WiW_7]. The pararegular T_0 uniformly λ-complete open continuous images of pararegular T_0 M-spaces are characterized as the pararegular T_0 β_c-spaces [WiW_7]. Preservation of the primitive base property, the primitive β_b-space property, and the primitive β_c-space property, for appropriate separation by open continuous mappings satisfying appropriate formulations of primitive uniform completeness can be obtained [Wi_{10}]. The regular T_0 q-spaces are the open continuous images of regular T_0 M-spaces [Na_3], [Wi_7]. This extends to pararegular T_0 q-spaces also. The Hausdorff spaces of point-countable type are the Hausdorff open continuous images of Hausdorff paracompact p-spaces [WiW_7], [Wi_5].

From characterization theorems of Section 5 above one can obtain various conditions of sufficiency for the range of an (appropriately) uniformly monotonically complete open continuous image of (an appropriately separated) space having a base of countable order and for β_b-, β_c-spaces to satisfy the condition of developability, be a paracompact space, be a θ-refinable p-space. Similarly, the theorem of Pasinkov-Worrell [Pa], [WoW_{12}] on the Čech completeness of the Hausdorff open continuous images of Čech complete spaces can be generalized.

Properties A and A' are generalizations of θ-refinability and absolute θ-refinability which do not imply $\delta\theta$-refinability and which can be used to broaden considerably certain of these results. If R is a completely regular T_0 range of a uniformly λ-complete open continuous mapping of a pararegular T_0 β_b-space S, then R is a p-space [WiW_{17}],[Wo_{25}]. If, additionally, S is a λ_b-space and R has Property A', then R is Čech complete. If R is a regular T_0 uniformly λ-complete open continuous image of a β_c-space and R has Property A, then R is a $\omega\Delta$-space (definition in [Bo]) [WiW_{14}].

The concept of a monotonically complete base may be viewed as a generalization of a classical characterization of compactness of a space S as equivalent with the condition that if T is a monotonic collection of a closed subset of S, there exists a point of S common to all members of T_1 [Mo],[WiW_6]. In base of countable order theory, the equivalence for regular spaces S having a base of countable order of the existence of a monotonically complete base for S and a countably monotonically complete base for S may be viewed as an analogue of the equivalence of compactness and countable compactness for Moore spaces [WiW_6]. From this viewpoint, uniformly monotonically complete mappings may be seen as generalizations of compact mappings. For regular spaces having bases of countable order, uniformly countably monotonically complete mappings may be seen as generalizing compactness of a mapping. From this standpoint, and from a background of the general problem of characterizing the spaces of Arhangel'skiǐ's class MOBI [Arh] we note the following characterizations from base of countable order theory, which generalize to the β_b-, λ_b-, β_c-, and λ_c-spaces:

Theorem. The minimal class C of Hausdorff spaces which contains (1) all metrically topological complete spaces, (2) each perfect image of each of its members, and (3) each open continuous Hausdorff image of each of its members is the class of basically complete spaces [WiW_6].

Theorem. The minimal class C of pararegular T_0 spaces which contains (1) all metrizable spaces, (2) each perfect image of each of its members, and (3) each open continuous countably monotonically complete pararegular T_0 open continuous image of each of its members is the class of pararegular T_0 spaces having bases of countable order [WiW_6].

7. Closed continuous mappings of developable spaces and of spaces having bases of countable order

R.L. Moore introduced the concept of an upper semicontinuous collection of closed point sets [Mo_2], [Mo_4] formulated in a way applicable to spaces satisfying the first axiom of countability. For these spaces with appropriate separation these collections correspond to closed continuous mappings of the spaces, as is well known [E]; and this mapping concept applies to spaces in general and can be put into an equivalence of a more general definition of upper semicontinuous collection (cf. [E]). Moore, and various of his students, studied various upper semicontinuous collections, particularly certain ones whose elements are continua [see Mo_2]. One of the main questions he raised concerning upper semicontinuous collections, and which was settled by Worrell [Wo_2], [Wo_7] is whether Axioms 0 and 1, and Axioms 0 and 1_3, respectively, are preserved by upper semicontinuous decompositions into compact sets of spaces satisfying these respective pairs of axioms.

Worrell showed that developability of a topological space is preserved by every first countable closed continuous mapping of such a space [Wo_5]. He showed that the properties of being a developable complete Aronszajn space and of being a developable complete Moore space are respectively preserved by peripherally compact closed mappings, of such spaces [Wo_{10}]. Under support of the National Science Foundation at the University of Texas and subsequently, the U.S. Atomic Energy Commission at Sandia Laboratory, Worrell showed, in 1962, that if the continuum hypothesis is true there exist first countable closed continuous mappings of locally compact Moore spaces respectively onto regular T_0 spaces which are complete Aronszajn spaces but not complete Moore spaces and onto regular T_0 spaces not having the Baire category property (such a space, of course, is not a complete Aronszajn space) [Wo_8]. Such mappings do not preserve the absolute θ-refinability of their domains.

The techniques of the proofs of preservation of regular T_0 developability and of the property of being a developable complete Aronszajn space of [Wo_{18}] applied to preservation of the properties of being a complete Aronszajn space and certain related properties, namely various formulations of the base of countable order property for regular T_0 spaces [Wo_{16}], [Wo_6], [Wo_{19}]. Moreover, it was evident that these properties were respectively preserved by continuous decompositions, in the sense of R.L. Moore, of those spaces [Wo_6]. The question as to whether there was an analogue for these spaces of the above invariance theorem for Moore spaces concerning *first-countable* closed continuous mappings onto regular spaces was a problem which was resolved in the course of showing, on a basis of the continuum hypothesis, non-invariance of the property of being a developable complete Aronszajn space under the action of a closed continuous first countable mapping onto a regular space. Worrell showed that, if the continuum hypothesis is true, then every regular T_0 first countable space of power $\leq c$ is a closed continuous image of a locally compact complete Aronszajn space. With this class of counterexamples to the fore, the exposition of [Wo_7] was developed so as to give a coherent essentially Axiom T_1 treatment of the preservation by peripherally compact closed continuum mappings of the topological uniformization property (developability) and respective completeness properties of a monotonic formulation (reducing to the complete Aronszajn space property for the regular T_0 cases) and a Cauchy type finite intersection property (reducing to the complete Moore space property for the regular T_0 cover). A careful reading of [Wo_7] with this background in mind, along with certain of the content of the present exposition, may help to indicate why Burke's result on perfect mappings of primitive spaces [Bu_2] is regarded of such importance. It seemed to Worrell of importance at the time (1963), and later in retrospect, to demonstrate an underlying technical coherence in the demonstration of invariance of these monotonic and finite intersection completeness properties.

Great tension in the development of topology has been generated by overtly opposing tendencies in formulations of properties explicitly in reference to conditions of monotonicity and explicitly in reference to conditions of finite intersection. Out of that tension has been generated one of the main influences occasioning the development of generalized base of countable order theory, where certain provisional resolutions can be identified.

The thrust of the closed mapping results generalizes to the β_b-, λ_b-, β_c-, and λ_c-space uniformizations for the regular, and more generally for the pararegular, cases [WoW_8], [WiW_7]. In particular, the preservation of developability by first countable closed continuous mappings, generalizes to preservation of the β_b-space condition by closed continuous mappings of subparacompact regular T_0 β_c-spaces onto regular T_0 q-spaces. There is a lot of detail to be filled in here, but methods have been worked out for much of it. For perfect mappings of these spaces (β_b-, λ_b-, β_c- and λ_c-spaces), a general approach of separation of invariances into preservation of the topological uniformization and preservation of an appropriate paracompactness-like property has been to the fore [Wo_{16}],[Wo_{10}]. Burke's result on perfect mappings of spaces having primitive bases strongly reinforces the appropriateness of this emphasis on separation of invariances. Also in this connection, a result of equal importance is that of Junnila [Ju] on the preservation of θ-refinability (= submetacompactness) by closed continuous mappings. The preservation of subparacompactness by such mappings was established by Burke in [Bu_1]. In illustration, we make an application of Burke's result to a derivation of the preservation of the property of being a regular T_0 space having a point regular (uniform) base similar to one which can be made to derive the A.H. Stone-Morita-Hanai Theorem [S], [MH].

Theorem. If f is a perfect mapping of a regular T_0 space S having a point regular base, then f(S) has a point regular base.

Proof outlined. Since S has a primitive base, by Burke's theorem f(S) has a primitive base [Bu_2]. Since S is metacompact, f(S) is metacompact [Wo_{10}]. Since closed sets in S are inner limiting sets, closed sets in the closed image f(S) are limiting sets. Primitive T_1 spaces in which closed sets are inner limiting sets have bases of countable order [WoW_2] [WiW_{11}]. Every T_1 metacompact space having a base of countable order has a point regular base [WoW_2].

It should be clear how to confine Burke's and Junnila's results to get preservation of developability by perfect mappings.

Note. First countability of a closed mapping of a regular T_0 metacompact space implies that no element of the induced decomposition has a non-compact boundary [WoW_8]. Extension of Burke's result to peripherally compact mappings could thus allow replacement of the hypothesis that f is perfect in the hypothesis that f is closed and first countable.

8. Some theorems on compactness

One of the classical themes in topology is the study of conditions characterizing for compactness. A work in this tradition, which helps to set the tone of much subsequent research, is the classical mémoire of Alexandroff and Urysohn on compact spaces [AlU_2]. Here some of the most splendid themes are sounded with a richness of content. A central question is, as we may phrase it in retrospect, what the relations of certain forms of point-countability are to compactness and cardinality. As Mary Ellen Rudin has shown, there may be underlying links here with the Souslin problem [Ru_2].

In their article [WoW_1] Worrell and Wicke emphasize a technical distinction between σ-distributive point-finitude and σ-distributive point-countability. They state a covering theorem [WoW_1] a corollary to which is that the compact spaces are the weakly δθ-refinable countably compact spaces [WoW_1], [WiW_{15}]. They extend this corollary in [WoW_{10}] and [WiW_{19}] to a characterization relaxing weak δθ-refinability to weak $[\alpha,\beta]^r$-refinability in a sense of Hodel and Vaughan [HoV] demonstrating basically complete examples to show that $[\alpha,\beta]^r$-refinability does not in general imply δθ-refinability. In a general development of barycentric equivalences of certain $[\alpha,\beta]^r$-refinability type concepts, the characterization of isocompactness is extended [WiW_{19}], [WiW_{20}].

One immediate application of these results is to give some indication as to how far one must go in a direction of non-paracompactness of a space for the M-space concept of Morita to have a significance in comparison with the class of Hausdorff perfect preimages of metrizable spaces (equivalently, the Hausdorff paracompact p-spaces of Arhangel'skiĭ [Arh_3]). If a T_2 M-space S is <u>not</u> a paracompact p-space, then S is not weakly δθ-refinable; indeed, for appropriate meanings for α and β, S is not weakly $[\alpha,\beta]^r$-refinable [WiW_{19}], [WiW_{20}]. Moreoever, in order that a pararegular β_c-space S <u>not</u> be a β_b-space, it is necessary that S not be locally δθ-refinable, more generally, that for appropriate α's and β's, S not be locally $[\alpha,\beta]^r$-refinable. From a viewpoint of the local-implies-global properties of the β_b- and β_c-space uniformizations, a question which comes to the fore with a particular emphasis is to what extent the β_c-space uniformization for regular T_0 spaces enlarges on the class of regular T_0 β_b-spaces. The techniques of construction of [Wo_{11}] can be applied to exhibit large classes of countably compact collectionwise normal β_c-spaces which are nowhere locally β_b-spaces (for construction, see [WoW$_5$]). Indeed, certain partial compactifications of certain of these spaces suffice.

First countability, in association with other conditions, can in general place stringent limitations on the cardinality of a compact continuum [cf. Arh$_6$]. Uniformized first-countability of the primitive base property requires *second countability* and hence metrizability of a compact Hausdorff space [WoW$_2$], [WoW$_6$]. Homogeneity and hereditary topological equivalence of non-degenerate subcontinua are properties placing strong requirements on non-degenerate compact metrizable continua [Bi$_1$], [Moi$_1$], [He]. Heredity topological equivalence of all non-degenerate closed dense in themselves subsets of a compact metrizable space S dense in itself requires topological equivalence with a Cantor set. After an analysis of Wicke's proof of non-invariance of Čech completeness under the actions of open continuous mappings between collectionwise normal spaces [Wi$_1$], Worrell showed, in extension of it, that for each $\aleph > \aleph_0$, there exists an arc having a countably compact subspace S with a base of countable order and with topologically equivalent local and global structures such that if X and Y are two points of S, there exists a homeomorphism of S onto itself taking X into Y and *leaving no point fixed* [Wo$_{11}$]. Such a space S is collectionwise normal Indeed, these spaces are regularly refinable. By comparison, note that a normal locally separable developable complete Aronszajn space with equivalent local and global topological structure is metrizable. Consider also the Reed-Zenor metrization theorem [RZ].

An underlying objective of the work [Wo_{11}] was to probe the extent one can go, from a standpoint of freedom of restrictions on transfinite weight after $\aleph_0$, in the domain of topological uniformized first-countable structure, with appropriate topological completeness, and still retain certain of the topological properties classically regarded as important and associated with, or indeed virtually defining, nice spatial properties. It was taken as an hypothesis that whatever one might select for a mode of first countable topological uniformization *within a framework of first-countable* structure, countable compactness would be a virtually optimal form of completeness within the domain of that first-countable structure for Hausdorff spaces. We took the position in our approach to non-first-countable topological uniformization that the bridge from metrizability should make a transition through a satisfactory and coherent theory of topological uniformization and completeness of first-countable structure. The burden of handling connectivity satisfactorily from a standpoint of this degree of generality was replaced by that of obtaining a satisfactory treatment of completeness.

We make two tests of the hypothesis that, in the presence of an appropriate topological uniformization, countable compactness should imply completeness of the space. Among spaces of point-countable type, consider the completely regular T_0 p-spaces. Such a space (S,τ) is Čech complete if and only if S is an innerlimiting set of a regular T_0 locally countably compact space. If (S,τ) is a completely regular T_0 β_b-space, then (S,τ) is a λ_b-space if and only if S is a set of interior condensation of a locally countably compact, regular T_0 space.

There are now a number of interesting results that have been obtained for normal M-spaces (spaces defined by Morita as quasi-perfect inverse images of metrizable spaces [Mor_1]. A particularly elegant result is Ishii's theorem that quasi-perfect images of normal T_1 M-spaces are M-spaces [I]. One may also observe Nagata's interesting result that regular T_0 q-spaces are open continuous images of regular T_0 M-spaces [Na_3], (Compare [Wi_5]). The pararegular T_0 β_c-spaces are the uniformity λ-complete open continuous images of regular T_0 M-spaces (cf. [Wi_7],[WiW_7]). Uniform λ-completeness of a mapping may be viewed as a generalization of countable compactness of a mapping. See [Wi_7] for further discussion of the concept.

9. Some further remarks on completeness and topological uniformization

We have published some extensive remarks concerning criteria for completness in topological uniformization [WiW_7] [WiW_6]. There is a relaxation of regularity, pararegularity, which in the presence of Axiom T_0 is stronger than Axiom T_2, is defined and discussed in relation to these criteria. The countable productivity of certain uniformizations based on primitive sequences, for appropriate separation, is observed. Some relations of sets of interior condensation to certain complete subspaces generalizing Alexandroff's characterization of metric topological completeness [Al_1] are considered in a context of some classical Baire category type theorems. The latter type theorems are considered in relation to β_b- and β_c-space uniformizations, in their role of identifying certain respective β_b- and β_c-subspaces. Some theorems of this type involving preservation of certain paracompactness-like conditions were discussed in 1968 [WiW_4].

We state as two of the criteria for a satisfactory concept

of topological completeness that, for appropriate separation, the completeness property be preserved by perfect mappings and by open continuous mappings which are uniformly complete will respect to that concept. From this standpoint, let us consider the situation for certain mappings of various types of spaces.

To begin with, let us recall the existence of perfect mappings of first countable arcs onto non-first countable spaces. While every such image is a q-space, the non-invariance problem here is not resolved simply by enlarging the class of spaces under consideration to the regular T_0 spaces of point-countable type. There exist perfect mappings of Hausdorff paracompact spaces of point-countable type onto spaces which are not q-spaces [L], [V]. This non-invariance problem met in the domains of first countable structure should, in our view, be solved there with an appropriate first countable topological uniformization.

Next, let us consider regular T_0 spaces S satisfying the classical Baire category condition (the elements of every countable collection of open sets dense in S have a common part dense in S). The property, preserved by open continuous mappings, has some completeness content. For instance, it requires separability for a Moore space satisfying a Souslin-like hypothesis (compare [Ru]).It has some interesting relations to the concept of a pseudo-base [O]. But there exists a perfect mapping of a normal T_1 space having the Baire category property onto a space not having the property [AL].

Consider now the semimetric spaces, particularly recalling McAuley's example of a Hausdorff paracompact semimetrizable space which is not metrizable [McA]. It is evident that there exist open continuous mappings of complete metric spaces onto normal T_1 spaces which are nowhere locally semimetrizable (see Section 8). Moreover, there exist a perfect mapping of a semimetric space not preserving semimetrizability [He], [L].

There exists an open continuous mapping of a complete metric space onto a regularly refinable, countably compact T_1 space with topologically equivalent local and global structures which is nowhere locally a p-space and which is nowhere locally Čech complete. The countably compact spaces of [Wo_{11}] are such spaces. A perfect mapping of a Čech complete space onto a completely regular space preserves Čech completeness [HI]. But the closely related p-space uniformization is not preserved by every perfect mapping of a p-space onto a completely regular space R, even under the strengthened hypothesis that S and R are locally second countable scattered spaces with θ-bases [Wo_{22}]. Indeed, conditions under which the range R of a perfect mapping of a completely regular T_0 p-space is a p-space may be treated along the lines of the Pasynkov-Worrell theorem on the Čech completeness of Hausdorff open continuous images of Čech complete spaces [Pa], [Wo_{17}]. It suffices, if R is completely regular, that R satisfy the generalization of θ-refinability we have called Property A [WoW_9],[WiW_{14}]. Similarly, one may replace the hypothesis of paracompactness in the theorem of Pasynkov-Worrell with Property A', which generalizes absolute θ-refinability and is stronger than Property A, and relax the hypothesis of Čech completeness on the domain to the hypothesis that it is locally a Hausdorff β_b-space.

From the above and the definition of a ωΔ-space [Bo], it may be seen that this condition is not preserved by every perfect mapping of a space having the property onto a completely regular space. But if R is a regular T_0 space which has Property A and which is a perfect image of a para-regular β_c-space, then R is a $\omega\Delta$-space. Moreover, for appropriate separation the p-space and the $\omega\Delta$-space properties are respectively preserved by perfect mappings of these types of spaces which have Property A'.

There exist metacompact complete Aronszajn spaces R which not complete Moore spaces [CČN]. By a characterization theorem of [WiW_2] every such non-Čech complete space R is a compact open continuous image of a complete metric space. It follows readily that there exist compact open continuous mappings of complete M-spaces onto regular T_0 spaces which do not preserve the M-space property, for no such non-paracompact space R is a perfect preimage of a paracompact space. It can readily be shown that there exists an open continuous mapping of a complete metric space onto a normal T_1 non-M-space. So while quasi perfect images of normal T_1 M-spaces, as was noted above, are M-spaces, the situation for open continuous mappings reflects that for metrizability. Moreover, there exists a perfect mapping of a Hausdorff locally compact M-space onto a non-M-space [Mor_2].

The property of being a space having a topology preserving complete uniformity in the sense of A. Weil is not preserved by every perfect mapping of a space having the property [HI]. Moreover, in a strong sense, Hausdorff's theorem on the metric topological completeness of metrizable open continuous images of complete metric spaces does not generalize for uniform spaces in the sense of A. Weil: the spaces of [Wo_{11}], (referred to in Section 8) nowhere locally have topology preserving complete uniformities though they are uniform spaces which are open continuous images of complete metric spaces [WoW_5]. Indeed, in a strong sense the Baire category theorem does not hold for all spaces having topology preserving complete uniformities, as may be readily observed from the theorem that all Hausdorff paracompact spaces have topology preserving complete uniformities [Na_2]. From Ponomarev's characterization of the T_0 open continuous images of metrizable spaces [Po] it may thus be seen that every first-countable T_0 space is an open continuous image of a space having a topology preserving complete uniformity.

10. *Inverse image approaches to topological uniformization*

As has been remarked (Section 8), the hypothesis was made in our work on generalized base of countable order theory (at Sandia Laboratory) that the bridge from metrizability to appropriate non-first-countable topological uniformizations should have a transition through a satisfactory theory of topological uniformization within the domain of first-countable spaces. The question arose whether a transition from such a topological uniformization for first-countability would, in the essentials, properly simply reduce to considering the certain inverse continuous images of a space having a first-countable topological uniformization. The characterization of the Hausdorff paracompact p-spaces as the Hausdorff perfect preimages of metrizable spaces [Wi_5] gave particular pertinence to the question. In particular, the question occurred as to whether (for appropriate separation) the metacompact p-spaces can be characterized as perfect preimages of spaces having point regular (uniform) bases.

The answer to the question is in the negative even for the case of the class of locally compact, Hausdorff screenable, metacompact, first-countable spaces S of power c [Wo_{18}]. There exists such a space S such that there exists no Lindelöf continuous mapping of S onto a regular T_0 space having a base of countable order.

By contrast with the approach of defining non-first-countable structure with respect to certain inverse continuous images of spaces with first-countable topological uniformizations, notably as certain closed inverse continuous images, one obtains as theorems that (for appropriate separation) in cases where spaces are such inverse images (for appropriate mappings) they are respectively topologically uniformizable by β_b-, λ_b-, β_c-, λ_c-, primitive β_b-, and primitive β_c-sequences. Thus every regular perfect preimage of a regular space having a base of countable order is a regular β_b-space, and every regular quasi-perfect preimage of a regular space having a base of countable order is a regular β_c-space. Analogously, every regular perfect (respectively, quasi-perfect) preimage of a regular space having a primitive base is a regular space having a primitive β_b- (respectively β_c-) sequence. Every regular perfect (respectively, quasi-perfect) preimage, of a complete Aronszajn space is a regular β_b- (respectively β_c-) space.

There are some related theorems on certain inverse images of topologically complete spaces having certain of the uniformizations of generalized base of countable order theory. In particular, the following theorem of [Wo_{20}] generalizes with respect to β_b- (λ_b-) spaces and β_c- (λ_c-) spaces, respectively: Every regular T_0 space S having a base of countable order which is the domain of a uniformly monotonically complete continuous mapping whose range is a complete Aronszajn space has a monotonically complete base of countable order. This theorem can be applied to obtain the following variation for inverse images of Hausdorff's theorem cited above of [Ha]: A metrizable space S is metrically topologically complete if and only if there exists a uniformly monotonically complete continuous mapping of S onto a complete Aronszajn space. An analogue can be obtained for complete Moore spaces in which metrizability of the domain S is replaced by absolute θ-refinability, the base of countable order property, and regular T_0 separation. These results generalize for Čech complete paracompact spaces and Čech complete absolutely θ-refinable spaces.

Under support of the National Science Foundation in 1962, Worrell obtained certain generalizations of the following theorem: If there exists a light perfect mapping of the regular T_0 locally connected space S onto a complete Aronszajn space, then S is a complete Aronszajn space. (Clearly, this implies an extension of Aronszajn's arc theorem.) This result can be extensively generalized in formulations pertaining to sets of *light* interior condensation of locally connected spaces [Wo_{20}].

11. Extensions and generalizations of metrization theory in base of countable order theory

In the primitive base concept, which generalizes the Nagata-Smirnov hypothesis of a σ locally finite space [N

[Smi] metrization theory is extended so that a generalization of the inner limiting set condition and certain completeness conditions generalizing local countable compactness are separated in characterizations of metrizability [WoW_2], [WiW_{11}], [WoW_6], [WoW_7] and metric topological completeness, respectively. This carries forward the advance of the Arhangel'skiĭ metrization theorem [Arh_2] on the metrizability of paracompact Hausdorff spaces having bases of countable order, which can be interpreted as having an implicit completeness hypothesis in that of presence of the base of countable order property.

We extend the Arhangel'skiĭ metrization theorem (on metrizability of paracompact Hausdorff spaces having bases of countable order) with σ-distributive point-countability type, more generally, $[\alpha,\beta]^r$-refinability type, hypotheses and an identification of a role of completeness in primitive sequence properties. Certain local restrictions of cardinality, in the presence of which certain conditions of point-countability and abstractions of a $[\alpha,\beta]^r$-refinable type can be used to obtain paracompactness [Wo_1], are replaced with a global refinement condition formulated with respect to the concept of the Alexandroff-Urysohn metrization theorem [AlU_1] of a collection of sets being inscribed in a covering [Wo_{15}]. This is associated with theorems characterizing compactness in reference to weak $[\alpha,\beta]^r$-refinability type conditions and countable compactness [WoW_1], [WoW_{10}],[WiW_{19}],[WiW_{20}]. With paracompactness and compactness characterizations, theorems of [WoW_1], [WiW_2] can be applied to obtain certain implicit metrization theorms.

The thrust of general metrization theory--including the Urysohn theorem [U_1], the Alexandroff-Urysohn theorem [AlU_1], the Bing metrization theorems [Bi_2], the Nagata-Smirnov theorem [Na_1],

[Smi] and the above cited Arhangel'skiĭ metrization theorem [Arh_1], and various metrization theorems involving separability-type cardinality conditions, such as that of F.B. Jones [Jo_1]-- as of about 1967 generalizes for characterizations of the Hausdorff paracompact p-spaces [Wo_{14}]. One such theorem, obtained independently by Worrell and Lutzer, is a generalization of the Nagata-Smirnov theorem which characterizes the Hausdorff paracompact p-spaces [Wo_{14}], [BuEL]. As noted in Section 5, much of the content of [WoW_1] generalizes in characterizations of θ-refinable paracompact p-spaces (for appropriate separation). With use of these extensions to non-first-countable spaces and theorems on lightness of mappings and light sets of interior condensation [Wo_{20}] and diagonal theorems [Wi_8], [Wi_9] additional theorems on metrizability can be obtained.

A central theorem in the classical Bing metrization paper [Bi] is the metrizability of collectionwise normal Moore spaces. R.L. Moore emphasized the question as to whether, in a current terminology, for a regular T_0 developable space, the collectionwise Hausdorff property and normalcy are equivalent. Worrell showed that for locally separable, developable, regular T_0 spaces, indeed on a more general basis of properties related to the Hodel-Vaughn $[\alpha,\beta]^r$-refinability concept, the collectionwise Hausdorff property is equivalent with metrizability; and he incorporated this result in his dissertation [Wo_6], [Wo_{23}]. At Sandia Laboratories, he showed, without employing the continuum hypothesis, that there exists a collectionwise Hausdorff Moore space which is not normal [Wo_4] At Wisconsin, under the influence of the work of Mary Ellen

Rudin, F. Tall obtained a result in which he reported the consistency of the hypothesis that normalcy of a locally separable Moore space of power c implies metrizability [T]. Fleissner extended this result to a form having application regardless of the weight of a locally separable Moore space [F1]. Worrell's example shows that in the consistency result of Nyikos [Ny], one cannot validly replace normalcy of a Moore space with the collectionwise Hausdorff condition.

G.M. Reed brought Worrell's announcement [Wo_4] to the attention of various of Engelking's students during study in Warsaw under an exchange program of the U.S. and Polish Academies of Science. Employing Martin's axiom, Alster and Pol obtained an example of a collectionwise Hausdorff non-normal Moore space [AP]. After his return to the United States, Reed reviewed these results at a seminar at the University of Wisconsin. M.L. Wage, then a graduate student of Mary Ellen Rudin's, succeeded in confirming Worrell's announcement [Wo_4]. He subsequently published this corroboration [Wa] with acknowledgement of Worrell's announcement. Alster and Pol succeeded in confirming the announcement [Wo_4] of the metrizability of collectionwise Hausdorff locally separable Moore spaces [AP].

The normal Moore space problem, with particular reference to explorations and investigations influenced by the work of Mary Ellen Rudin, has been a source of immense stimulation to the field of set-theoretic topology. In the background there is the question: Suppose metrizability of a normal Moore space is a consistency result, and suppose that the importance presumably attributed to normalcy of a space is reflected into first-countable topological uniformization theory so as to encompass explicitly certain nonmetrizable spaces. Where does one look for such spaces, and how far must one go beyond the classical emphasis on paracompactness-like properties in order to find them?

Perhaps the answer is simpler than one typically would have thought. Perhaps one of the first requirements is to acknowledge that the "long line" is prototypical of natural topological structure [Wo_{17}].

12. The finite intersection property in base of countable order theory

It is provocative that R.L. Moore, who to the end of his career indicated or would seem to have indicated, an almost adamant insistence or preference for characterization of compactness in reference to a condition of monotonicity, expressed topological uniformization in his Axiom 1 as a finite intersection property. One is tempted to say that it does not make sense on the surface. Yet, whatever we may in retrospect regard as the basis in his mathematical intuition, there is a stratum of topological structure at which there is a coherence.

To plumb this stratum, we go beyond a naive focus on paracompactness-like properties. How hard this can be amid all the elegance to be found there, and, most especially, the absorbing early engagement with inner relations with set theory and mathematical logic! Beyond the positive integers, we look at the set of countable ordinals and listen to a prompting that tells us that this space is a gateway to a vast domain of topological structure and, that if it was closed to earlier structural conceptions, that did not make it pathological.

In a context of the study of continua, the primitive sequence uniformizations of base of countable order theory, and what we here refer to as generalized base of countable order theory, can be given equivalent monotonic and finite intersection formulations, analogously to the way compactness can be equivalently characterized [Mo_1],[KS], and [Re], respectively. To say that on the surface this possibility would likely hardly be suspected is an understatement. The realization in our work followed years of researches, and in an atmosphere in which the bias against the formulations expressed in reference to conditions of monotonicity was a strong counter-current to our efforts in publication.

At this stage, with a particular illumination afforded by Burke's theorem [Bu_2], we can see the primitive base concept as an underlying topological uniformization for metrizability of an extraordinary conceptual simplicity. In the form currently frequently used, the definition may be stated as follows: A primitive base for a space S is a sequence H_1, H_2,... of well-ordered open coverings for S such that if P is a point of an open set D, there exists some n such that the first member of H_n that contains P is a subset of D. The space of the countable ordinals with respect to the order topology has a primitive base, as do all σ-scattered spaces [WiW_{16}],[WiW_{18}]. One can equivalently define primitive base adding the conditions that (1) for each n, each member h of H_n contains a point belonging to no element of H_n preceding h and (2) if n < k and P is a point, the first member of H_k that contains P is a subset of the first member of H_n that contains P [WoW_4]. In our announcement to the American Mathematical Society in 1965, and in our subsequent presentation, we formulated the primitive base property as follows: There exists a sequence H_1, H_2,... of well-ordered collections of open sets such that if P is a point of an open set D, there exist positive integers n and k such that (1) some element n of H_n contains P and (2) the kth such element is a subset of D [WoW_4]. In this form, the θ-base property can be readily seen to be a special instance of a primitive base. There is, moreover, an economy in that the collections H_n are not required to cover S.

We characterized spaces having bases of countable order, for the essentially T_1 cases, as spaces which have primitive bases and in which closed sets are sets of interior condensation [WoW_4], [WiW_{11}]. We proved that Hausdorff compact spaces with primitive bases are second countable. This is a special case of the theorem that a locally countably compact regular T_0 space has a base of countable order if and only if it has a primitive base.

Putting the primitive base property in the primitive sequence form of the second definition above, and adding the hypothesis that if h_1, h_2,... is a primitive representative of H_1, H_2,... there exists a point common to the primitive part $p(h_n, H_n)$ of each h_n with respect to H_n, we obtained a formulation of a primitively complete base. We showed that a regular T_0 space is a complete Aronszajn space if and only if it has a primitively complete base [WoW_4], [WiW_{11}]. We applied the concept of a complete primitive base to obtain characterizations of metrically topologically complete spaces [WoW_6]

[WoW_7] recognizing the existence of monotonically complete paracompact Hausdorff spaces having primitive bases which are not metrizable. In particular, the "Michael line" is such a space; indeed, there exist Lindelöf non-second countable regular spaces having primitive bases [Be]. A property equivalent to the base of countable order property, for essentially T_1 spaces, was obtained by relaxing the requirement of the above formulation that there exist a point common to the sets $p(h_n,H_n)$ to the requirement that if there exists a point common to the sets h_n, then there exists a point common to the sets $p(h_n,H_n)$ [WoW_4], [WiW_{11}].

Primitive completeness of a primitive base H_1, H_2,... which is a primitive sequence satisfies a condition of monotonicity as respects primitive parts; if h belongs to H_n, h' belongs to H_k, and $p(h,H_n)$ intersects $p(h',H_n)$ then p is a subset of $p(h',H_n)$ or $p(h',H_n)$ is a subset of $p(h,H_n)$. Any collection K of such sets $p(h',H_n)$ which has the finite intersection property is both monotonic and countable. One can obtain an equivalent formulation of primitive completeness for a primitive space by using the sequence formulation of the first definition above and requiring for collections K of sets $p(h,H_n)$ with the finite intersection property that there exist a point common to all the members of K. Such a sequence H_1, H_2,... may be thought of as identifying a concept of *primitive basis compactness.*

Similar equivalences of finite intersection formulations can be made for the β_b-, λ_b-, β_c- and λ_c-space definitions and primitive β_b- and primitive β_c-space formulations and for sets of interior condensation. Moreover, Burke's theorem extends to preservation of the primitive β_b-space and primitive

β_c-space properties by perfect mappings, respectively, of regular primitive β_b-spaces and regular primitive β_c-spaces.

13. *Some other characterizations of topological uniformizations of generalized base of countable order theory and related results*

J. Chaber early expressed an interest in base of countable order theory. Chaber, Čoban, and Nagami showed equivalences of certain of the definitions of Arhangel'skiĭ and of Wicke and Worrell with certain formulations made with respect to sieves [CČN]. Chaber has published other papers concerning applications and extensions of primitive sequences [Ch_1],[Ch_2].

Phillips has published papers on completeness in base of countable order theory, some of which involve primitive structure theory [Ph_1]-[Ph_5].

In the course of his paper [Mi_4], Michael obtains some results on completeness in base of countable order theory. For the regular cases, certain of the pertinent formulations, are equivalent to certain of the uniformizations reviewed here.

Fletcher and Lingren [FL] point out some interesting relations of the concept of a θ-space to that of a primitive base which were a stimulus for further developments found in [Wi_{13}]. [Wi_{14}], [Wi_{12}].

Arhangel'skiĭ formulated a p-space definition with respect to Wallman compactifications [Arh_3]. The heuristic point is, as we interpreted it, strongly in accord with our views on separation in topological uniformization. Some of the main results in base of countable order theory (in the first-countability uniformizations) can be obtained on an essentially

T_1 basis. In some instances, results can be put on what is' *implicity* an essentially T_0 basis as is the case of Ponomarev's characterization [Po] generalized for pseudometrizable domains.

We characterized regular T_0 β_b-spaces in an equivalence of the class of certain T_0 hyperspaces formed by their compact sets with the class of T_0 uniformly monotonically complete open continuous images of metrizable spaces [WoW$_4$]. The regular T_0 λ_b-spaces were similarly characterized in an equivalence of the class of certain T_0 hyperspaces formed by their compact sets with the class of T_0 open continuous images of metrically topologically complete spaces [WoW$_4$].

Concepts of absolute θ-refinability, absolute subparacompactness, absolute metacompactness can be formulated with respect to embeddings in locally countably compact regular T_0 spaces.

There are a number of our results, and those of others, on paracompactness-like properties, which we have not reviewed here or have not reviewed in detail. Included in this category are certain results on countable paracompactness and normalcy which we consider to have a considerable importance and a nontrivial bearing (compare [Ru$_2$] and the Reed-Zenor metrization theorem [RZ]). Included in this category also are certain theorems on cardinal restrictions on preimages of various mappings and on extensions of mappings (see [WiW$_2$]) and other results [vDW], [Wi$_{10}$], [Wi$_{12}$]-[Wi$_{14}$].

Some more general theorems on open continuous mappings than stated here can be obtained for inductively open mappings [WiW$_2$].

H. Junnila has developed many interesting results on submetacompactness and related topics. He reported on certain of this work during a research fellowship at Ohio University in 1978. See [Ju] for references.

Bennet and Lutzer obtained an equivalence for θ-bases and quasi-developability [BeL].

Burke and Lutzer presented a survey of the area of generalized metric spaces in 1975 [BuL].

An example especially pertinent to weak covering properties is in [vDW].

14. Generalized base of countable order theory and continua

Definitions of arcs, simple closed curves, and certain other continua are made in Chapter I of R.L. Moore's *Foundations of Point Set Theory* on a basis of compactness, separation, and connectivity. In their freedom from stipulated conditions of first-countability or weight, these definitions find application in generalized base of countable order theory.

In the Preface of the revised edition of Moore's *Foundations*, Moore calls particular attention to a result of F.B. Jones [Jo_2] from which it follows that *compact continuum* can validly replace *simple closed curve* in the statement of Axiom 5 of the first edition [Mo_4]. Jones' axiom, called Axiom 5 in the revised edition, states: If A and B are two points every region that contains A contains a compact continuum that separates A from B. Moore remarks that he considered Jones' improvement to be a major one. From the viewpoint of generalized base of countable order theory, this indeed would se seem an appropriate statement.

Sufficiently careful consideration of Čech's definition of topological completeness will show serious limitations in its applicability to the flat space S obtained as a Cartesian product of the "long line" with itself, provided one wishes genuinely non-trivial extensions of concepts of flatness given by Moore's Axioms 0,1,2,3,4 and 5 in which completeness replaces Moore's Axiom 1 with basic completeness [WiW_5] resulting set of axioms, for appropriate interpretations of region, are satisfied by subspaces of S which are not Čech complete, through locally they are metrically topologically complete spaces satisfying Axioms 0,1,2,3,4 and 5. The equivalence of completeness, in the sense of a relative topology satisfying Axiom 1, with that of an inner limiting set of spaces doing so, which has a prime place in the Preface to [Mo_4] has a counterpart in the equivalence for point sets of a basically complete space S of basic completeness and being a set of interior condensation of S. Moreover, as has been developed in the preceding section, these properties can be put in finite-intersection form.

The above remarks take on resonance against a background of the work of Schoenflies, Jordan, Moore, F.B. Jones, and others who have made fundamental contributions to the study of topological properties of two-space and certain generalizations (as in the case of R.L. Moore and F.B. Jones). We do not even attempt here any exhaustive survey of the background researches in this area leading to the present ready application. Neither do we feel that mention of their applicability should be omitted, particularly when certain of the work, notably that of F.B. Jones, has figured prominently in the background.

Moore's Axiom 2 introduces local connectivity and non-degeneracy of the components of the space. In retrospect, from a standpoint of generalized base of countable order theory, juxtaposition of axioms seems remarkable. One can characterize locally connected spaces satisfying Axiom 1 with use of absolute θ-refinability and certain continuous mappings as has been indicated (implicitly) above. Local connectivity has a strong bearing on first-countable topological uniformization: A regular T_0 locally connected space satisfies Moore's Axiom 1 (with respect to an interpretation of "region" as "open set") if and only if it is absolutely θ-refinable and each point belongs to an open set which is the domain of a light perfect mapping onto a complete Aronszajn space as its range [Wo_{20}]. Indeed, a regular T_0 locally connected space S has a base of countable order if and only if each point belongs to an open set which is the domain of a light perfect mapping whose range has a base of countable order [Wo_{20}]. Moreover, if S is a regular T_0 locally connected β_b-space then S has a base of countable order if and only if each point belongs to an open set which is the domain of a light continuous mapping whose range is a regular T_0 space having a base of countable order [Wo_{25}]. One can relax the hypothesis that S be a β_b-space to the hypothesis that S is a β_c-space if S is locally normal.

It is clear that these theorems on lightness of mappings, as well as certain of Wicke's diagonal theorems, with Aronszajn's theorem [Aro] imply a number of additional arc theorems as corollaries. There are technical problems that are present in relaxing the separation hypothesis from the stipulation that the spaces be regular T_0 to the hypothesis that they are Hausdorff. It is possible to extend the arc theorem to the class of pararegular T_0 locally connected, basically complete spaces [WiW_3], [WiW_6].

15. Summary

What has been said reduces in its quintessence to the simplicity of statement of the primitive base definition. The latent combinatorial content has the classical theory of metrization as one of its aspects. This formulation was achieved under the support of the research administration of Sandia Laboratories and under the guiding mathematical influence of R.L. Moore and P.S. Alexandroff. Supplementary to the record here of its development are the abstracts in the Notices of the American Mathematical Society from 1962 to the present written by the authors singly or in combination.

References

[AL] J.M. Aarts and D.J. Lutzer, *Completeness properties designed for recognizing Baire spaces*, Dissertationes Math. 116 (1974), 1-48.

[Al_1] P.S. Alexandroff, *Sur les ensembles de la première classe et les espaces abstraits*, C.R. Acad. Sci. Paris 178 (1924), 185-187.

[Al_2] ________, *On the metrization of topological spaces*, Bull. Acad. Polon. Sci., Sér. Sci. Math. Astr. Phys. 8 (1960), 135-140. (Russian)

[Al_3] ________, *On some results concerning topological spaces and their continuous mappings*, Proc. Symp. Gen. Topology, Prague (1961), 41-54.

[AlU_1] P.S. Alexandroff and P.S. Urysohn, *Une condition necessaire et suffisante pour qu'une classe (L) soit une classe (D)*, C.R. Acad. Sci. Paris 177 (1923), 1274-1276.

[AlU_2] ________, *Mémoire sur les espaces topologiques compacts*, Verh. Akad. Wetensch. Amsterdam 14 (1929).

[AP] K. Alster and R. Pol, *Moore spaces and collectionwise Hausdorff property*, Bull. Acad. Polon. Sci., Sér. Sci. Math. Astr. Phys. 23 (1975), 1189-1192.

[Arh_1] A.V. Arhangel'skiĭ, *On open and almost open mappings of topological spaces*, Dokl. Akad. Nauk SSSR 147 (1962), 999-1002. (Russian)

[Arh_2] A.V. Arhangel'skiǐ, *Certain metrization theorems*, Uspehi Mat. Nauk 18 (1963) no. 5 (113), 139-145. (Russian).

[Arh_3] ________, *On a class of spaces containing all metric and all locally bicompact spaces*, Mat. Sb. 67 (1965), 55-85. (Russian)

[Arh_4] ________, *Bicompact sets and the topology of spaces*, Trudy Mosk. Mat. Obšč. 13 (1965), 3-55. (Russian)

[Arh_5] A.V. Arhangel'skiǐ, *Mappings and spaces*, Uspehi Mat. Nauk 21 (1966) no. 4 (130), 133-184. (Russian)

[Arh_6] ________, *The cardinality of first-countable bicompacta*, Dokl. Akad. Nauk SSSR 187 (1969), 967-970; English translation in Sov. Math. Dokl. 10 (1969), 151-154.

[Aro] N. Aronszajn, *Über die Bogenverknüpfung in topologischen Raümen*, Fund. Math. 15 (1930), 228-241.

[Au_1] C.E. Aull, *A generalization of a theorem of Aquaro*, Bull. Austral. Math. Soc. 9 (1973), 105-108.

[Au_2] ________, *Quasi-developments and $\delta\theta$-bases*, J. of Lond. Math. Soc. 9 (1974), 197-204.

[Ba] R. Baire, *Sur les fonctions de variables réelles*, Ann. di Math. 3 (1899), 1-123.

[Be] H.R. Bennett, *On quasi-developable spaces*, Gen. Top. Appl. 1 (1971), 253-262.

[BeL] H.R. Bennett and D.J. Lutzer, *A note on weak θ-refinability*, Gen. Top. Appl. 2 (1972), 49-54.

[Bi_1] R.H. Bing, *A homogeneous indecomposable plane continuum*, Duke Math. J. 15 (1948), 729-742.

[Bi_2] ________, *Metrization of topological spaces*, Can. J. Math. 3 (1951), 175-186.

[Bo] C.J.R. Borges, *On metrizability of topological spaces*, Can. J. Math. 20 (1968), 795-804.

[BuEL] D.K. Burke, R. Engelking, and D. Lutzer, *Hereditarily closure-preserving collections and metrization*, Proc. Amer. Math. Soc. 51 (1975), 483-488.

[BuL] D.K. Burke and D.J. Lutzer, *Recent advances in the theory of generalized metric spaces*, in *Topology: Proceedings of the Memphis State University Conference*, Marcel Dekker, Inc., New York, 1976, 1-70.

[Bu_1] D.K. Burke, *On subparacompact spaces*, Proc. Amer. Math. Soc. 23 (1969), 655-663.

[Bu_2] ________, *Spaces with a primitive base and perfect mappings*, preprint.

[C] G. Cantor, *Uber eine eigenschaft des Inbegriffes aller reellen algebraischen Zahlen*, J. Reine Angew. Math. 77 (1874), 258-262.

[C_1] E. Cech, *Sur la dimension des espaces parfaitement normaux*, Bull. Int. Acad. Tcheque Sci. 33 (1932), 38-55.

[$\check{C}_2$] ________, *On bicompact spaces*, Ann. Math. 38 (1937), 823-844.

[CČN] J. Chaber, M.M. Čoban and K. Nagami, *On monotone generalizations of Moore spaces, Cech complete spaces and p-spaces*, Fund. Math. 84 (1974), 107-119.

[Ch_1] J. Chaber, *On point-countable collections and monotonic properties*, Fund. Math. 94 (1977), 209-219.

[Ch_2] ________, *Primitive generalizations of σ-spaces*, preprint.

[D] A.S. Davis, *Indexed systems of neighborhoods for general topological systems*, Amer. Math. Monthly 68 (1961), 886-893.

[vDW] E.K. van Douwen and H.H. Wicke, *A real, weird example*, Houston J. Math. 3 (1977), 141-152.

[E] R. Engelking, *General topology*, Warsaw, 1977.

[Fl] W.G. Fleissner, *When normal implies collectionwise Hausdorff: consistency results*, Doctoral Thesis, Univ. Calif., Berkeley, 1974.

[FL] P. Fletcher and W.F. Lindgren, *θ-spaces*, Gen. Top. Appl. 9 (1978), 139-153.

[dG] J. deGroot, *Subcompactness and the Baire category theorem*, Indag. Math. 25 (1963), 761-767.

[Ha] F. Hausdorff, *Über innere Abbildungen*, Fund. Math. 23 (1934), 279-291.

[H] R.W. Heath, *Semi-metric and related spaces*, in *Proceedings of Topology Conference at Arizona State University*, Tempe, Ariz., 1967, 153-161.

[He] G.W. Henderson, *Proof that every compact decomposable continuum which is topologically equivalent to each of its nondegenerate subcontinua is an arc*, Ann. Math. 27 (1960), 421-428.

[HenI] M. Henriksen and J.R. Isbell, *Some properties of compactifications*, Duke Math. J. 25 (1958), 83-106.

[Ho] R.E. Hodel, *Spaces defined by sequences of open covers which guarantee that certain sequences have cluster points*, Duke Math. J. 39 (1972), 253-263.

[HoV] R.E. Hodel and J.E. Vaughan, *A note on [a,b]-compactness*, Gen. Top. Appl. 4 (1974), 179-189.

[I] T. Ishii, *On closed mappings and M-spaces II*, Proc. Japan. Acad. 43 (1967), 757-761.

[Jo_1] F.B. Jones, *Concerning normal and completely normal spaces*, Bull. Amer. Math. Soc. 43 (1937), 671-677.

[Jo_2] ________, *Concerning R.L. Moore's Axiom 5*, Bull. Amer. Math. Soc. 44 (1938), 689-692.

[Ju] H.J.K. Junnila, *On submetacompactness*, Topology Proceedings 3 (1978), 375-405.

[Ko] G. Köthe, *Die Quotiènträume eines linearen vollkommenen Raümes*, Math. Zeit 51 (1947), 17-35.

[KS] K. Kuratowski and W. Sierpiński, *La théoréme de Borel-Lebesgue dans la théorie des ensembles abstraits*, Fund. Math. 2 (1921), 172-178.

[L] D.J. Lutzer, *Semimetrizable and stratifiable spaces*, Gen. Top. Appl. 1 (1971), 43-48.

[McA] L.F. McAuley, *A relation between perfect separability, completeness, and normality in semi-metric spaces*, Pac. J. Math. 6 (1956), 315-326.

[Mi_1] E.A. Michael, *A theorem on semi-continuous set-valued functions*, Duke Math. J. 27 (1959), 647-651.

[Mi_2] ________, *The product of a normal space and a metric space need not be normal*, Bull. Amer. Math. Soc. 69 (1963), 375-376.

[Mi_3] ________, *A note on closed maps and compact sets*, Israel J. Math. 2 (1964), 173-176.

[Mi_4] ________, *Complete spaces and triquotient maps*, Ill. J. Math. 21 (1977), 716-733.

[Moi_1] E.E. Moise, *An indecomposable plane continuum which is homeomorphic to each of its nondegenerate subcontinua*, Trans. Amer. Math. Soc. 63 (1948), 581-594.

[Moi_2] ________, *A note on the pseudo-arc*, Trans. Amer. Math. Soc. 63 (1949), 57-58.

[Mo_1] R.L. Moore, *On the most general class L of Fréchet in which the Heine-Borel-Lebesgue theorem holds true*, Proc. Nat. Acad. Sci. 5 (1919), 206-210.

[Mo_2] R.L. Moore, *Concerning upper semi-continuous collections of continua which do not separate a given continuum*, Proc. Nat. Acad. Sci. 10 (1924), 356-360.

[Mo_3] ________, *Concerning separability*, Proc. Nat. Acad. Sci. 25 (1942), 56-58.

[Mo_4] ________, *Foundations of point set theory*, rev. ed. Amer. Math. Soc. Coll. Pub. XIII. 1962.

[MH] K. Morita and S. Hanai, *Closed mappings and metric spaces*, Proc. Japan Acad. 32 (1956), 10-14.

[Mor_1] K. Morita, *Products of normal spaces with metric spaces*, Math. Ann. 154 (1964), 365-382.

[Mor_2] ________, *Some properties of M-spaces*, Proc. Japan Acad. 43 (1967), 869-872.

[Na_1] J. Nagata, *On a necessary and sufficient condition of metrizability*, Journ. Inst. Polyt. Osaka City Univ. 1 (1950), 93-100.

[Na_2] ________, *On topological completeness*, Journ. Math. Soc. Japan 2 (1950), 44-47.

[Na_3] ________, *Mappings and M-spaces*, Proc. Japan Acad. 45 (1969), 140-144.

[Ny] P.J. Nyikos, *The normal Moore space problem*, Topology Proceedings 3, no. 2, (1978), 473-494.

[O] J.C. Oxtoby, *Cartesian products of Baire spaces*, Fund. Math. 49 (1961), 157-166.

[Pa] B.A. Pasynkov, *On open mappings*, Dokl. Akad. Nauk SSSR 175 (1967), 292-295; English translation in Sov. Math. Dokl. 8 (1967), 853-856.

[Ph_1] T.M. Phillips, *Completeness in Aronszajn spaces*, Proc. Top. Conf. Univ. North Carolina at Charlotte, Academic Press, New York, 1974, 457-465.

[Ph_2] ________, *Extending continuous functions*, Proc. Top. Conf. Memphis State Univ., Marcel Dekker, Inc., New York 1976, 229-237.

[Ph_3] ________, *A note on monotonic orthobases*, Proc. Amer. Math. Soc. 65 (1977), 150-154.

[Ph_4] ________, *Centered bases, nested bases and completability of Aronszajn spaces*, Can. J. Math. 30 (1978), 1331-1335.

[Ph_5] ________, *Primitive extensions of Aronszajn spaces*, Pac. J. Math. 77 (1978), 233-242.

[Po] V.I. Ponomarev, *Axioms of countability and continuous mappings*, Bull. Acad. Polon. Sci. Sér. Sci. Math., Astr. Phys. 8 (1960), 127-133. (Russian)

[RZ] G.M. Reed and P.L. Zenor, *Metrization of Moore spaces and generalized manifolds*, Fund. Math. 91 (1976), 203-210.

[Re] F. Reisz, *Stetizkeitsbegriff und abstrakte Mengenlehre*, Atti IV Contr. Internat. Mat. Roma 2 (1908), 18-24.

[Ru_1] M.E. Rudin (Estill), *Concerning abstract spaces*, Duke Math. J. 17 (1950), 317-327.

[Ru_2] M.E. Rudin (Estill), *Countable paracompactness and Souslin's problem*, Can. J. Math. 7 (1955), 543-547.

[SZ] M. Sion and G. Zelmer, *On quasi-metrizability*, Can. J. Math. 19 (1967), 1243-1249.

[Sm] Yu M. Smirnov, *On metrization of topological spaces*, Uspehi Mat. Nauk 6 (1951) no. 6 100-111; English translation: Amer. Math. Soc. Transl. Ser. 1, 8 (1962), 63-77.

[Smi] J.C. Smith, *A remark on irreducible spaces*, Proc. Amer. Math. Soc. 57 (1976), 133-139.

[Ste] L.A. Stern, *A direct proof that a linearly ordered space is hereditarily collectionwise normal*, Proc. Amer. Math. Soc. 24 (1970), 727-728.

[S] A.H. Stone, *Metrizability of decomposition spaces*, Proc. Amer. Math. Soc. 7 (1956), 690-700.

[St] M.H. Stone, *Applications of the theory of Boolean rings to general topology*, Trans. Amer. Math. Soc. 41 (1937), 375-481.

[T] F.D. Tall, *Set-theoretic consistency results and topological theorems concerning the normal Moore space conjecture and related problems*, Thesis, Univ. Wisconsin, 1969.

[Tu] J.W. Tukey, *Convergence and uniformity in topology*, Ann. of Math. Studies 2, Princeton, 1940.

[Ty] A. Tychonoff, *Über die topologische Erweiterung von Raümen*, Math. Ann. 102 (1930), 544-561.

[U_1] P.S. Urysohn, *Über die Metrisation der kompakten topologischen Raüme*, Math. Ann. 92 (1924), 275-293.

[U_2] ________, *Über die Mächtigkeit der zusammenhängenden Mengen*, Math. Ann. 92 (1924), 275-293.

[V] J.E. Vaughan, *Spaces of countable and point-countable type*, Trans. Amer. Math. Soc. 151 (1970), 341-351.

[Wa] M. Wage, *A collectionwise Hausdorff non-normal Moore space*, Can. J. Math. 28 (1976), 632-634.

[We] N. Wedenisoff, *Sur les espaces metriques comlets*, Journal de Mathematiques, Ser. 9, 9 (1930), 377-381.

[Wey] H. Weyl, *Die Idce der Riemannschen Fläche*, Leipzig, 1913.

[Wi_1] H.H. Wicke, *The non-preservation of Cech completeness by an open continuous mapping between collectionwise normal T_1 spaces*, Notices Amer. Math. Soc. 13 (1966), 601.

[Wi_2] ________, *The regular open continuous images of complete metric spaces*, Pac. J. Math. 23 (1967), 621-625.

[Wi_3] ________, *Concerning an arc theorem*, Notices Amer. Math. Soc. 14 (1967), 290.

[Wi_4] ________, *On a theory of topological structure involving principles of uniformization*, Kempner Colloquium Lecture, University of Colorado, 1967.

[Wi_5] H.H. Wicke, *On the Hausdorff open continuous images of Hausdorff paracompact p-spaces*, Proc. Amer. Math. Soc. 22 (1969), 136-140.

[Wi_6] ________, *Base of countable order theory and some generalizations*, Proc. Houston Topology Conf. 1971, 76-95.

[Wi_7] ________, *Open continuous images of certain kinds of M-spaces and completeness of mappings and spaces*, Gen. Top. and Its Appl. 1 (1971), 85-100.

[Wi_8] ________, *On spaces whose diagonal is a set of interior condenstaion*, Notices Amer. Math. Soc. 19 (1972), A-657.

[Wi_9] ________, *Primitive structures and diagonal conditions*, Notices Amer. Math. Soc. 22 (1975), A-587.

[Wi_{10}] ________, *Complete mappings in base of countable order theory*, in *Set-Theoretic Topology*, ed. George M. Reed, Academic Press, 1977, 383-412.

[Wi_{11}] ________, *Monotonic β-spaces and θ-diagonals*, Notices Amer. Math. Soc. 25 (1978), A-328.

[Wi_{12}] ________, *A functional characterization of primitive base*, Proc. Amer. Math. 72 (1978), 355-361.

[Wi_{13}] ________, *A generalization of θ-space and point-countable base*, Notices Amer. Math. Soc. 25 (1978), A-668.

[Wi_{14}] ________, *Using θ-space concepts in base of countable order theory*, Topology Proceedings 3(1) (1978), 267-285.

[WiW_1] H.H. Wicke and J.M. Worrell, *Open continuous mappings of spaces having bases of countable order*, International Congress of Mathematicians, Moscow, August 18, 1966. Abstract in *Abstracts of Brief Scientific Communications*, Int'l Cong. Math., Moscow, 1965, p. 14.

[WiW_2] ________, *Open continuous mappings of spaces having bases of countable order*, Duke Math. J. 34 (1967), 255-272; errata 813-814.

[WiW_3] ________, *On a class of spaces containing Arhangel-skii's p-spaces*, Notices of the Amer. Math. Soc. 14 (1967), 687.

[WiW_4] ________, *Quasi-hereditary properties, Baire category, and non-first countable structure*, Notices Amer. Math. Soc. 15 (1968), 162.

[WiW_5] ________, *On the open continuous images of paracompact Cech complete spaces*, Pac. J. Math. 37 (1971), 265-275.

[WiW_6] ________, *Topological completeness of first countable Hausdorff spaces I*, Fund. Math. 75 (1972), 209-222.

[WiW_7] ________, *Completeness and topological uniformizing structures*, Proc. Second Pittsburgh Int'l Conf. Gen. Top. and Its Appl. 1972, Springer-Verlag, New York, 1974, 557-585.

[WiW_8] ________, *On topological structures defined by primitive sequences*, International Congress of Mathematicians, Vancouver, B.C., Canada, August 1974.

[WiW_9] ________, *A characterization of primitive bases*, Proc. Amer. Math. Soc. 50 (1975), 413-450.

[WiW_{10}] ________, *The hereditary Lindelöf property, primitive structures, and separable metrizability*, Notices Amer. Math. Soc. 22 (1975), A-425.

[WiW_{11}] ________, *A characterization of spaces having bases of countable order in terms of primitive bases*, Can. J. Math. 27 (1975), 1100-1109.

[WiW_{12}] ________, *Primitive structures in general topology*, Studies in Topology, Proc. Conf. held at Charlotte, N.C. 1974, Academic Press, New York, 1975, 581-599.

[WiW_{13}] ________, *The local-implies-global characteristics of primitive sequences*, in Proc. Memphis State Univ. Top. Conf., Marcel Dekker, New York, 1976, 269-282.

[WiW_{14}] ________, *The concept of a θ-refinable embedding*, Gen. Top. Appl. 6 (1976), 167-181.

[WiW_{15}] ________, *Point-countability and compactness*, Proc. Amer. Math. Soc. 55 (1976), 427-431.

[WiW_{16}] ________, *Scattered spaces of point-countable type*, Proc. Fourth Prague Top. Symposium 1976, Prague, 1977, 513-516.

[WiW_{17}] ________, *Topological completeness of first countable Hausdorff spaces II*, Fund. Math. 41 (1976), 11-27.

[WiW_{18}] ________, *Spaces which are scattered with respect to collections of sets*, Top. Proc. 2 (1977), 281-307.

[WiW_{19}] ________, *A covering property which implies isocompactness II*, Top. Proc., to appear.

[WiW_{20}] ________, *Certain generalizations of point-countable refinements*, talk, Int'l Top. Conf., Moscow, USSR, 1979.

[Wo_1] John M. Worrell, Jr., *Concerning scattered point sets*, Doctoral dissertation, University of Texas, Austin, 1961. University Microfilms, Inc.

[Wo_2] ________, *Concerning upper semicontinuous collections of mutually exclusive closed and compact point sets*, invited talk, Sandia Laboratory, Albuquerque, New Mexico, April, 1962.

[Wo_3] ________, *Upper semicontinuous decompositions of spaces satisfying an axiom of N. Aronszajn*, Dept. of Math. Res., Sandia Laboratories, 1963 (in review).

[Wo_4] ________, *Isolated sets of points in nonmetrizable spaces*, Notices Amer. Math. Soc. 11 (1964), 250.

[Wo_5] ________, *Upper semicontinuous decompositions of developable spaces*, Notices Amer. Math. Soc. 11 (1964), 250.

[Wo_6] ________, *Decompositions of spaces having bases of countable order*, Notices Amer. Math. Soc. 11 (1964), 773-774.

[Wo_7] ________, *Upper semicontinuous decompositions of developable spaces*, Proc. Amer. Math. Soc. 16 (1965), 485-490.

[Wo_8] ________, *On boundaries of elements of upper semicontinuous decompositions I*, Notices Amer. Math. Soc. 12 (1965), 219.

[Wo_9] ________, *A characterization of metacompact spaces*, Portugaliae Mathematica 25 (1966), 171-174.

[Wo_{10}] ________, *The closed continuous images of metacompact topological spaces*, Protugaliae Mathematica 25 (1966), 175-179.

[Wo_{11}] ________, *On compact spaces and Cech completeness*, Notices Amer. Math. Soc. 13 (1966), 644.

[Wo_{12}] ________, *The paracompact open continuous images of Cech complete spaces*, Notices Amer. Math. Soc. 13 (1966), 858.

[Wo_{13}] ________, *Some properties of full normalcy and their relation to Cech completeness*, Notices Amer. Math. Soc. 14 (1967), 555.

[Wo_{14}] ________, *Concerning the paracompact p-spaces of Arhangel'skii*, Notices Amer. Math. Soc. 14 (1967), 949.

[Wo_{15}] ________, *On collections of domains inscribed in a covering of a space in the sense of Alexandroff and Urysohn*, Protugaliae Mathematica 26 (1967), 405-420.

[Wo_{16}] ________, *Upper semicontinuous decompositions of spaces having bases of countable order*, Portugaliae Mathematica 26 (1967), 493-504.

[Wo_{17}] ________, *A comparison of the spacial conceptions of R.L. Moore and A.V. Arhangel'sii*, invited talk, University of Houston Topology Conference, March 1968.

[Wo_{18}] ________, *On continuous mappings of metacompact Cech complete spaces*, Pac. J. Math. 30 (1969), 555-562.

[Wo_{19}] ________, *Characterizations of certain subspaces*, Notices Amer. Math. Soc. 16 (1969), 687-688.

[Wo_{20}] ________, *Lightness of mappings and the topological theory of metrization*, Notices Amer. Math. Soc. 17 (1970), 976-977.

[Wo_{21}] ________, *Some remarks concerning the Nagata-Smirnov Theorem*, invited talk, Pittsburgh Int'l Conf. Gen. Top. and Its Appl., Pittsburgh, June 1970.

[Wo_{22}] ________, *A perfect mapping not preserving the p-space property*, mansucript submitted 1970.

[Wo_{23}] ________, *Metrization of locally separable spaces*, Notices Amer. Math. Soc. 18 (1971), 264.

[Wo_{24}] ________, *Characterizatons of Cech complete developable spaces*, invited talk, Ohio University, Athens November 1971.

[Wo_{25}] ________, *Sequential refinements*, talk, Conf. Gen. Metrizable Spaces, Ohio University, December 1975.

[Wo_{26}] ________, *Locally separable Moore spaces*, Set-Theoretic Topology, Academic Press, Inc., George M. Reed, ed., 1977, 413-436.

[WoW_1] J.M. Worrell, Jr. and H.H. Wicke, *Characterizations of developable topological spaces*, Can. J. Math. 17 (1965), 820-830.

[WoW_2] ________, *Concerning spaces having bases of countable order*, Notices Amer. Math. Soc. 12 (1965), 343-344.

[WoW_3] ________, *Non-first-countable topological structure*, Notices Amer. Math. Soc. 14 (1967), 935.

[WoW_4] ________, *Concerning certain hyperspaces*, Notices Amer. Math. Soc. 16 (1969), 303.

[WoW_5] ________, *Extension of a result of Dieudonné*, Proc. Amer. Math. Soc. 25 (1970), 634-637.

[WoW_6] ________, *A central metrization theorem I*, Notices Amer. Math. Soc. 20 (1973), A-355.

[WoW_7] ________, *A central metrization theorem II*, Notices Amer. Math. Soc. 20 (1973), A-381.

[WoW_8] ________, *Perfect mappings and certain interior images of M-spaces*, Trans. Amer. Math. Soc. 181 (1973), 23-25.

[WoW_9] ________, *Perfect mappings and certain interior images of M-spaces*,

[WoW_9] ________, *Remarks on a property of θ-refinable spaces*, Notices Amer. Math. Soc. 21 (1974), A-622.

[WoW_{10}] ________, *A covering property which implies isocompactness I*, Proc. Amer. Math. Soc., to appear.